Bioengineering

Mirjana Pavlovic

Bioengineering

A Conceptual Approach

Springer

Mirjana Pavlovic
Department of Computer and Electrical Engineering
Florida Atlantic University
Boca Raton, FL, USA

ISBN 978-3-319-10797-4 ISBN 978-3-319-10798-1 (eBook)
DOI 10.1007/978-3-319-10798-1
Springer Cham Heidelberg New York Dordrecht London

Library of Congress Control Number: 2014949238

Printed on acid-free paper

Springer is part of Springer Science+Business Media (www.springer.com)

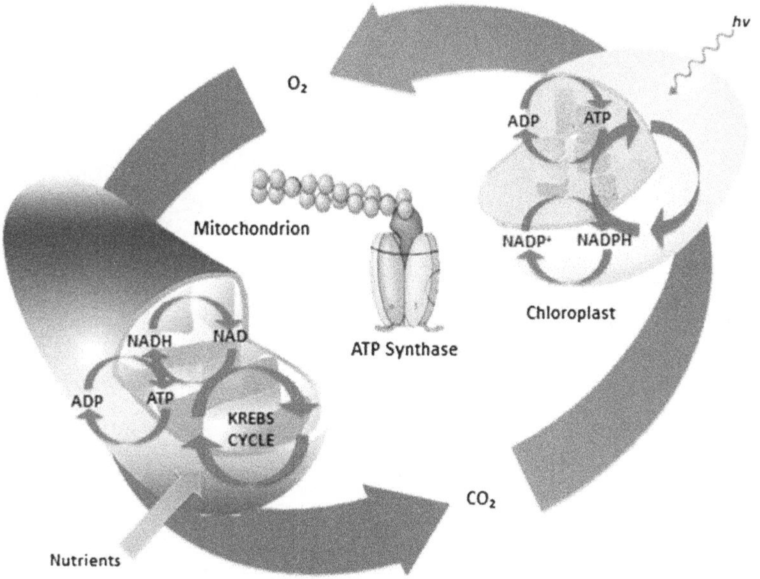

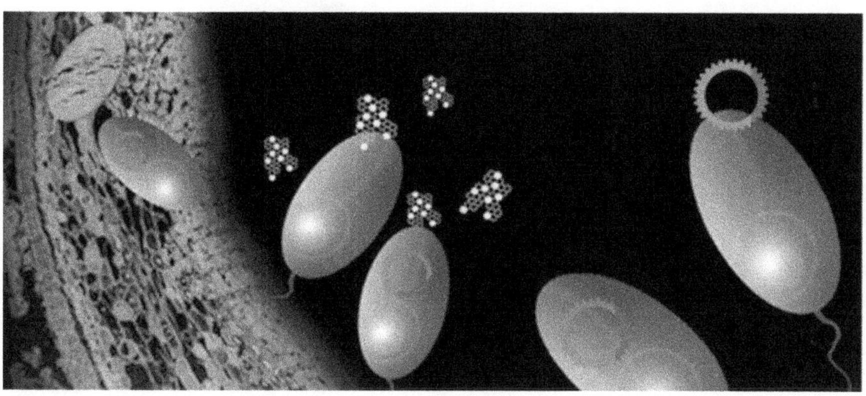

Illustrated by John Mayfield, undergraduate DIS student at FAU

This book is written in memory of the shadows of my parents who taught me that giving is the highest expression of power.

To MOM and DAD with love and unforgettable memories.

Thank You Note

This book is product of love and enthusiasm for the rapidly growing field of science which involves integration of different disciplines, something that I have sensed as a need at a very early stage of my road less travelled. In trying to develop the particular subjects/topics/courses at Florida Atlantic University (FAU) within a bioengineering group I have established significant and friendly relationships with a lot of people which I owe gratitude for this book design, and publication, and hopefully, its life in the future. Those are Dr. Zvi Roth, who has initiated the program and stood by me when it was the most difficult, Drs. Nurgun Erdol and Borko Furht, Chairmen and big fans of modernization and development of integrated programs, Dr. Maria Larrondo Petrie, with her encouraging, supportive, and warm friendship, Dr. Hanqi Zhuang who always believed in me, and most of my colleagues from Department of Computer Science and Electrical Engineering, at FAU. My graduate and DIS students and their passion for bioengineering, their work and research that they have done with me or other mentors, were also strong, supportive, inspiring, and driving forces during this long journey toward the light. Quite unexpectedly, a young man with infinite patience and talents, undergraduate DIS/research student, John Mayfield, was capable of following my thoughts and ideas giving his tremendous input in illustrating this fascinating field: a combination of nature and human work. He used some existing visualizations as models and guides for each of his visual elaborations. And finally, all of my friends and family members, especially my extremely constructively helpful brother, deserve to be mentioned within this list for encouraging me to get into this adventure. I do hope it will show up useful to those who the book is purposely written for.

John Mayfield

Abstract

The book reflects the critical principles and basic concepts in bioengineering. It integrates the biological, physical, and chemical laws and principles enlightening bioengineering as emerging, novel, complex approach with deep roots in the fundamental science. It is a concise review on the critical topics in this field including both: biological/medical and engineering aspects to it. It should be kept in mind yet, that the book is not bioengineering itself, but rather the introduction to this subject, with essential purpose to introduce those who do not have necessary background, to fundamental biological and physiological principles, that are significantly implicated in bioengineering. Therefore, the physical/chemical properties of cells, the natural design and function of tissues and organs, along with the main principles of molecules of life existence, composition, conformation, and interplay within different physiological scenarios are described and explained. They are used as the fundament for complex cellular and tissues/organs physiological functions such as function of heart, neuronal, skeletal muscle, and other cells and tissues: lungs, overall circulation, liver, gastrointestinal tract, and kidneys. The emerging concepts of nanotechnology, drug delivery, biomaterials, scaffolds, biomagnetism, and regenerative/cellular therapy are outlined, emphasized, and their status of development and progress is evaluated. Molecular aspects of life communication and molecular aspects of bioengineering as a fundamental approach in this field are interrelated and therefore compared in order to give an insight into fundamental, structural dimension of this approach and its brilliant natural or scientific solutions. The leading breakthrough personalities and events are mentioned where appropriate, and their impact on scientific development of this field, emphasized. The author has combined her own laboratory experience and data with those of others in order to give the book, both: monograph and scientific-book character. The book is written by Dr. Mirjana Pavlovic, M.D., Ph.D., who is teaching these subjects/courses for engineers and science students, and is highly recommended as a helpful tool along with any textbook.

Preface

Science is organized knowledge.

Herbert Spencer (1820–1903)

Biological engineering or bioengineering is the application of concepts and methods of biology to solve real-world problems related to the life sciences and/or the application thereof, using engineering's own analytical and synthetic methodologies and also its traditional sensitivity to the cost and practicality of the solution arrived at. In this context, while traditional engineering applies physical and mathematical sciences to analyze, design and manufacture inanimate tools, structures and processes, biological engineering uses primarily the rapidly developing body of knowledge known as molecular biology to study and advance applications of living organisms. In a word, biological engineering is based as well as classical engineering upon: chemistry, electricity, mechanics, magnetism and life science/medical principles.

What is the Difference Between Bioengineering and Biomedical Engineering?

Bioengineering: *biological engineering*, *biotechnological engineering*, or *bioengineering* (including *biological systems engineering*) is the application of concepts and methods of physics, chemistry, mathematics, and computer science to solve problems in life sciences, using engineering's own analytical and synthetic methodologies and also its traditional sensitivity to the cost and practicality of the solution(s) arrived at [1–2]. In this context, while traditional engineering applies physical and mathematical sciences to analyze, design, and manufacture inanimate tools, structures, and processes, biological engineering uses the same sciences, as well as the rapidly developing body of knowledge known as molecular biology to study many aspects of living organisms. Thus, biological engineering *is a science-based discipline founded upon the biological sciences in the same way that chemical engineering, electrical engineering, and mechanical engineering are based upon chemistry, electricity and magnetism, and classical mechanics, respectively* [3].

Biological engineering can be differentiated from its roots of pure biology or classical engineering in the following way. Biological studies often follow a *reductionist approach* in viewing a system on its smallest possible scale which naturally leads toward tools such as *functional genomics*. Engineering approaches, using classical design perspectives, are constructionist, building new devices, approaches, and technologies from component concepts. Biological engineering utilizes both kinds of methods in concert, relying on reductionist approaches to identify, understand, and organize the fundamental units which are then integrated to generate something new. In addition, *because it is an engineering discipline, biological engineering is fundamentally concerned with not just the basic science, but the practical application of the scientific knowledge to solve real-world problems in a cost-effective way.*

Although engineered biological systems have been used to manipulate information, construct materials, process chemicals, produce energy, provide food, and help maintain or enhance human health and our environment, our ability to quickly and reliably engineer biological systems that behave as expected is at present less well developed than our mastery over mechanical and electrical systems [1].

The differentiation between biological engineering and *overlap with biomedical engineering can* be unclear, as many universities now use the terms "bioengineering" and "biomedical engineering" interchangeably. However, according to Prof. Doug Laufenberg of *MIT*, biological engineering (like biotechnology) has a *broader base* which applies engineering principles to an enormous range of size and complexities of systems ranging from the molecular level—molecular biology, biochemistry, microbiology, pharmacology, protein chemistry, cytology, immunology, neurobiology, and neuroscience (often but not always using biological substances)—to cellular and tissue-based methods (including devices and sensors), whole macroscopic organisms (plants, animals), and up increasing length scales to whole ecosystems. Neither biological engineering nor biomedical engineering is wholly contained within the other, as there are *non-biological products for medical needs and biological products for nonmedical needs* [2].

ABET, the US-based accreditation board for engineering B.S. programs, makes a distinction between biomedical engineering and biological engineering; however, *the differences are quite small.* Biomedical engineers must have life science courses that include *human physiology* and have experience in performing measurements on living systems while biological engineers must have *life science courses* (*which may or may not include physiology*) and experience in making measurements not specifically on living systems. Foundational engineering courses are often the same, and include thermodynamics, fluid and mechanical dynamics, kinetics, electronics, and materials properties.

How Bioengineering Relates to Areas such as Stem Cell Research?

They are fundamentally interrelated, since stem cells are known to be the building blocks of entire organism, the "blank chips" with great potential to Trans-differentiate into different tissues, and so regenerate, repopulate, and recruit new cells in order to heal the process caused by the initial tissue damage [3]. Here we are in the tissue engineering area, the subarea of biomedical engineering, where stem cell application is still debatable in some respect, but the results of which are also encouraging. The great breakthrough is the discovery and use of adult stem cells, which can be found and taken out of the human body and used either for classical transplantation or tissue reparation when necessary. There is a considerable advance in Computer Aided Tissue Engineering (*CATE*), where the dimensions of tissue damage can be determined, and tissue samples designed by the use of stem cells and scaffolds (the supportive structures made from biocompatible biomaterial), which are enabling stem cells to differentiate and grow in accordance with original tissue architecture, leading toward complete and perfect reparation. It is also strengthen by *ink-jet printing system*, where the stem cell patterns are layered by dispensing them through notorious ink-jet cartridge [3]. Stem cells have the capability of self-renewal, expansion under hypoxic conditions, and multipotency-capacity to differentiate into many directions dependent on the conditions. There are even trials with cells of an old organ which behave like stem cells when introduced into damaged one. Stem cell researchers explain that those cells already know their environment and are well instructed; in fact they memorize how to arrange and to what extent to grow. This approach is developed by Dr. *Anthony Atala* and known as "*transplantation without a donor*." A great success of stem cell application is especially noticed in the disease known as *osteogenesis imperfecta*, where the bones in children are extremely fragile, and when applied in early stage of child development they can dramatically improve their future life. I am personally collaborating with two groups from Europe, and they have very good results with application of autologous adult stem cells in acute myocardial infarction and other ischemic diseases.

What are the Discipline's Main Subareas? Is it OK to Specialize in Only One of These Areas?

They are really numerous, and I think that each is equally important since either bioengineering or biomedical engineering has so many subdisciplines which are interrelated and it is difficult to make strict distinctions. In fact, the heart of these two disciplines is *integrative thinking* and as such, involves the ideas for the solutions that are coming from life scientists and engineers at the same time. The first such "crossing over" happened between *Alexander Fleming*, who has discovered Penicillin but did not have the possibility to expand its production, and *Howard Florey*, who was a pharmacologist (chemical engineer) and who invented technology for Penicillin production using Fleming's frozen samples [2]. Today, for example, for a good Rational Vaccine Design (RVD) you need the interaction of bioinformatician and immunologist in order to do it well. The first one will do the data mining and necessary mathematical transformations in order to find the best possible candidate for the vaccine, while another will lead the bioinformatician through the field of immunology known as vaccination and finally check it experimentally in the wet-lab. So, the hypothesis is tested and either confirmed or rejected. Yes, it is OK to specialize in only one of these areas if you understand that the teamwork is the *essential* request for successful bioengineering solution.

What are the Typical Jobs that Engineers Perform in Industry?

Biological engineers or *bioengineers* are engineers who use the principles of biology and the tools of engineering to create usable, tangible, economically viable products. Biological engineering employs knowledge and expertise from a number of pure and applied sciences, such as mass and heat transfer, kinetics, biocatalysts, biomechanics, bioinformatics, separation and purification processes, bioreactor design, surface science, fluid mechanics, thermodynamics, and polymer science. It is used in the design of medical devices, diagnostic equipment, biocompatible materials, renewable bioenergy, ecological engineering, and other areas that improve the living standards of societies. In general, biological engineers attempt to either mimic biological systems to create products or modify and control biological systems so that they can replace, augment, or sustain chemical and mechanical processes. Bioengineers can apply their expertise to other applications of engineering and biotechnology, including genetic modification of plants and microorganisms, bioprocess engineering, and biocatalysis.

Because other engineering disciplines also address living organisms (e.g., prosthetics in mechanical engineering), the term biological engineering can be applied more broadly to include agricultural engineering and biotechnology. In fact, many old agricultural engineering departments in universities over the world have rebranded themselves as *agricultural and biological engineering* or *agricultural and biosystems engineering*. Biological engineering is also called bioengineering by some colleges and biomedical engineering is called bioengineering by others, and is a rapidly developing field with fluid categorization. The main fields of bioengineering, and therefore, the typical jobs that they can find may be categorized as:

- *Bioprocess Engineering*: Bioprocess Design, Biocatalysis, Bioseparation, Bioinformatics, Bioenergy.
- *Genetic Engineering*: Synthetic Biology, Horizontal Gene Transfer.
- *Cellular Engineering*: Cell Engineering, Tissue Culture Engineering, Metabolic Engineering.
- *Biomedical Engineering*: Biomedical Technology, Biomedical Diagnostics, Biomedical Therapy, Biomechanics, Biomaterials.
- *Biomimetics*: The use of knowledge gained from evolved living systems to solve difficult design problems in artificial systems.

How is the Market for Fresh Graduates? What are the Typical Salaries?

This is developing field in a rapid expansion, so the market is open to fresh graduates, either at universities, hospitals, or industries. The typical salaries are: $45,000–$55,000 and within a year can reach even $60,000.

What are the Hot Research and Development Topics?

One of the greatest is growing organs from patient's own tissue. A very good example of that is the bladder. Clinical trial is going on to collect the data. Great "hit" is drug delivery through particular vectors, the surface of which has the molecules that bind to specific receptors on damaged tissues. In that way, drug delivery is targeted toward only damaged tissue (cancer, inflammation, etc.) and the medication affects only sick cells without touching normal ones. This enables precise dosage and individual targeted therapy. The bioinstrumentation has brought up also incredible solutions such as eradication of cancer cells by using golden nanoparticles in combination with laser technique. Gene therapy has raised the hope in treatment of hemophilia. Almost unbelievable, but true, the mouse eye is developed to the certain point in one experimental trial. The development of mouse micro-brain is one of the greatest challenges in the development of this field.

What are the Long-Term Challenges and Future Directions?

Since the very first use of stem cells in bioengineering, they have been used with hope that they can have anti-ageing and life-improvement effect. Is the longevity the ultimate goal? For those who really live in that hope I think that, as a human race with defined life we cannot live much longer than we do. But as long as we leave, we should have a good quality of life. And that for sure, will be better, and therefore also, somewhat longer. So, let us say that it is the ultimate goal and in my vision that is on its way to be achieved. It does not mean, of course, that stem cells are the answer to every question. Their use has also its disadvantages and limitations dependent on the scenario in question.

What are the Academic Prerequisites (Science and Math, Software Tools, etc.) and What is the Key Academic Bottlenecks En Route to Graduation?

In my experience, at least here, at FAU I have found that students with good understanding of basic sciences (math, chemistry, and physics) even without any biological experience can "conquer" biological knowledge to that extent that they feel very comfortable in becoming independent in their work. Especially if they are scientifically oriented and therefore, very resourceful, they can surprise you pleasantly with problem solving and creativity skills. Both are important for bioengineering and their own growth. My students were amazingly interested in what they were doing and therefore their knowledge was/is exceptionally solid.

What are Typical Topics for Senior Design Projects?

I would say: nanotechnology, rational vaccine design, gene therapy, stem cell application, bioinstrumentation, etc.

How Much of the Engineer's Work is Done at the "Systems Level" and How Much at the "Individual Device Level"?

It is really hard to say. I do believe that it goes in parallel, since both directions are challenging and necessary to be developed, and as long as we as humans are different, so there will be those who are interested in one and those who have an interest in another direction. In that sense, both directions will be and I think they are, developed with great enthusiasm and intellectual investment. An especially important application is the analysis and cost-effective solution of problems related to human health, but the field is much more general than that. For example, biomimetics is a branch of biological engineering which strives to understand how living organisms, as a result of the prolonged trial-and-error processes known as evolution, have solved difficult problems in the past, and to find ways to use this knowledge to solve similar problems in artificial systems [4]. On the other hand, systems biology seeks to utilize the engineer's familiarity with complex artificial systems, and perhaps the concepts used in "reverse engineering," to facilitate the difficult process of recognition of the structure, function, and precise method of operation of complex biological systems [1, 4, 5, and 6].

Boca Raton, FL Mirjana Pavlovic

References

1. Saltzman MW: Biomedical Engineering. (2009), Cambridge University Press, New York. 2009. ISBN: 978-0-521-840099-6 (hardback)
2. Pavlovic M (Ed), Balint B: Stem cells and Tissue engineering, Springer New York (2013). ISBN:978-1-4614-5505-9 (eBook)
3. Berger SA, Goldsmith W, Lewis ER. (Eds): Introduction to Bioengineering (1996), Oxford University Press, Oxford New York. ISBN:0-19-856516 (Hbk)
4. Vunjak-Novakovic G, Scadden DT: Biomimetic platforms for human stem cell research. (2011) Cell Stem Cell, 8:252–261.
5. Hall GE, Guyton AC, Textbook of Medical Physiology. (2011) Philadelphia, PA, Sounders Elesevier, ISBN:978-1-4160-4574-8
6. Pavlovic M, Balint B: Stem Cells and Tissue Engineering. (2013) Springer Briefs in Electrical and Computer Engineering, Springer NY, ISBN 978-1-4614-5504-2, pp. i–xii, 1–153

Contents

Chapter 1
Cell Content and Basic Construction

The complexity of the simplest known type of cell is so great that it is impossible to accept that such an object could have been thrown together suddenly by some kind of freakish, vastly improbable, event. Such an occurrence would be indistinguishable from a miracle.

Michael Denton (1943–)

This first chapter is mostly, informative. It is either recapitulation of your previous knowledge about cell basics or quite a new field for you, dependent on your education, so far. Having in mind that it is a developing field from bioengineering point of view, you should have your solid visual picture of the cell in your mind. Visualize whatever is possible, especially what is invisible by naked eye! We all know that if we don't see something it does not mean that "entity" does not exist, or will not emerge once we find the tools to detect it.

M. Pavlovic, *Bioengineering*, DOI 10.1007/978-3-319-10798-1_1

Cell Content and Basic Construction

Breakthrough from visible to invisible with primitive compound microscope: Zacharias Jansen (1565-1588), Anton de Leeuwenhoek (1632-1723) and Robert Hooke (1635-1703)

Introduction: Cell Compartmentalization

This book will help you with terminology and meaning of the terms, since it will facilitate development of a vocabulary after every chapter that you can very efficiently use to either fortify your knowledge or confirm and memorize it. But, do not only memorize. Always think about the questions beyond the scope and try to find solutions or look for them. Nobody knows answers to all questions—neither me. It is in human nature—to be limited and reach the individual plateau in final personal evolution. We are all different and our plateau levels are different. However, fear not, thrust yourself and go on! Go on with the questions. You will make a great move if you ask an intriguing question; you have a chance to change the world with answer. Fundamentally, maybe. And maybe that will be the question that you will want to answer within your research work. Don't be shy to ask for that possibility, since you might lose it if you don't. Do always what you really wish and like, since otherwise you will end up doing what somebody else want and you might not neither want nor like it. Try to avoid that personal catastrophe, since you live in the country of great opportunities.

We know that understanding of the life is tightly linked to the understanding of its cellular and molecular structures and their function, as well as genetic code in each species. The essential breakthroughs in development of biology are done by Charles Darwin and Gregor Mendel in nineteenth century. Darwin has proposed his **theory of evolution through natural selection and species adaptation** as an underlying mechanism for survival, while Mendel proposed the **concept of inheritance based on chromosomal interplay during cell division with mathematical precision**. Both have open the door for further consideration of chromosomal structure which with time escalated into DNA discovery and confirmation of its structure and determining its function in inheritance.

Breakthrough: Charles Darwin (1809-1882) and Gregor Mendel (1822-1884): Natural selection with adaptation as underlying mechanism of evolution, and principle of inheritance as interplay of chromosomes combined during fertilization (Mendelian rules of genetics)

Today, we know that people live longer than they did in the past. Overall life expectancy has increased from 50 to almost 80 (1900–2000).

1). The growth and expansion of biomedical engineering is a critical factor in this extension of life and improvement of health. So, what is the essence of terrestrial life? If you look into the most active organelle in living cells (animal and plant): mitochondria and chloroplast you will conclude that it is exchange of the matter (CO_2 produced by animal cells and O_2 produced by plant cells) and energy-light or ATP molecule (cell energy currency) synthesized in the cell.

In order to reach efficient solutions to the problems linked to life, biologist and bioengineer must work together. They do that through many projects in which biology, medical, physics; chemistry and mathematical knowledge are integrated. Although still difficult to define, bioengineering is revolutionary touching biological sciences in terms of focusing research toward very specific and precise outcomes [1–3]. Bioengineering captures a spectrum of different disciplines which all together function harmonically when needed to comfort the requirements (Biomolecular engineering, Biochemical engineering, Biotechnology, Nanotechnology, Biomaterials, Biomechanics, Bioinstrumentation, etc.). By definition, bioengineering is engineering that is applied to Life science, while biomedical engineering is focused specifically to human health [1]. Yet, the borderlines are not so strict due to one important thing: they are mostly based on cell structure and function, or physiology. Therefore, the basic knowledge of the cell is necessary whether you want to study one or another (Fig. 1.1).

The life starts with the cell. It is either unicellular organism *per se*, or the physiological unit of multicellular organism [3–10]. It is an open system, border-lined with membrane [5]. The exchange of matter and energy is taking place through. Cell function is based upon existence of cellular organelles (compartmentalized part of the cell). It starts with nucleus as the biggest and significant from reproductive (life maintaining) point of view, and after that many others including: cilia, flagella and microfilaments as the smallest ones. Cell membrane is the semipermeable membrane with proteins immersed into phospholipid's bilayer. It communicates through pores (mechanical, passive transport) and/or ion channels and different carriers, and receptors integral proteins (active transport). Non-compartmentalized part of the cell, in which the organelle are immersed, is called cytosol or cytoplasm.

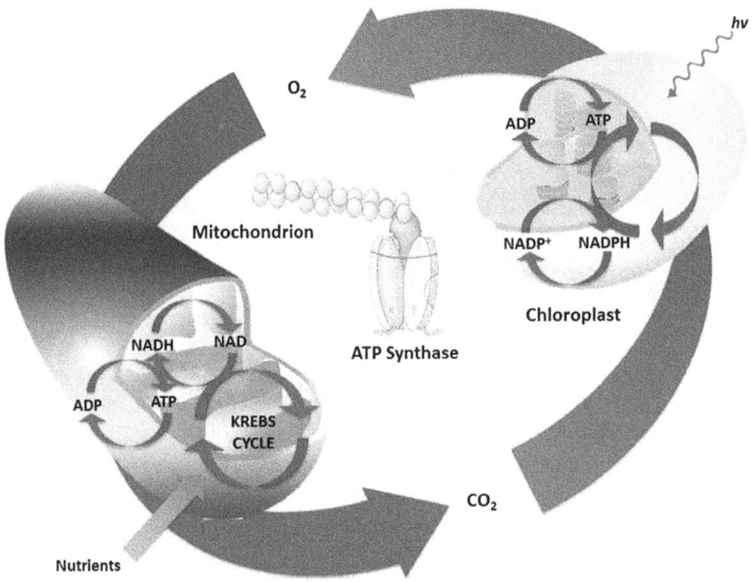

Fig. 1.1 The principle of the balance of the life: communication between plants and animals

Cell has many functions, dependent on its final differentiation stage: It is a solvent, since it contains water and salts. It is also storage of enzymes, the basis for signaling mechanisms, reproduction (nucleus), cell respiration and aerobic glycolysis (mitochondria), ATP synthesis (mitochondria and cytosol), protein synthesis (RER), protein package (SER), translation (ribosomes), package of proteins and biosynthesis of complex carbohydrate matter (Golgi apparatus), hydrolysis-degradation of other proteins (lysosomes), catalysis of the conversion of toxic hydrogen peroxide into water and oxygen (peroxisomes, etc.). Many cells and especially different cell types are involved in formations of tissues and organs. There are different tissues and organ systems in the body each of which is designed by nature to perform different function. Their work is orchestrated by control mechanisms in the body that keep the organism in the state of either dynamic equilibrium or homeostasis not allowing the body to get out of that energy state and molecular order. Complexity of organism, especially human, is possible thanks to stem cells which are during gametogenesis, embryogenesis and fetal period of life differentiating into distinctive, specialized tissues, and organ cells [3].

Biological systems can form populations, and communities. Group of living and non-living systems interacting on the same landscape is giving the ecosystem. A major geographical parts characterized by a particular type of flora and fauna is called **biome**. It is deeply influenced by **climate**.

An International conservation designation given by UNESCO is a **Biosphere** [3]. They are created to promote and demonstrate a balanced relationship between humans and the biosphere.

Hierarchy of life is again, designed by nature. We are learning about principles of design and usually precisely put that into laws articulated usually by mathematical equations. Each equation carries a certain message that can be translated into speaking language. It is not surprising at all, since we are aware that the cell is composed of atoms and complex biological molecules that have their physicochemical properties, which can be mathematically defined. Furthermore, the entire organism with its special organ's physiology is another aspect of the nature's design. That is overall physiology of the body which can also show to be founded on many of mathematical principles such as *Bernoulli's equation*, *Henderson–Hasselbalch's equation*, *Flick's Law*, *Bohr's effect*, *diffusion coefficient*, *osmotic pressure*, *oncotic pressure*, *filtration pressure*, *blood pressure*, *blood vessel resistance*, *viscosity*, *viscoelasticity*, etc. [6–10].

Bioengineering Aspects to Cell Compartmentalization

- The skills of engineer and life scientist are complementary. To convert the premises of molecular biology into new processes to make new products requires the INTEGRATION of these skills.
- And this INTEGRATION is the ultimate goal of bioengineering, by which the gap between two fields will be bridged.

References

1. Saltzman, M.W.: Biomedical Engineering. Cambridge University Press, New York (2009). ISBN 978-0-521-840099-6
2. Pavlovic, M., Balint, B.: Stem Cells and Tissue Engineering. Springer Briefs in Electrical and Computer Engineering. Springer, New York (2013), pp. i–xii, 1–153. ISBN 978-1-4614-5504-2
3. Berger, S.A., Goldsmith, W., Lewis, E.R. (eds.): Introduction to Bioengineering. Oxford University Press, Oxford (1996). ISBN 0-19-856516
4. Johnson, A.T.: Biology for Engineers, 1st edn. Taylor & Francis, London (2011). ISBN 0-19-856516
5. Andjus, R.K.: General physiology and biophysics. Modules 1–7. Center for Multidisciplinary Studies. University of Belgrade, Belgrade, Serbia (2002)
6. Domach, M.M.: Introduction to Biomedical Engineering. Prentice Hall, Upper Saddle River (2009). ISBN 10: 0-13-602003-8
7. Palsson, B.P., Bhatia, S.N.: Tissue Engineering. Prentice Hall, Englewood Cliffs (2003). ISBN 0130416967
8. Bronzino, J.D. (ed.): The Biomedical Engineering Handbook, vol. 1 & II, 2nd edn. CRC, Boca Raton (2000). ISBN 0-8493-0461-X
9. Rashidi, H.H., Buehler, L.: Bioinformatics Basics. Applications in Biological Science and Medicine. CRC, Boca Raton (2000). ISBN 0-8493-2375-4
10. Jones, N.C., Pevzner, P.A.: An Introduction to Bioinformatics Algorithms. Massachusetts Institute of Technology, Cambridge (2004). ISBN 0-262-10106-8

Chapter 2
The Advanced Architecture of the Cell

Learn from yesterday, live for today, hope for tomorrow. The important thing is to not stop questioning.

Albert Einstein (1879–1955)

This chapter will introduce you with the detailed cell construction: elements, molecules, forces and bonds between them, macromolecules and their functions in the cells as well as movable, working molecules that are maintaining cell energetic level, being capable of performing specific functions. This entire book is giving you the overall picture of organ-tissue functions starting from the skin (integumentary system) that wraps the body as the organ which is separating the body toward external environment as well as the internal organs the functions of which are interrelated. In order to understand it, it is necessary to understand how the nature has designed the construction of the cell, or what does cellular architecture look like?

M. Pavlovic, *Bioengineering*, DOI 10.1007/978-3-319-10798-1_2

The Advanced Architecture of the Cell

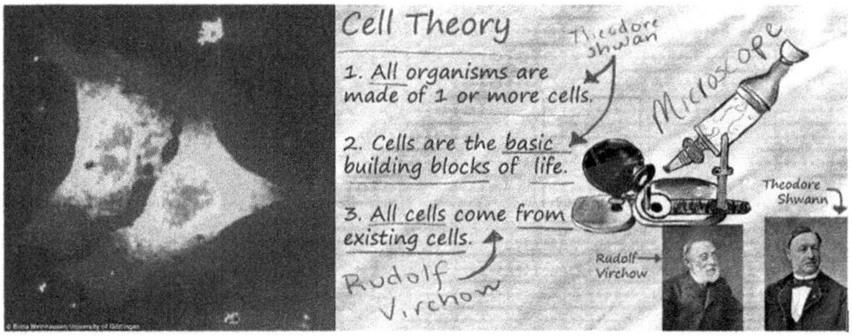

Recent breakthrough: Prof Sarah Köster with a team of German's researchers have managed to find a way of looking inside cells in their natural state and have produced the world's first X-ray of a single living cell (pictured)

Early breakthrough (19th century): Shwan, Schliden, and Virchow: The theory of the cell based upon magnifying glass and light microscope findings (invented by A. Leuwenhoek and R. Koch)

The structure of the cell gives definitely the chemical context to life, since cell is water–based solution with elements, molecules, and macromolecules dissolved within it. This immediately answers the question why bioengineer should have to understand chemistry, since if he wants to solve the problem by improving some function in the living system; he has to understand **chemical laws** and **processes**. This aspect of integral thinking is involved in designing new molecules for treating diseases, such as:

- Liposomes/Doxorubicin and other drug delivery systems
- Non-viral gene therapy
- Plaques removal from Alzheimer's disease (AD)
- Creating artificial devices
- Nuclear Magnetic Resonance (NMR) spectra, of important biomolecules, etc.

Examples of applied chemistry in bioengineering are numerous, and they are growing in number every day, especially polymers, for drug delivery systems as we shall see later on.

A matter consists of chemical elements either in pure form or in combination called compounds [1]. Compound is bigger and heavier than element. Therefore, chemistry is fundamental to understanding the life, since life is built up of the matter. What is the matter that makes life so specific? There are about **25** chemical elements essential to life among **92** known as naturally occurring [1]. They are organized in Periodic Chart of elements and designated by symbols or letters. Biologically, the most important are carbon, oxygen, hydrogen, and nitrogen from which **about 96 % of living matter** is built up. There are other biologically important elements such as (Ca, P, K, S, Na, Cl, and Mg) that make up remaining **4 %** of an organism's weight. So called trace elements also need to be present in very low concentrations but are of vital importance (Br, Cr, Co, Cu, F, I, Fe, Mn, Mo, Se, Si, Sn, V, and Zn). Elements can compose the compounds in a fixed ratio, and with different properties than the elements alone have (Nalco is different than either Na or Cal). Dependent on the types of bonds and the size of molecules, the compounds can be **micro** (water and salt bonds-inorganic matter) and/or **macromolecules** (C–C or CONH bonds-organic matter) [2].

The behavior of the element is determined by the structure of the atoms that are building the element. The atom is the smallest possible unit of matter that retains the physical and chemical properties of its element. Atoms are tiny particles, not visible by naked eyes. Atoms of the same element share similar chemical properties. Atoms are made up of subatomic particles, the three of which—the most stable are: neutrons (no charge, neutral), protons (positive charge), and electrons (negative charge). If an atom is electrically neutral, the number of protons equals the number of electrons, which yields an electrostatically balanced charge. They are further divided in smaller particles known as subatomic, or elementary particles. Today we know that hundreds of elementary particles have been discovered (neutrino, mesons, moons, positrons).

These elementary particles are made up of extremely small particles called quarks. But the quarks, even so small, have their own organization. According to Gel Man's system of symmetry, there are quarks and antiquarks, matter and antimatter (of the opposite charge). There are 4 different kinds of quarks which are = 2/3, –1/3, –1/3, = 2/3 that of the electron charge. The quarks combine to make different elementary particles. Each meson, for example, can be conceived as the union of quark and antiquark. Knowing these entities is necessary for example for development of nanotechnology-nanoparticles that can improve some function in the body or be of diagnostic or curative importance.

In atom, e.g. element, number of protons is constant while the number of neutrons can vary. All atoms of an element have the same atomic number (number of protons in an atom of particular element). In a neutral atom the number of protons is equal to number of electrons. The number of protons and neutrons in an atom is known as mass number. The mass of proton and the mass of neutron are both **about 1**. The atom of an element which has the same atomic number but different mass number is called **isotope** (same number of protons but different of neutrons). Some isotopes are radioactive. These are instable, with the spontaneously decaying nucleus, emitting subatomic particles and/or energy as **radioactivity**. The use of isotopes in biomedical sciences is of great importance for radioactive labeling of substances in many assay designs, determination of the age of fossils, etc. [1].

Electrons are also important from many aspects, especially valence electrons (electrons in the outermost energy shell-valence shell since they tend to fill incomplete valence shells by interacting with other atoms. This is the reason for creating chemical bonds—attractions that hold atoms together. Examples of bonds are: covalent, ionic, metallic, hydrogen, and van der Waals.

Coulomb Forces: A Simplified View of Bonding

The bonds between atoms hold a molecule together. But what causes bonding? Two atoms form a bond only if their interaction is energetically favorable, that is, if energy—heat, for example—is released when the bond is formed. Conversely, breaking that bond requires the input of the same amount of energy.

The two main causes of the energy release associated with bonding are based on Coulomb's law of electric charge:

1. Opposite charges attract each other (electrons are attracted to protons).
2. Like charges repel each other (electrons spread out in space).

Each atom consists of a nucleus, containing electrically neutral particles, or neutrons, and positively charged protons (Fig. 2.1). Surrounding the nucleus are negatively charged electrons, equal in number to the protons so that the net charge is zero. As two atoms approach each other, the positively charged nucleus of the first atom attracts the electrons of the second atom; similarly, the nucleus of the second atom attracts the electrons of the first atom. As a result, the nuclei are held together by the electrons located between them.

This sort of bonding is described by **Coulomb's law**: Opposite charges attract each other with a force inversely proportional to the square of the distance between the centers of the charges.

$$\text{Attracting force} = \text{constant} \bullet \frac{(+)\text{charge} \bullet (-)\text{charge}}{\text{distance}^2}$$

This attractive force causes energy to be released as the neutral atoms are brought together. This energy is called the **bond strength**.

When the atoms reach a certain closeness, no more energy is released. The distance between the two nuclei at this point is called the **bond length** (Fig. 2.1). Bringing the atoms closer together than this distance results in a sharp *increase* in energy. Why? As stated above, just as opposite charges attract, like charges repel.

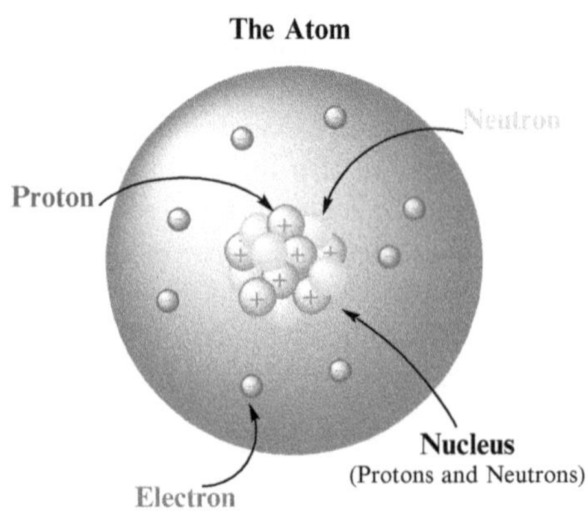

Fig. 2.1 The scholastic model of the atom: nucleus and shells

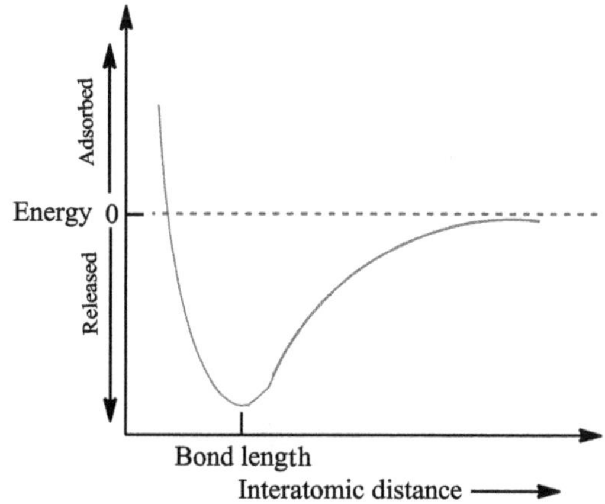

Fig. 2.2 The changes in energy, E, that result when two atoms are brought into close proximity. At the separation defined as bond length, maximum bonding is achieved

If the atoms are too close, the electron–electron and nuclear–nuclear repulsions become stronger than the attractive forces. When the nuclei are the appropriate bond length apart, the electrons are spread out around both nuclei, and attractive and repulsive forces balance for maximum bonding. The energy content of the two-atom system is then at a minimum, the most stable situation (Fig. 2.2).

Covalent bonds are chemical bonds between the atoms formed by sharing a pair of valence electrons (Fig. 2.3). They are strong and good example is H_2. In the molecule, the **nuclei are shielded** from each other by the **two** electrons. In the molecule there is an **electrostatically stable configuration** for the 2 negatively and 2 positively charged particles (the electrons and the protons).

An alternative to this type of bonding results from the complete transfer of an electron from one atom to the other. The result is 2 charged *ions*: 1 positively charged, a *cation*, and 1 negatively charged an *anion* (Fig. 2.4). Again, the bonding is based on coulombic attraction, this time between two ions (Fig. 2.4).

The coulombic bonding models of attracting and repelling charges shown in Figs. 2.4 and 2.5 are highly simplified views of the interactions that take place in the bonding of atoms. Nevertheless, even these simple models explain many of the properties of organic molecules.

Ionic bond formed between ions (positive-cations and negative anions) by the electrostatic attraction after a complete transfer of an electron from the donor atom to an acceptor (change of transfer). These bonds are strong in crystals but fragile in the water. Ionic compounds are called **SALTS** (Fig. 2.5).

The molecular polarity is determined by the position of polar and non-polar covalent bonds. Strength of these bonds is given in the table (expressed as energy (GA). The strongest is covalent and the weakest van der Waals forces.

Bonding of the atoms within a molecule (H–H) where the line represents a pair of shared electrons is known as **structural formula**. Formula which indicates the number and type of atoms, but does not reveals the structure is known as **molecular formula. Molecules** are building up of two or more atoms held together by covalent bonds. **Compounds** are the substances composed of two or more elements combined **in a fixed ratio and can have covalent or ionic bonds**.

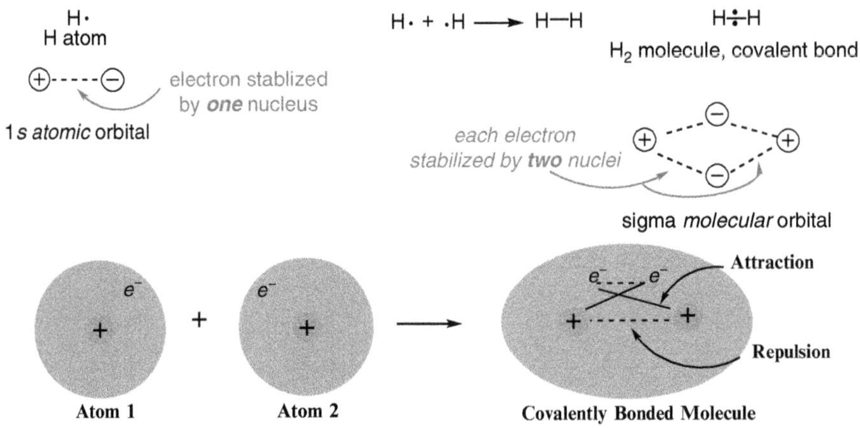

Fig. 2.3 Covalent bonding. Attractive (*solid-line*) and repulsive (*dashed-line*) forces in the bonding between two atoms. The large *spheres* represent areas in space in which the electrons are found around the nucleus. The small circled plus sign denotes the nucleus

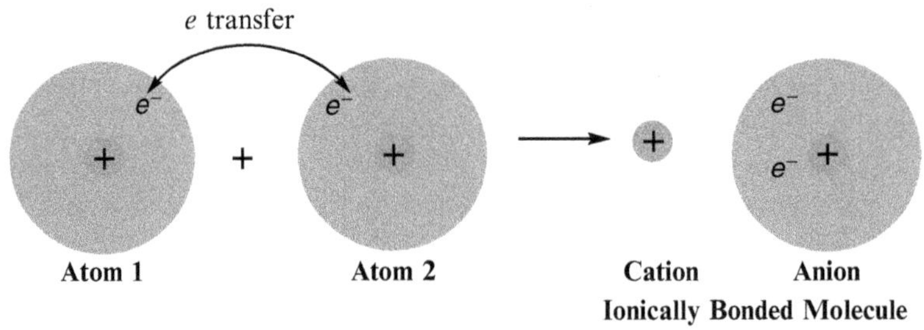

Fig. 2.4 Ionic bonding. An alternative mode of bonding results from the complete transfer of an electron from atom 1 to atom 2, thereby generating two ions whose opposite charges attract each other

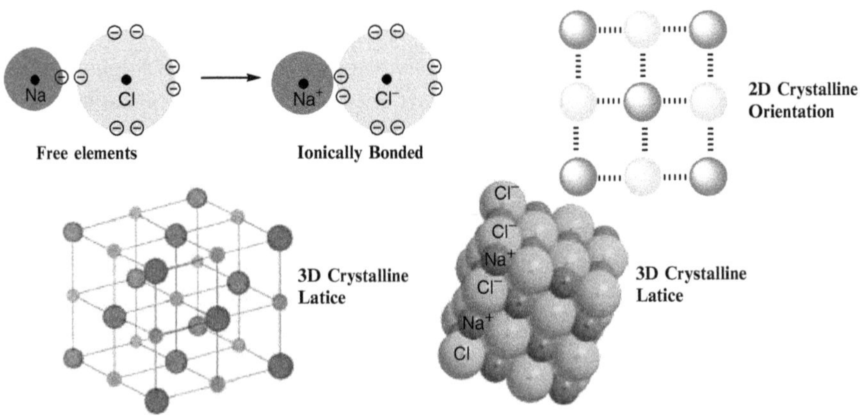

Fig. 2.5 Two different views of crystalline sodium chloride

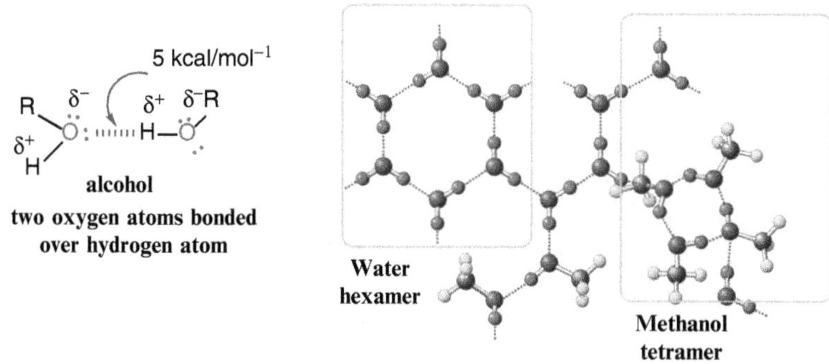

Fig. 2.6 Hydrogen bond between alcohol and water molecules

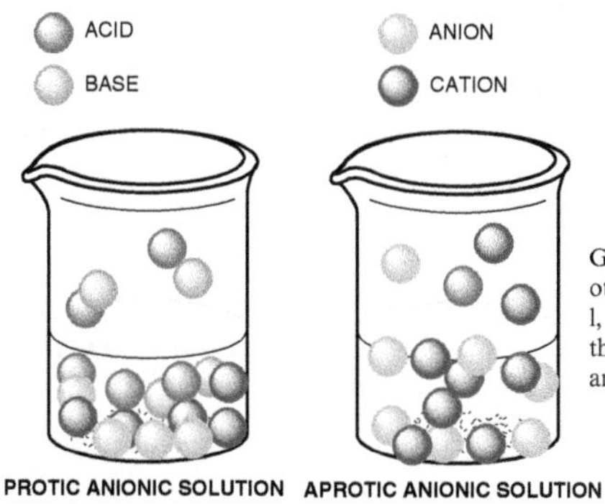

Green cicles represent anions, blue circles represent cations, other colored circles represent neutral molecules; l, liquid phase; g, gaseous phase. For the gaseous phase over the aprotic ionic liquid, the representation is purely schematic and has no implication for the actual degree of aggregation.

Fig. 2.7 Solutions: protic and aprotic. For the protic ionic liquids, a dynamic equilibrium exists between the ionic and dissociated forms: [BH]+X-(l) B (l)+HX (l) B (g)+HX (g)

Metallic bond: Metallic bonding is the type of bonding found in metallic elements. This is the electrostatic force of attraction between positively charged ions and delocalized outer electrons.

Hydrogen bond: formed by the charge attraction when a hydrogen atom covalently bonded to one electronegative atom is attracted to another electronegative atom. There is no **orbital overlap** as it is in covalent bond, so its strength is about ten time weaker than that of covalent or ionic bonds (Figs. 2.6 and 2.7).

However, they are very important in fixing properties such as:

- Solubility,
- Melting points
- Boiling points

in determining the form and stability of crystal structures. Therefore, they play a crucial role in biological systems.

van der Waals Forces

These are weak interactions that occur between atoms and molecules that are very close together and result from charge asymmetry in electron clouds [3]. These forces are responsible for the condensation of the gases into liquids, and the freezing of liquids into solid. Functional groups determine the type and strength of these interactions. There are several types of intermolecular interactions. Thus, ionic compounds contain oppositely charged particles held together by extremely strong electrostatic interactions. These ionic interactions are much stronger than the intermolecular forces present between covalent molecules (Fig. 2.8).

But, even though CH_4 has no net dipole, at any one instant its electron density may not be completely symmetrical, resulting in a temporary dipole. This can induce a temporary dipole in another molecule. The weak interaction of these temporary dipoles constitutes van der Waals forces (Fig. 2.9).

All compounds exhibit van der Waals forces. The surface area of a molecule determines the strength of the van der Waals interactions between molecules. The larger the surface area, the larger the attractive force between two molecules, and the stronger the intermolecular forces (Fig. 2.10).

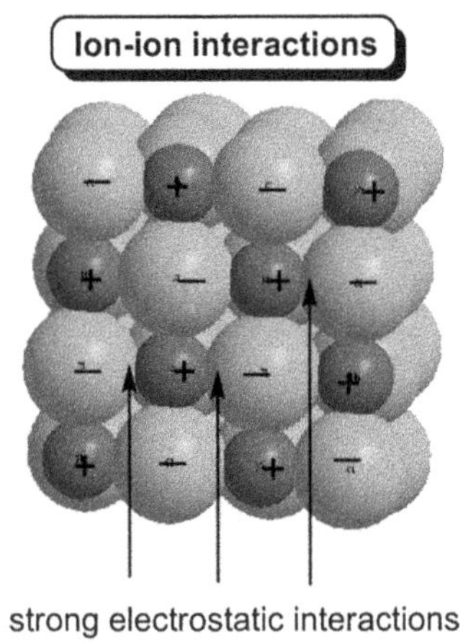

Fig. 2.8 Strong ion–ion electrostatic interactions in crystallic form of sodium chloride

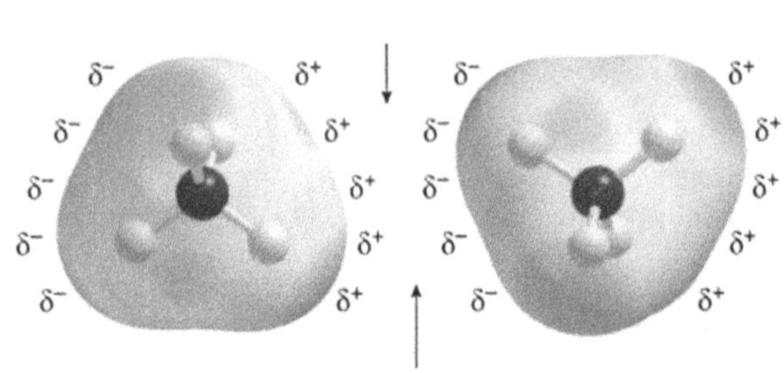

Fig. 2.9 van der Waals interactions between two CH_4 molecules. Unsymmetrical electron density creates a temporary dipole

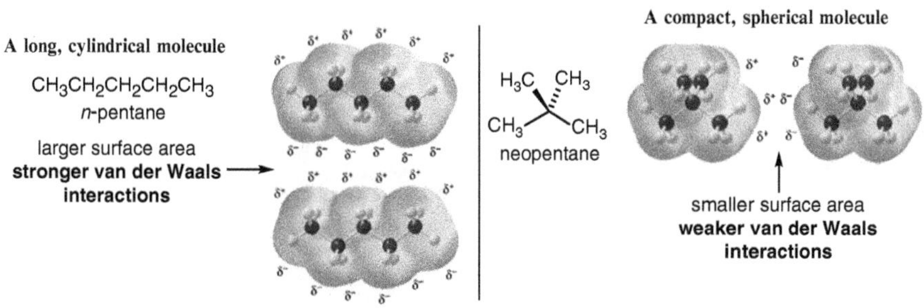

Fig. 2.10 Surface area and van der Waals forces

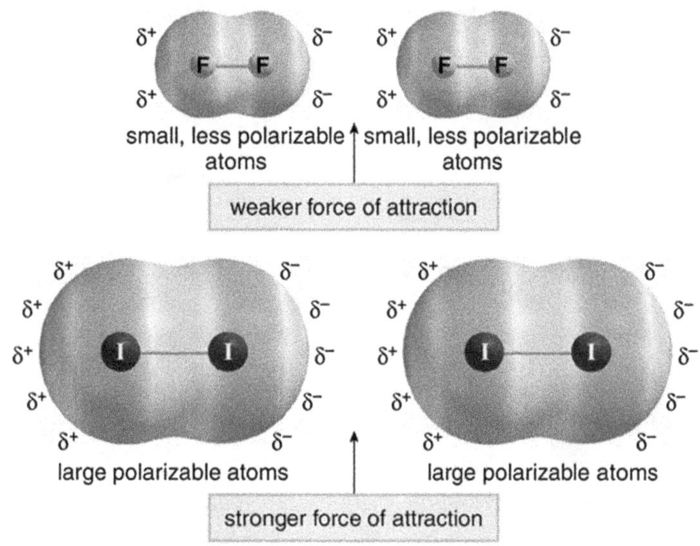

Fig. 2.11 Weaker and stronger forces of attraction affected by polarizability between smaller (fluorine) and larger atoms (iodine)

van der Waals forces are also affected by **polarizability**. Polarizability is a measure of how the electron cloud around an atom responds to changes in its electronic environment. Thus, larger atoms, like iodine, which have more loosely held valence electrons, are more polarizable than smaller atoms like fluorine, which have more tightly held electrons. Thus, two F_2 molecules have little attractive force between them since the electrons are tightly held and temporary dipoles are difficult to induce (Fig. 2.11).

A molecule's biological function is related to the shape. The chemical reactions make and break chemical bonds (Fig. 2.12).

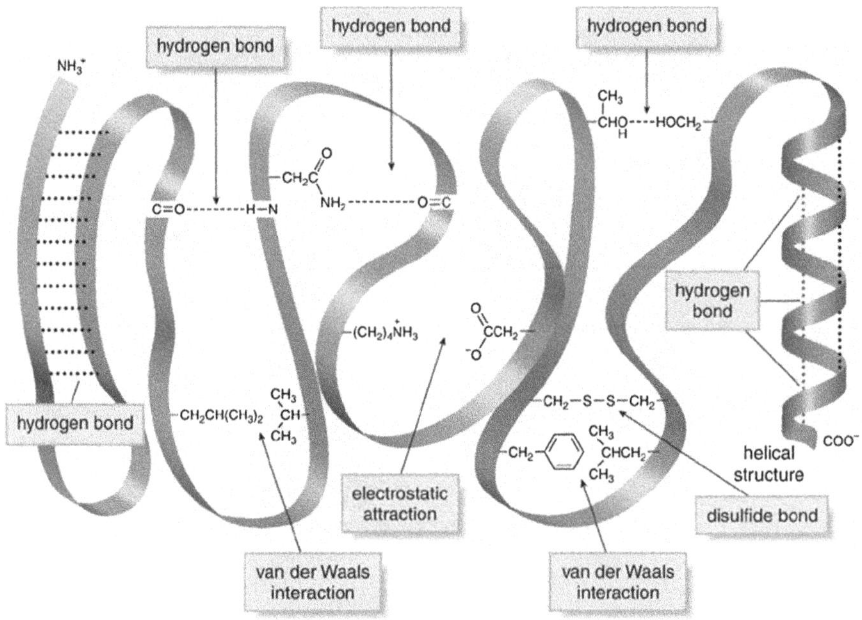

Fig. 2.12 The different stabilizing interactions in secondary and tertiary protein structure

Cell Energy, Kinetics, Electrolytic Dissociation and Acid–Base Equilibrium

Cell activities as well as body activities require expenditure of energy. Human gains energy trough the food that they eat. This energy is stored or expended to sustain life. All of the chemical reactions in our body result in utilization or accumulation of the energy [4]. It is important to separate the possibility of reaction occurring from the **rate** at which the reaction will proceed. These concepts are related, but distinct. The role of **enzymes** (proteins which are specialized to serve as biocatalizators, to speed up the chemical reactions in the cells), is essential.

The energy currency of the cells is ATP, organic compound which occurs as an intermedier in metabolism and thanks to three phosphorus groups the last of which has the weakest covalent bond, can be hydrolyzed to ADP and Pi releasing energy needed for cellular processes. Energy is always released from chemical bonds and required for their formation. From thermodynamic point of view, the overall heat of formation (enthalpy) is a measure of the order, the amount of energy that is either consumed or released when for example the water is formed, and is called ΔHf (25 °C and 1 atm) = ΔH of formation. Heats of formation can be used to calculate the enthalpy changes of other reactions. Negative indicates exothermic (released E) and positive-(consumed E) endothermic reaction. The entropy of the system is the measure of disorder in the system or the amount of energy in the system that cannot be used to work. For any change in the state of the system, a change in entropy or ΔH can be calculated.

Gibbs free energy (**G**) is related to both entropy (**S**) and enthalpy (**H**). It is actually a measure of the potential energy of the system, which is a function of enthalpy and entropy.

$$\Delta \mathbf{G} = \Delta \mathbf{H} - \mathbf{T} \Delta \mathbf{S}$$

The value of ΔG can be used to predict whether a reaction is favorable/spontaneous under the given conditions. Most biological reactions have a positive ΔG, so they do not occur spontaneously. How do the unfavorable reactions proceed? They require an INPUT of energy, which most often come from the braking of a high-energy phosphate bond found in as special biochemical called ATP. ATP hydrolysis is energetically favorable while ATP synthesis is not, but it occurs thanks to respiratory chain of events and proton-motive force or respiratory chains created during cell respiration. Three essential events are occurring on the inner mitochondrial membrane during mitochondrial respiration:

1. Separation of charges (negative inside—electrons—and positive outside—protons)
2. Synthesis of endogenous water
3. Synthesis of ATP from ADP and Pi on ATP-ases/synthases of the inner mitochondrial membrane (Fig. 2.13).

During cell metabolism, the pH (negative logarithm of concentration of hydrogen ions) is changing toward basic (higher than 7 which is neutral) or acid level (lower then 7). Acids are molecules that release protons (H^+) when added to aqueous solutions, while bases are the molecules that release hydroxyl ions (OH^-) when added to acid solution. Deviation of a patient's blood pH from its normal value (which is 7.4, near neutral) is always a sign of serious illness. Variations of pH are found either in different part of the body or within cellular compartments due to their different metabolic activities based upon enzymatic content. The equilibrium constant for the reaction is expressed as:

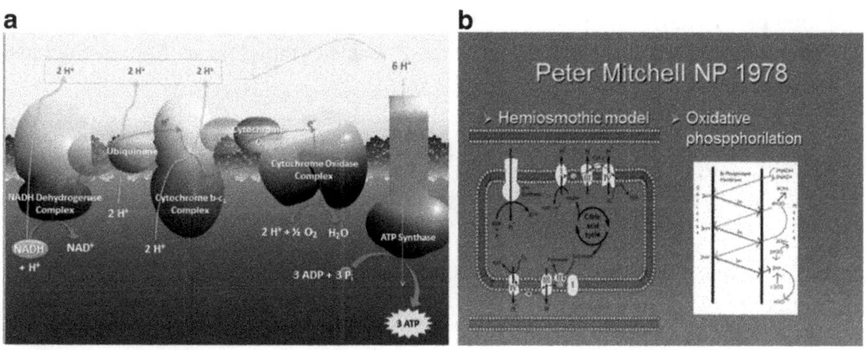

Fig. 2.13 (**a**, **b**) Proton-motive force of the respiratory chain and essential events on the inner mitochondrial membrane during its formation

The Henderson-Hasselbalch Equation

$$pH = pK_a + \log\frac{[A^-]}{[HA]}$$

- *Lawrence Joseph Henderson* (1878–1942) was a talented biochemist, among many other titles, who spent most of his career at Harvard. He was responsible for developing the components of the equation after studying equilibrium reactions that took place within blood as a result of respiration (specializing in "fatigue"). His equation was incomplete without a solid calculations going into it.
- *Karl Albert Hasselbalch* (1874–1962) was a chemist who studied pH closely. He also studied blood and reactions that took place with oxygen, to put in the simplest of terms. He eventually modified Henderson's equation by putting mathematical logs into it creating a solid relationship. The Henderson-Hasselbalch equation can be used to prepare buffer solutions and to estimate charges on ionizable species in solution, such as amino acid side chains in proteins. Caution must be exercised in using this equation because pH is sensitive to changes in temperature and salt concentration in the solution being prepared.

Dissociation constant: the equilibrium constant for the decomposition of a complex ion into its components in solution. The smaller the value of K, the lesser the dissociation of the species in solution. This value varies with temperature, ionic strength, and the nature of the solvent.

Buffers

1. Buffer solutions consist of either: a weak acid and salt of its conjugate base (e.g. HF and NaF) or: a weak base and the salt of its conjugate acid (e.g. NH_3 and NH_4Cl). Buffer solutions are resistant to pH change despite small additions of acid or base. Buffer systems are very important in living systems (e.g. constant blood pH is vital).
2. When H^+ is added to a buffered solution, it reacts completely with the weak base present:

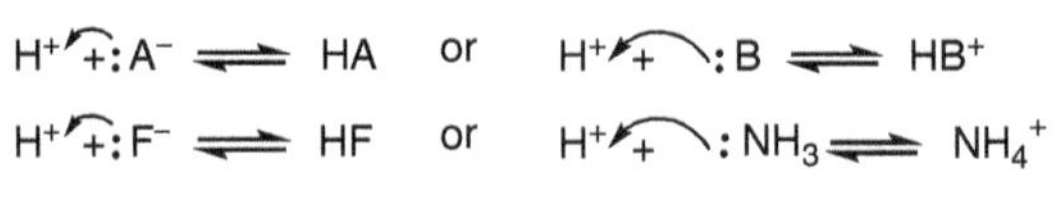

3. When OH^- is added to a buffered solution, it reacts completely with the weak acid present:

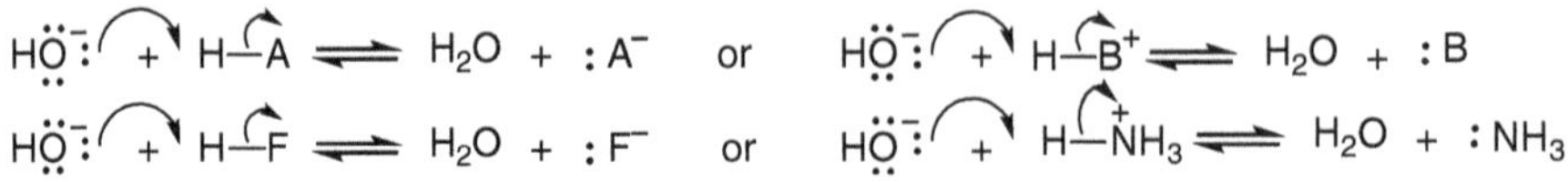

4. Steps (2) and (3) are stoichiometry problems: In step (2), H^+ is completely consumed, leaving excess A^- (or B). In step (3), OH^- is completely consumed, leaving excess HA (or BH^+). You must determine which species remain and how much of each remains in solution. Once $[A^-]$ and [HA] are calculated, the pH of solution can be calculated from the Henderson-Hasselbalch equation:

$$\mathrm{pH} = \mathrm{p}K_a + \log\frac{[\mathrm{A}^-]}{[\mathrm{HA}]}$$

5. The pH of a buffered solution will be determined by the ratio $[A^-]$/ [HA], (or [B]/ $[BH^+]$). As long as this ratio remains constant, the pH remains constant. This will be case if **[HA] and $[A^-]$, (or [B] and $[BH^+]$) are large relative to $[H^+]$ and $[OH^-]$**. Optimum buffering occurs when $[A^-]=[HA]$. In this case, the ratio $[A^-]$/ [HA] is most resistant to pH change when H^+ or OH^- is added.

$$\mathrm{pH} = \mathrm{p}K_a + \log\frac{[\mathrm{A}^-]}{[\mathrm{HA}]} \quad \text{If} \left[\mathrm{A}^-\right]=\left[\mathrm{HA}\right] \text{ then } \frac{[\mathrm{A}^-]}{[\mathrm{HA}]}=1 \text{ and } \mathrm{pH} = \mathrm{p}K_a$$

The pK_a of the weak acid selected for the buffer should be as close as possible to the desired pH.

Macromolecules of Life

Those are the key classes of molecules that are constructed to LINKING small molecules. These are mostly members of a more general class of chemicals called polymers which are large molecules formed by bonding of many smaller chemicals, called monomers, into one long molecule. Because of their large size they are called macromolecules [5].

1. **Nucleotides**
   The monomer of nucleic acid polymers is called a nucleotide. They are composed of pentose, inorganic phosphate and organic base. Dependent on pentose (ribose or deoxyribose) and composition of the bases, they will make DNA (deoxyribose) or RNA (ribose and Uracil instead of Tymine).
2. **Nucleotide basis** (Cytosine, Guanine, Thymine, and Uracil)
3. **Nucleic acids** (DNA and RNA)
   DNA is double-stranded in a form of double-helix; the process of two single strands of DNA assembling into double-stranded DNA is called **hybridization**. Not every pair of DNA strands can form a double helix. Hybridization can only occur if the two strands have **complementary sequence**. Complementary strands are "mirror images": each strand contains the same information (although the strands are not identical) but they are mirror images prepared in special way.

First, the complementarity strand is pointed into different direction: if one strand is arranged phosphate to pentose, phosphate to pentose facing upward, then the complementary strand is perfectly **predictable**. The basis on the complementary strand match a particular pattern: A goes with T, C goes with G, G does with C, and T goes with A. These matches—often called **base pairings**—are determined by **hydrogen bonding interactions** between the nucleotides. It is the hydrogen bonding of complementary base-pair matching that holds the two DNA strands together in a stable double helix.

4. **Proteins and how they are made**?
   Proteins are produced by chemical reactions that are directed by DNA. One of the main functions of DNA in our cells is to provide the **information blueprint** for synthesis of the proteins that our cell will need. They are synthesized according to **central dogma of molecular biology defined by Watson and Crick**: replication, translation, transcription: the language of the bases sequence in DNA during replication is transcribed by mRNA and translated into the amino-acid sequence of the protein synthesized on ribosomes in the cytosol. During translation, the information on mRNA is translated, through series of reactions, into a linear sequence of amino acids that will become a protein. They can be constitutive proteins of the cell, carriers, messengers, enzymes, antibodies, enzymes.
5. **Carbohydrates**:
   Major source in human diet. According to size: monosaccharides (ribose and deoxyribose) disaccharides (sucrose) and polysaccharides (starch and cellulose). $(CH_2O)n$, where *n* is the number of carbon atoms in the molecule, although there are the exceptions to this rule.
6. **Lipids** are not polymers, but fairly large molecules built from a combination of other simple units.
   Triglycerides, phospholipids, and steroids: hydrophobic, hydrophilic, and amphyphillic.
7. **Natural and Synthetic Polymers**
   Large molecules composed of multiple identical or similar units (monomers) linked by covalent bonds. DNA and RNA are natural polymers of nucleotides. Polypeptide is natural linear organic polymer consisted of a large number of amino acid residues bonded together through peptide (CONH) bond into a chain. Polysaccharide is a biological macromolecule composed of monosaccharide subunits. **Polymerization** is the chemical process of making of a polymer from a collection of monomers. Polyvinylchloride is synthetic organic polymer used in biomedical purposes.

Emphasizing Bioengineering Aspects to Advanced Architecture of the Cell

The German scientists used an X-ray beam to scan the internal nanostructure of the cells, but only blasted them for 0.05 s at a time to avoid damaging the living cells too quickly [6]. Their method produced images so clear that nanometer-scale

a

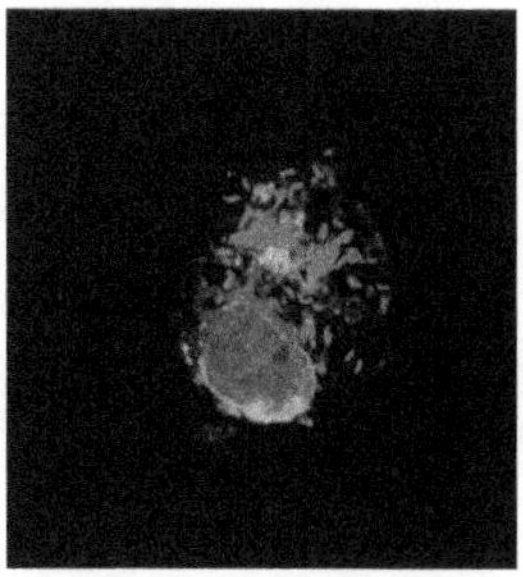

b

Cell Line	(6R)DDATHF	Aminopterin (pmol/h/mg prot.)	Tetrahydrofolate
CCRF-CEM	750	1280	595
CCRF-CEM R16	352	447	276

Fig. 2.14 (**a**) Fluorescent labeling of COX enzyme activity (*green*) during macrophage phagocytosis of Bacillus Chalmette Gerene (author's unpublished data). (**b**) Activity of enzyme FPGS for different folate and antifolate substrates expressed as pmol/h/mg of protein-published [7]

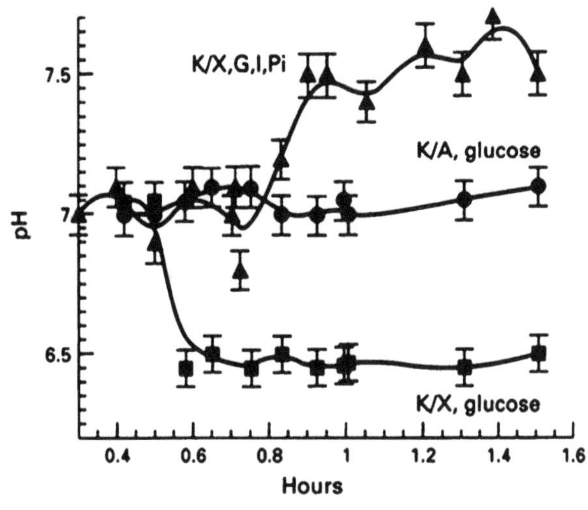

Fig. 2.15 The effect of ketamine/xylazine and ketamine/acepromazine combinations on 9 L glioma pH after injection of glucose as measured by ^{31}P NMR. The effect of insulin on 9 L glioma pH following i.p. injections of ketamine/xylazine and glucose/Pi. Each variable represents data accumulated from three separate tumour-bearing rats [8]

structures are visible. The researchers studied living and chemically fixed cells using the 'nan diffraction technique' and when they compared the images of the cells, the new X-rays prove that the chemical fixing process makes big changes to tiny 30–50 nm structures in the cell [6]. While the scientists have not speculated on what the new technique could mean for medical and scientific research, it will make it possible for experts to study living cells at high resolution and understand a living cell' inner mechanics better.

A very new bioengineering aspect emerges from that quite new level, which can detect living molecules in their natural location and movement.

The author's work on detection of enzyme activity at fluorescence level and biochemical level are presented in Figs. 2.14a, b [7]. On the other hand, the author's experience in "wondering throughout the cell" and looking for pH changes in transplanted neural tumor (glioma) under the skin of rat in trying to reach pH changes that will increase tumor's radiosensitivity, measured, monitored and detected using NMR, are presented with result published in British Journal of Cancer in 1996 [8] (Fig. 2.15).

References

1. Brown, L., Holme, T.: Chemistry for Engineering Students. Mary Finch, Books/Cole, Cengage Learning, Belmont (2011). ISBN 13:078-1-4390-4791-0
2. Johnson, A.T.: Biology for Engineers, 1st edn. Taylor & Francis, London (2011). ISBN 13: 978-1420077636
3. Andjus, R.K.: General Physiology and Biophysics. Modules 1–7. Center for Multidisciplinary Studies, University of Belgrade, Belgrade (2002). ISBN 86-80109-11-8
4. Hall, G.E., Guyton, A.C.: Textbook of Medical Physiology. Sounders Elesevier, Philadelphia (2011). ISBN 978-1-4160-4574-8
5. Shuler, M.L., Karg, F.: Bioprocess Engineering Basic Concepts, 2nd edn. Prentice Hall, PTR, Upper Saddle River (2002). ISBN 0-13-081908-5
6. http://www.dailymail.co.uk/sciencetech/article-2570300/The-X-ray-living-cell-Breakthrough-lets-scientists-study-cancer-minute-detail. html#ixzz2upd14slJ
7. Pavlovic, M., Leffer, J., Russello, O., Bunni, M.A., Beardsley, G.P., Priest, D.G., Pizzorno, G.: Altered transport of folic acid and antifolates through the carrier mediated reduced folate transport system in a human leukemia cell line resistant to (6R) 5,10-Dideazatetrahydrofolic acid (DDATHF). In: Chemistry and Biology of Pteridines and Folates. Advances in experimental medicine and biology, vol. 338, pp. 775–778. Springer, New York (1993)
8. Pavlovic, M., Wroblewski, K., Manevich, Y., Kim, S., Biaglow, J.E.: The importance of choice of anaesthetics in studying radiation effects in the 9L rat glioma. Br J Cancer **74**(Suppl. XXVII), S222–S225 (1996)

Chapter 3
Cell Physiology: Liaison Between Structure and Function

Nothing in life is to be feared, it is only to be understood.
Now is the time to understand more, so that we may fear less.

Marie Curie (1867–1934)

Cell is the basic functional unit in the body. There are about 200 different (specialized) types of cells in human body, although each is genetically the same. Yet, they are different in size, shape and function due to the fact that not all the genes and not the same set of genes are functional or being used in each particular cell type (gene selectivity). Despite this diversity of cell composition and function, most cells in the body have the same structural organization. There are between 50 and 200 trillion of cells in the body of an average person (estimated) and they are constantly being dividing, metabolizing, working, dying and being replaced by integrated mechanisms. Therefore, structure, morphology, and function are tightly coupled in the cell giving to each specific cellular entity unique and distinguishable features.

Yuri Ovchinnikov (1934–1988), Peter Mitchell (1920–1992) and Paul Boyer (1918–) breakthrough in transport and energy storage across the cell and inner mitochondrial membrane (ATP and GTP formation and energy storage and release; ionophores)

M. Pavlovic, *Bioengineering*, DOI 10.1007/978-3-319-10798-1_3

Cell Physiology: Structure and Function

Cell Structure and Function

Physiology is defined as the science that treats of the functions of the living organism and its parts, and of the physical and chemical factors and processes involved [1–5]. With respect to that. There are two essential, distinguishable groups of cells that appear either as unicellular or multicellular organisms.

In **prokaryotic cells** (typically small, **about 1 μ or more**), which lack the membrane separated nucleus (but not DNA content needed for replication), cytoskeleton and cytoplasmic organelles, there is a **rigid cell wall**, maintaining their shape [3, 4]. In bacteria, *peptidoglycan*, a polysaccharide that cross links and add to cell stability is present in the wall. Gram-positive bacteria have predominantly a peptidoglycan wall, while Gram-negative bacteria have two walls: a thinner, *inner wall*, containing peptidoglycan and *an outer lipopolysaccharide* layer. Despite small size these organisms are biochemically diverse, with a rapid doubling time. Biomedical engineers often use them for production of **recombinant proteins**.

Eukaryotic cells (fungi, algae, protozoa, plants and animals) are typically larger, **about 10 μ and more**; have a more complex structure with the plasma membrane consisting of lipid bilayer, separating the intracellular from extracellular space. Although with a membrane boundary, this is still an *open system* since it communicates with external world either through pores or different forms of *active and passive transport* of molecules and ions [4–7] (Fig. 3.1).

Plant cells have walls to give them structure. Animal cell membrane has elastic and fluid properties. Around the cell is extracellular matrix (**ECM**), produced by

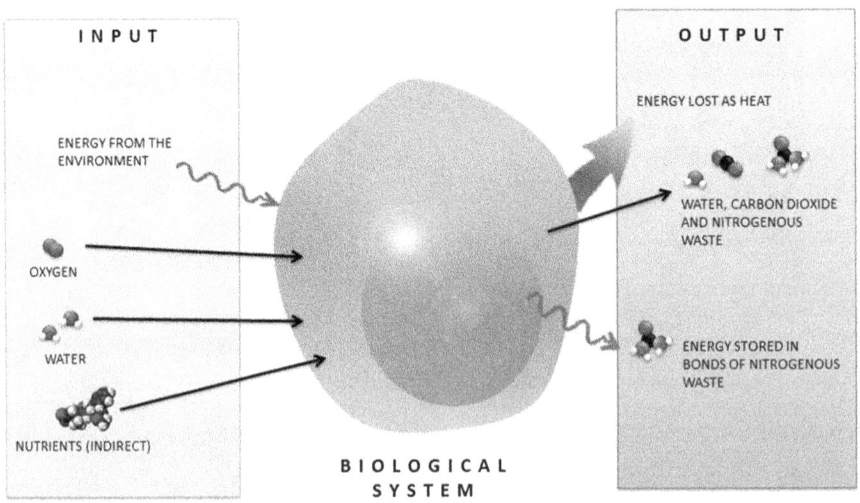

Fig. 3.1 Biological systems and energy exchange

cells, which holds them together and allows them to form tissues [3, 7, 8]. Specialized structural molecules are secreted locally by cells and assembled to form **a scaffold** that supports cell attachment, spreading, proliferation, migration, and differentiation. Cells influence the chemistry of their surrounding ECM by direct secretion of molecules, but they also modify the physical characteristics of the matrix locally by releasing of the enzymes, which can digest or stabilize the gel matrix, or by application of physical forces, can physically rearrange the gel components [6–12]. The composition and organization of macromolecules of the ECM helps determine the tissue structure and physical properties (examples: the soft cartilage in nose and ear, the **basal lamina sheet** underlying epithelial cells (and secreted by epithelial cells) in the intestine, and tendons, which attach muscles to the bone. Bone contains a mineral-rich ECM. Proteoglycan, collagen and elastin are the special *structural molecules* of ECM, while *special adhesives* are: fibronectin and laminin [3, 11]. Collagen provides strength, while elastin provides elasticity. Fibronectin serves as a cross-linker between collagen and GAGs, while laminin also contains binding sites for cell attachment and promotion of neurite growth [11, 12].

Cell membrane and basic functions: Cell membrane is a **semipermeable**, lipid bilayer with proteins immersed into it having very different functions [3, 4]. This plasma membrane restricts the movement of the water into and out of the cell. The plasma membrane is described as *a fluid mosaic* ("model of fluid mosaic") composed of lipids, carbohydrates, and proteins. In lipid bilayer, amphyphylic phospholipids are arranged with their hydrophobic tails pointing toward the interior of the membrane and their hydrophilic heads exposed to adjacent water phases, serve as the main elements of the membrane.

Biomedical engineers have also used lipids to construct devices (for instance: lyposomes—as drug carriers).

Transport across the cell membrane might capture quite a few mechanisms [3, 5]:

- ***Diffusion*** is the spontaneous movement of particles from an area of high concentration to the area of low concentration.
- ***Passive transport*** (***diffusion***): is the process by which water and small uncharged molecules, such as oxygen (O_2) and carbon dioxide (CO_2) pass through the plasma membrane.
- ***Active transport*** (***facilitated diffusion***): process of moving a molecule from an area of low concentration on one side of the membrane to the area of high concentration on the other (against the concentration gradient).
- ***Osmosis***: The diffusion of water through semi-permeable membrane. It is a physical process in which a solvent moves, without input of energy, across a semi-permeable membrane (permeable to the solvent).
- **Solvent**: A solvent is a liquid, solid, or gas that dissolves another solid, liquid, or gaseous solute, resulting in a more complex solution [2]. The most common solvent in everyday life is water. Most other commonly-used solvents are organic chemicals. These are called organic solvents, but not the solute, separating two solutions of different concentrations.

- **Solutions**: are the mixtures of solutes and solvents in different proportions giving different final concentrations in chemical systems. In a physiological, cellular systems the solutions can be classified with respect to normotonic solution of the cell (isotonic, hypotonic, and hypertonic) solutions. There are a couple laws that can be applied to the solutions in the cell, such as:
- ***Fick's first Low of Diffusion***:

$$Jx = D\,dc/dx$$

where Jx is the rate of diffusion, D is difference in concentration, and dc/dx is a membrane thickness.

- ***Van't Hoff's equation for osmotic pressure***:

$$\Pi = \boldsymbol{RT\Sigma c_i}$$

where the osmotic pressure, Π, in dilute solutions is $\Pi = RT\Sigma c_i$, R is the universal gas constant, T is the absolute temperature, and c_i is the molar concentration of solute i.

Many molecules do not diffuse through lipid bilayers. There are accessory molecules (constitutive molecules) in membrane bilayer that regulate the transport of the molecules that do not pass freely through the lipid bilayer [4].

Facilitated Transport via Transporters

Glucose transport protein shuttles glucose molecule through the hydrophobic membrane bilayer which is impermeable for glucose. This is possible through conformational changes of the protein stimulated by binding of glucose to the protein. Facilitated transport protein as glucose transporter is present in the cell membrane in limited number. Both facilitated transport and simple diffusion depend on the concentration gradient; net solute transport always occurs from high to low concentration [4, 5].

Active transport systems are similar to facilitated transport systems; both involve the participation of trans-membrane proteins that bind a specific solute. In primary active transport systems, however THE ENERGY IS PROVIDED (most often by the hydrolysis of ATP) to drive a conformational change in the transporter that leads to solute transport. (Na^+/K^+ pump or ATP-ase activity is moving both anions "up" their gradients). Na^+ is extracellular and K^+ intracellular cation [4, 5] (Fig. 3.2).

Secondary active transport systems also move solutes "up" their concentration gradients, but they gain the energy from the different source (co-transporters and exchangers) (Fig. 3.3).

These conditions establish membrane potential, which a the systems start working create action potential, characteristic with Na ions entering the cell and K ions out the cell. High Na influx will cause the spike, which will after repolarization establish the resting potential again. Action potentials are very important in excitable tissues such as heart and brain [4, 5].

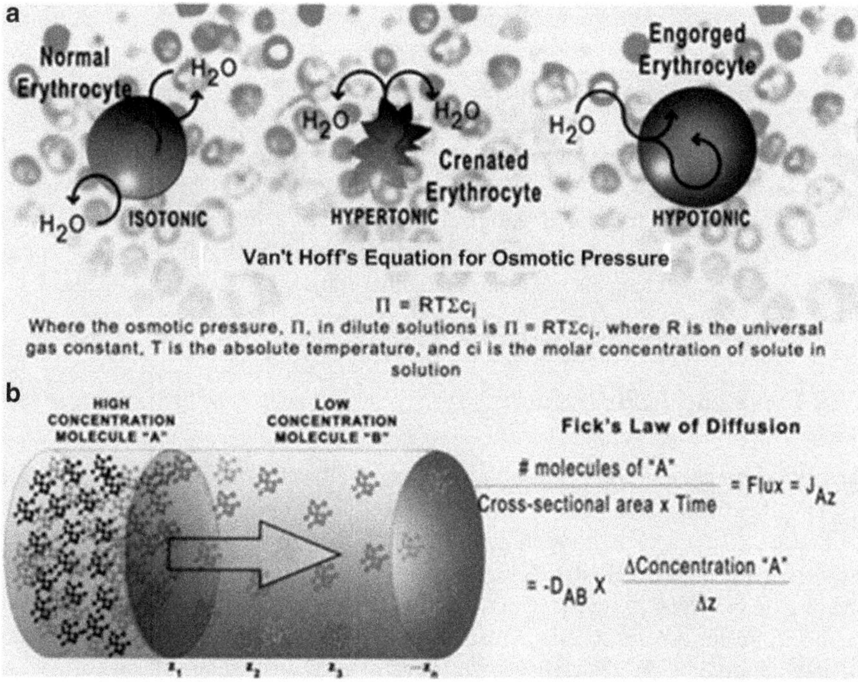

Fig. 3.2 (**a**) and (**b**), Osmotic pressure (**a**) and Fick's Law of diffusion (**b**). Ion transport, membrane and action potentials (excitable tissues)

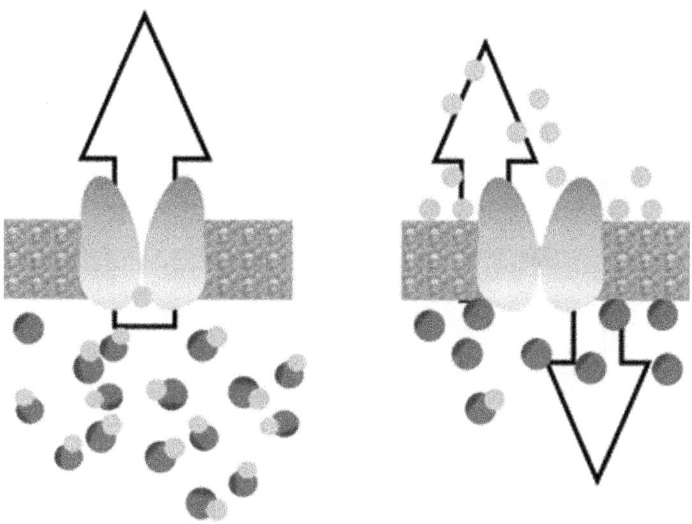

Fig. 3.3 Facilitated ion transport via transporters through mitochondrial membrane

Membrane and Action Potential

Cells and microorganisms exquisitely sensitive to changes in their ionic environment, making ion movement across membranes profoundly important in physiology. Ion movement across the membranes is regulated by specialize membrane proteins called ion channels [4, 5]. They are selective in many aspects to the size, charge, binding tip protein surface and stabilization of the non-hydrated ions. Some channels are voltage regulated (gated); others are regulated by binding ligands to the membrane [4]. Ion channels are faster than other transport systems [3–5] (Fig. 3.4).

Cell Cycle and Cell Division

The fundamental purpose of life and fundamental cell activity is **reproduction** [5–7]. All cells go through division and death. Cells divide in an orderly sequence of events. *Cell cycle* is continuously repeated daughter cell progression which begins in the phase called **G1**, because it represents the gap between Mitosis (M phase) and DNA synthesis (S phase). Cells can exit the cell cycle and remain in an indefinite period of rest called **Go**; neurons are suspendered in Go for the lifetime of an individual [4, 5]. Thus, they are referred to as non-dividing cells [4].

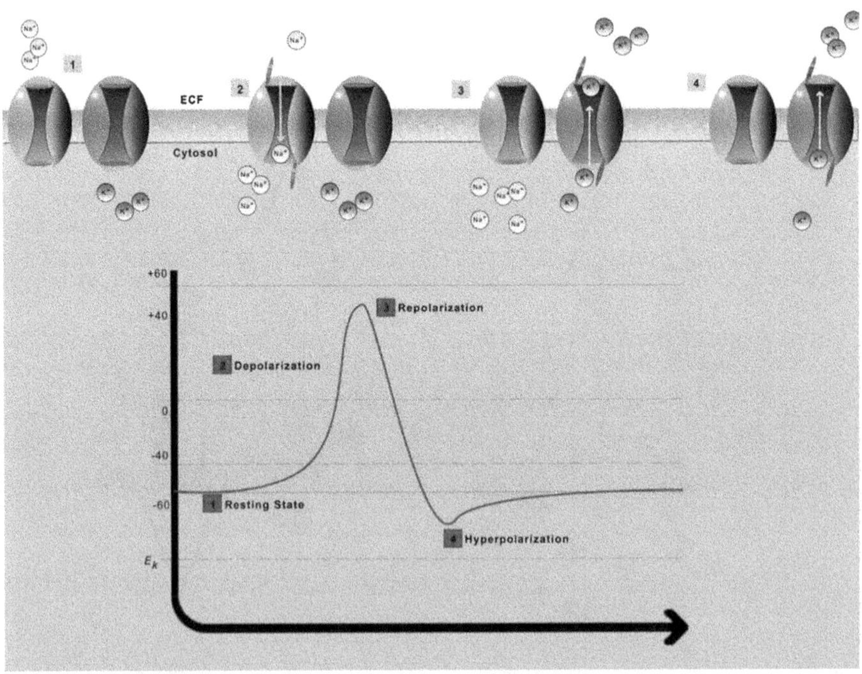

Fig. 3.4 Processes of depolarization and repolarization over cell membrane in excitable tissues

Types of Cell Division: Meiosis/Mitosis

Difference between meiosis and mitosis is that meiosis takes place in gametes (*germ-line cells*) and it is a division with reduction of chromosome number in both mother's and father's cells, while mitosis occurs in tissues other than gametes (*somatic*) and during mitosis, the cell's genome is replicated and the cells split once giving two new identical cells [13–17]. However, meiosis involves one genome replication and two cells are splitting, so that four haploid cells are produced from one parent cell. It occurs in two main stages: Meisosis I and Meisosis II. Chromosomes exist in pairs. Each partner in the pair has the basic shape, structure, and contains complimentary genes. There are 23 pairs or 46 chromosomes in humans. A **homolog** refers to the corresponding partner in a pair [13–17].

Importance of Mitosis

1. Each **homolog** carries information about the same hereditary trait, though the information may vary: dominant, recessive (blond, brown,) etc.
2. A single pair of homologous chromosomes may carry information for several thousand hereditary traits.
3. A female has 23 pairs of homologous chromosomes. This includes 22 pairs of **autosomes**, and 1 pair of **sex chromosomes (XX)** (**sex chromosomes identical**).
4. A male has 22 pairs of homologous chromosome (autosomes), and 1 pair of non-homologous chromosomes—**sex chromosomes** (**XY**) (**sex chromosomes non-identical**).
5. During period of time between cell divisions, DNA is in the form of **chromatin**. When chromatin condenses, chromosomes become visible. Chromosome is actually 10,000 times longer than when coiled and condensed. Condensing is necessary to fit in nucleus and separate in cell division.
6. Chromosome structure: two **chromatids** (sister chromatids) attached by **centromere**.

Mitosos is important for repair of somatic cells of individual young or adult organism, since every moment cells are naturally dying in the body and need to be replaced by new ones with the same function. It is process by which the nucleus of the cell is divided into two nuclei, each with the same number and kinds of chromosomes as parent cell. Cytokinesis is process by which the cytoplasm divides, forming two distinct cells.

Importance of Meiosis

1. Meiosis is significant for development of a new organism. Development starts with a **zygote** with a *2n* of 46. Development is due to mitosis and differentiation.

2. The zygote begins with one member of each pair of chromosomes from the father and one member of each pair from the mother.
3. **Meiosis** assures that the genetic code is passed on to next generation. It assures that the gametes are prepared with the ***n*** number of chromosomes (23) so that sexual reproduction and fertilization will result in the ***2n*** number (46) having in each pair of chromosomes one father's and one mother's alel.

Meiosis is division of gametes-not somatic cells and it also provides for variation due to a reshuffling of the genes—genetic recombination (chiasma)—crosssing over of two non-sister chromosomes. In that way, genes can be exchanged and variation of the gametes is increased. By pairing haploid number of chromosomes (half from mother and half from father), the zygotis growing into embryo and a new individual with all the cells at one place in the – → blastocyst and later on, in uterus.

There is an **Interphase** before meiosis–similar to mitosis. It means, that preceding meiosis chromosomes are replicated. Result is two sister **chromatids** attached by a **centromere. Centrioles** also get reproduced.

A. **Meiosis I**—Reduction division

- **Prophase I**: Chromosomes condense, **synapsis**—pairing of homologous pairs, **chiasma** and **crossing over**, **tetrads** appear, centriole pairs separate, spindle microtubules for between them, chromosomes begin to migrate to equator (metaphase plate). Longest stages—90 % of meiosis.
- **Metaphase I**: Homologous pairs (unlike single chromosomes in mitosis) line up on metaphase plate. **Spindle fibers** from poles attach to a single chromosome of the pair.
- **Anaphase I**: **Spindle apparatus** moves chromosomes towards poles. Chromatids remain attached. Homologous pairs are pulled apart and towards opposite poles.
- **Telophase I**: Chomosomes reach poles. Each pole has monoploid (haploid number). **Cytokineses** occurs and two daughter cells are formed. Result of meiosis I is two daughter cells with *n* number of chromosomes (Fig. 3.5).

B. **Meiosis II**: A mitotic-like process

- **Prophase II**: Note–there is little or no interphase before prophase II. A spindle apparatus forms. Chromosomes are pulled to **metaphase plate**.
- **Metaphase II**: Similar to **mitosis**. Individual chromosomes line up on equator.
- **AnaphaseII**: Centromeres split and chromatids are pulled apart, becoming separates chromosomes and move towards opposite poles.
- **Telophase II**: Nuclei form and cytokinesis occurs. Result is now four daughter cells with *n* number.

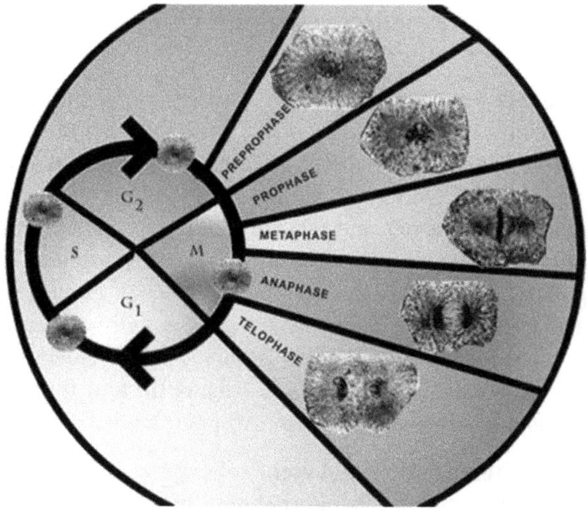

Fig. 3.5 Mitosis/mitotic-like process

Meiosis in Males: Spermatogenesis

Spermatogenesis occurs in the seminiferous tubules within the testes, and takes about 74 h: 300 million are produce per day. Once ejaculated they live about 48 h within female reproductive tract. It begins at the onset of **puberty** and continues throughout male's life. **Spermatagonia** in seminiferous tubules of testes become **primary spermatocytes**, which enter meiosis.

At the end of **Meiosis** I there are two **secondary spermatocytes**. Telophase of Meiosis II results in four **spermatids** of equal size and cellular material. A **flagellum** (**tail**) forms and a **protein cap** which contains enzymes necessary to penetrate the coat of the **ovum**. The top of the head is called the **acrosomal cap** which contains the enzymes. The result is a mature **spermatozoon**.

Male Puberty

At onset of puberty (sexual maturity) between ages 11–14 the **anterior pituitary** starts to secrete the gonadotropic hormones—**FSH** and **LH**. Release is controlled by **hypothalamus. FSH** acts on seminiferous tubules to initiate spermatogenesis. **LH** assists **seminiferous tubules** to develop mature sperm. **Testosterone** (**androgen**) is produced from **cholesterol** or Acetyl coenzyme A.

Meiosis in Females: Oogenia/Oogenesis

Oogenia begins in the embryo as early as the 3rd month. At birth female is born with most or all of primary oocytes—100,000 or more. **Primary oocyte** continue to develop at onset of puberty. Several primary oocytes continue each month in **menstrual cycle**. Only one survives per month to become ovum. Cells in **ovaries** that give rise to oocytes and ova are called **oogonia**. After 3rd month oogonia develop into **primary oocytes**. These cells enter **prophase I** but do not continue until puberty. **FSH** stimulates continuation. Result of meiosis I is one **secondary oocyte** and the **1st polar body**, which is essentially discarded nuclear material. It may develop further to **two polar bodies**. After **Meiosis II** the secondary oocytes produces two cells of unequal size—the **ootid and the second polar body**. The ootid eventually becomes the **ovum**. All polar bodies disintegrate and die.

Cell differentiation and stem cells: process through which the cell is becoming specialized. Stem cells are building blocks for tissues and organs [8]. The main characteristics are: pluripotency, self-renewal, transdifferentiation, and plasticity. Therefore, they are used in regenerative therapy [8–12] (Fig. 3.6).

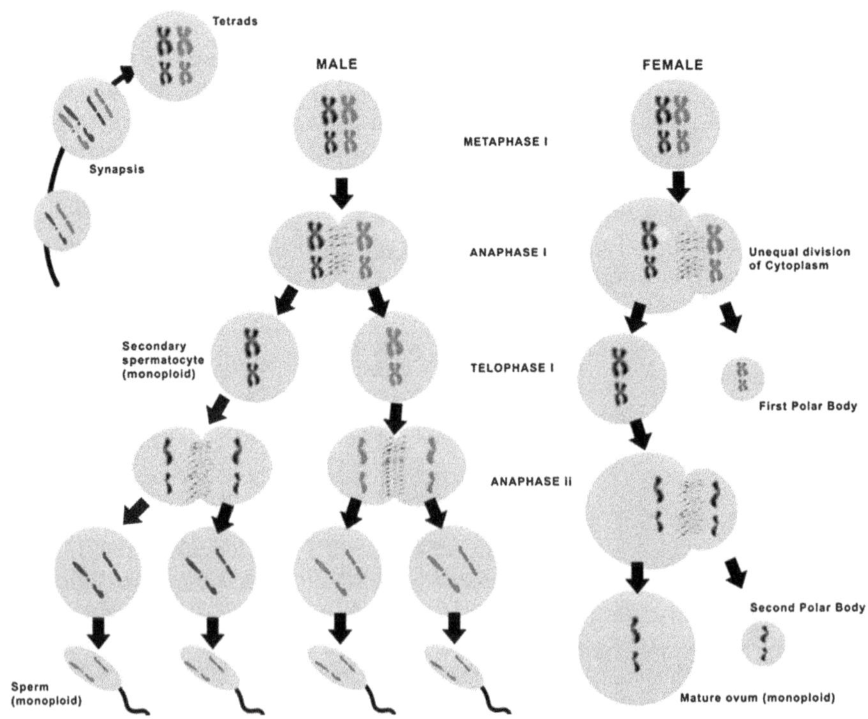

Fig. 3.6 Mitosis in male and female

Cell death: Apoptosis, autophagy, and necrosis. Cell death can occur through three mechanisms: **apoptosis, autophagy**, and **necrosis.** *Apoptosis*, or programmed cell death, results in controlled cell shrinkage and nuclear fragmentation via the action of caspases, as well as an anti-inflammatory cytokine release. In contrast, *necrosis* signals via RIPK1 signal molecule (RIP1), leading to cell swelling, lyses, and a pro-inflammatory cytokine release. *Autophagy* destroys the cell's damaged proteins and organelles via an intracellular catabolic process in the lysosome. Multiple physiological processes require the removal of specific cells by a controlled cell-death program. For example, tissue remodeling activates apoptosis, whereas energy metabolism and growth regulation responses rely on autophagy. Developmental processes often activate apoptosis, while bodily injuries or infection more commonly induce *necrosis*. The molecular mechanisms behind these cell death pathways overlap, and can be co-activated during some cellular functions. **Apoptosis** and **necrosis** both signal through the *death domain receptors* FAS, TNFRSF1A (TNFR1), and TNFRSF10A (TRAIL-R), while **autophagy** and **apoptosis** share BCL_2 family members as key players.

Emphasizing Bioengineering aspects to Cell Physiology with an accent on the knowledge of cell division, new micro-techniques, and bioinformatics approach

Knowledge of the cell mitotic and meiotic cycles is necessary from especially two points of view and two main breakthroughs in engineering the very specific cell division areas:

1. **In vitro fertilization and** (Fig. 3.7)
2. **Therapeutic cloning** (Fig. 3.8)

Furthermore, many companies (QIAGEN, for example) provide a **broad range of assay technologies for cell death research** that enable analysis of gene expression and regulation, epigenetic modification, genotyping, and signal transduction pathway activation. Solutions optimized for cell death studies include PCR array, miRNA, siRNA, mutation analysis, pathway reporter, chromatin IP, DNA methylation, and protein expression products [13–20].

Feature selection techniques have become an apparent need in many bioinformatics applications. In addition to the large pool of techniques that have already

Fig. 3.7 In vitro fertilization

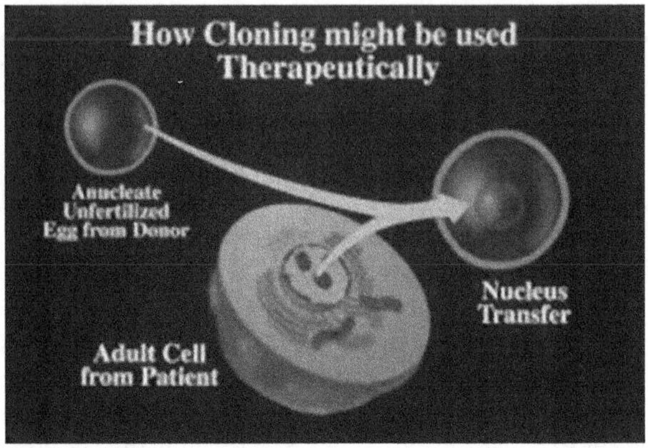

Fig. 3.8 Therapeutic cloning

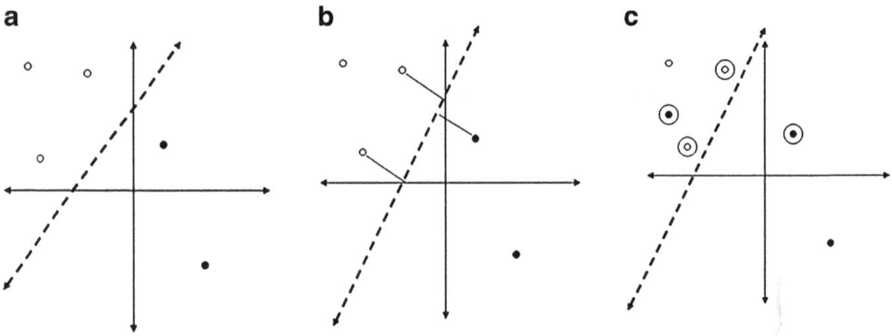

Fig. 3.9 Support Vector Machine: (**a**) separating hyperplane (**b**) maximum margin hyperplane (**c**) soft margin (from author's collaborative study, published [20]

been developed in the machine learning and data mining fields, specific applications in bioinformatics have led to a wealth of newly proposed techniques [20]. In a recent article, we make the interested reader aware of the possibilities of feature selection, providing a basic taxonomy of feature selection techniques, and discussing their use, variety and potential in a number of both common as well as upcoming bioinformatics applications. Thus, using *bioinformatics tools* we have realized that MicroRNAs (miRNAs) may serve as **diagnostic** and **predictive** biomarkers for cancer, rapidly and uncontrolled dividing cell conglomerate. A number of biomarkers for tissue-specific cancer have been confirmed by our techniques [20]. In addition, seven miRNAs have been newly identified by our methodology as possible important biomarkers for hepatocellular carcinoma or breast cancer, pending wet lab confirmation. In this paper [20], these biomarkers were identified from miRNA expression data sets by combining **multiple feature selection techniques** (i.e., creating an ensemble), and then classified by different learners. In general, creating a subset of features by selecting only the highest ranking features (miRNAs) improved upon results generated when using all the miRNAs, and the ensemble approach out performed individual feature selection methods [20]. Furthermore, a subset of features created by using a Support Vector Machine wrapper was found to generally outperform the ensemble subset.

References

1. O'Toole, M.T.: Miller-Keane Encyclopedia and Dictionary of Medicine, Nursing, and Allied Health. Saunders Elsevier, Philadelphia (2003)
2. Brown, L., Holme, T.: Chemistry for Engineering Students. Mary Finch, Books/Cole, Cengage Learning, Belmont (2011). ISBN 13:078-1-4390-4791-0
3. Johnson, A.T.: Biology for engineers. Taylor & Francis, Boca Raton (2011). ISBN 978-1420077636
4. Andjus, R.K.: General physiology and biophysics [in Serbian]. Modules 1–7. Center for multidisciplinary studies. University of Belgrade, Belgrade, Serbia (2002)
5. Hall, G.E., Guyton, A.C.: Textbook of medical physiology. Sounders Elsevier, Philadelphia (2011). ISBN 978-1-4160-4574-8
6. Shuler, M.L., Kargi, F.: Bioprocess engineering: basic concepts, 2nd edn. Prentice Hall, PTR, Upper Saddle River (2002). ISBN 0-13-081908-5
7. Saltzman, M.W.: Biomedical engineering. Cambridge University Press, New York (2009). ISBN 978-0-521-840099-6
8. Pavlovic, M., Balint, B.: Stem Cells and Tissue Engineering. Springer Briefs in Electrical and Computer Engineering. Springer, New York (2013), pp. i–xii, 1–153. ISBN 978-1-4614-5504-2
9. Berger, S.A., Goldsmith, W., Lewis, E.R. (eds.): Introduction to bioengineering. Oxford University Press, New York (1996). ISBN 0-19-856516
10. Domach, M.M.: Introduction to biomedical engineering, 2nd edn. Prentice Hall, Upper Saddle River (2009). ISBN 0-13-602003-8
11. Palsson, B.P., Bhatia, S.N.: Tissue Engineering. Prentice Hall, Englewood Cliffs (2003). ISBN 0130416967
12. Bronzino, J.D. (ed.): The Biomedical Engineering Handbook, vol. 1 & II, 2nd edn. CRC, Boca Raton (2000). ISBN 0-8493-0461-X
13. Rashidi, H.H., Buehler, L.: Bioinformatics basics. Applications in biological science and medicine. CRC, New York (2000). ISBN 0-8493-2375-4
14. Jones, N.C., Pevzner, P.A.: An Introduction to Bioinformatics Algorithms. Massachusetts Institute of Technology, Cambridge (2004). ISBN 0-262-10106-8
15. Huang, Y.L.X., Shao, X.F., Huang, Q., Deng, M.Y., Wu, L., Zeng, Q.L., Shao, J.Y.: MicroRNAmiR-21 over expression in human breast cancer is associated with advanced clinical stage, lymph node metastasis and patient poor prognosis. RNA **14**(11), 2348–2360 (2008)
16. Zhao, J.J., Lin, J., Yang, H., Kong, W., He, L., Ma, X., Coppola, D., Cheng, J.Q.: MicroRNA-negatively regulates Estrogen receptor and is associated with tamoxifen resistance in breast cancer. J. Biol. Chem. **283**, 31079–31086 (2008)
17. Zheng, Y., Kwoh, C.K.: Cancer classification with microRNA expression patterns found by an information theory approach. J. Comput. **1**(5), 30–39 (2006)
18. Zhu, S., Si, M.L., Wu, H., Mo, Y.Y.: MicroRNA-21 targets the tumor suppressor gene tropomyosin 1(TPM1). J. Biol. Chem. **282**, 14328–14336 (2007)
19. Zhu, S., Wu, H., Wu, F., Nie, D., Sheng, S., Mo, Y.Y.: MicroRNA-21 targets tumor suppressor genes in invasion and metastasis. Cell Res. **18**(3), 350–359 (2008)
20. Kotlarchyk, A., Khoshgoftaar, T.M., Pavlovic, M., Zhuang, H., Pandya, A.S.: Identification of microRNA biomarkers for cancer by combining multiple feature selection techniques. J. Comput. Meth. Sci. Eng. **11**(5–6), 283–298 (2011)

Chapter 4
Genomics

> *Science is not only compatible with spirituality; it is a profound source of spirituality.*
>
> **Carl Sagan** (1934–1996)

Genomics is the science of understanding structure, function, and evolution of genomes in all forms of life. It is an application of genomic data using biotechnologies, to solve challenging problems of biology and medicine. Genomics in Bioinformatics have concentrated on few of the model organisms, and the analysis of regulatory systems. Structural genomics discusses structures of the protein products of genomes, by revealing similarities to proteins of known structure. The current interest of research community lies in getting complete genome sequences for different organisms as a Landmark achievement of completion of HGP–Human Genome Project, which sequence all the present genes of human DNA.

Genomics: What Was Behind Human Genome Project?

Genomics is the science of understanding structure, function, and evolution of genomes in all forms of life. It is an application of genomic data using biotechnologies, to solve challenging problems of biology and medicine.

M. Pavlovic, *Bioengineering*, DOI 10.1007/978-3-319-10798-1_4

Conceptual breakthrough: Double helix DNA and Central dogma of Molecular Biology: James Watson, Frances Crick and Rosalind Franklin (1953). X-ray crystallography revealed structural and indicated functional aspects of DNA presented in central Dogma:

DNA → mRNA → protein

Genomics in Bioinformatics have concentrated on few of the model organisms, and the analysis of regulatory systems and can be seen as: **structural**, **functional**, and **comparative genomics** with respect to focus of research within this field.

Structural genomics discusses structures of the protein products of genomes, by revealing similarities to proteins of known structure. The current interest of research community lies in getting complete genome sequences for different organisms. Completion of HGP-Human Genome Project, which sequence all the present genes of human DNA, although extremely important for research development, is still just locating the position of the human genes, not defining their function and which needs more complex approaches. Apart from this, completely sequenced genomes comprehensive data is freely available for the whole researcher's community on GOLD-Genomes On-Line Database; whereas NCBI Entrez Genome project gives organism specific searchable databases for complete and incomplete genomes, achievement of great significance for researchers in that field.

Functional genomics approaches involving the use of large-scale and/or high-throughput methods to understand function and expressions of major biomolecules [1–6]. Identification of genes causing disease conditions and complex traits, including responses to drugs and other xenobiotics comprise as the most significant area of functional genomics research [7–9]. All kind of post genome sequencing analysis such as **mutation** or **polymorphism**, and understanding gene and protein function & interaction at molecular level is relevant from the pharmacogenomics or proteomics point of view. *Uni-Prot/Swiss-Prot* is a curated protein sequence database with high level of annotation description of the function of a protein, its domains structure, post-translational modifications, variants, integration with other databases and minimum level of redundancy (Fig. 4.1).

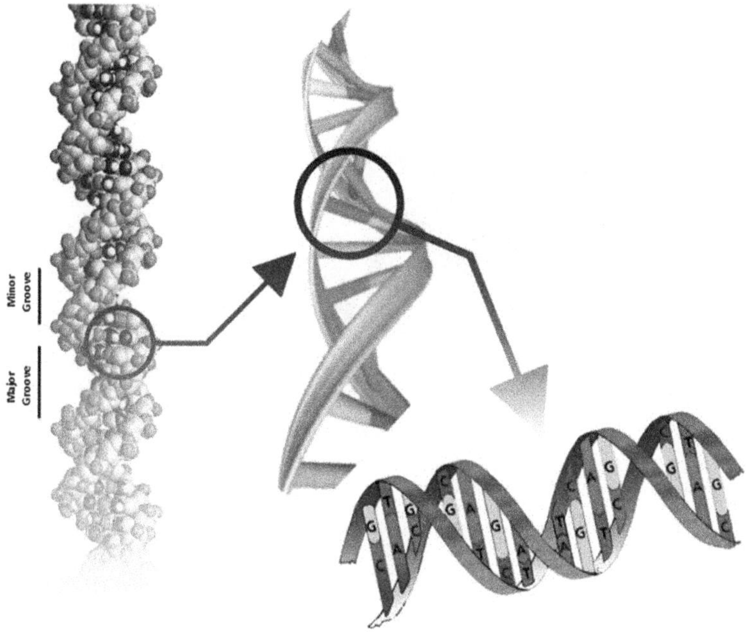

Fig. 4.1 The structure and conformation of DNA

Comparative genomics plays a significant role in the post-human genome sequencing era, to transmit the annotated genome sequence to the physiological functions of a cell. In silico metabolic pathway reconstruction, metabolic pathway comparison, pathway based analysis of expression data, using software such as Pathway based Analysis and Gene Expression analysis Tools and KEGG-Kyoto Encyclopedia of Genes and Genomes system, scientists are able to validate functional annotation, identification of novel genes and their protein products [10–15].

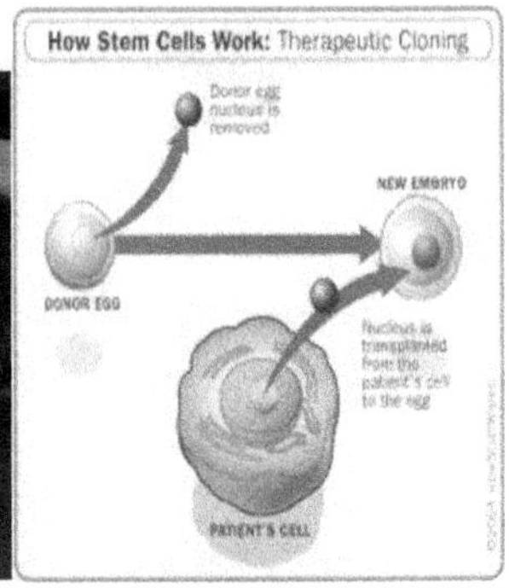

Breakthrough: in vitro fertilization (IVF): Patrick Steptoe and Robert Edwards, NP 2010	Breakthrough in DNA amplification (cloning): Kary Mullis, PCR 1983, NP 1993	Only a handful of the labs in the world are currently using SCNT techniques in human stem cell research: Harvard Stem Cell Institute, the University of California San Francisco, the Oregon Health&Science University Stemgen (La Jolla, CA) and possibly Advanced Cell Technology are currently researching a technique to use somatic-cell nuclear transfer to produce embryonic stem cells. In the United Kingdom, the Human Fertilization and Embryology Authority has granted permission to research groups at the Roslyn Institute and the Newcastle Centre for Life. SCNT may also be occurring in China.

Emphasizing Bioengineering Aspects to Genomics:

Emphasizing Bioengineering Aspects to Genomics

DNA replication is the process of producing two identical replicas from one original DNA molecule [4, 6]. This biological process occurs in all living organisms and is the basis for biological inheritance. This is excellent example of molecular bioengineering: highly intelligent integration of biology and development of new technology in terms of processing and engineering **DNA sequencing** as a fundamental tool in biological and medical research. DNA molecules contain the heritable genetic information in all living organisms and encode all the proteins in our body [14, 15]. Therefore, determination of DNA sequence is useful in basic biological research, evolutionary biology, as well as the applied biological fields, such as diagnostic or forensic research [15]. **High-throughput DNA sequencing is essential for personalized medicine**. To achieve this dream, the price of genome sequencing should be dramatically decreased to a level that most people can afford.

Despite the refinements of **Sanger sequencing**, the current genome sequencing cost remains formidable. Therefore, revolutionary advances in DNA sequencing technology are demanded. To overcome the limitations of the current sequencing technologies, a variety of new DNA sequencing methods have been investigated with the aim of eventually realizing the goal of the $1,000 genome, including sequencing by synthesis (SBS). Thus, the certain companies built upon current state-of-the-art sequencing technologies such as SBS to develop novel proof-of-principle technologies for high-throughput DNA sequencing; demonstrate a general platform for high sensitivity biomolecular detection; and briefly study DNA processing protein (e.g., helicase) functions at the atomistic level.

As a summary of the resulting work on HGP and sequencing and cloning, the development and continuous improvement with respect to three fundamental technical achievements has evolved:

1. First, **a new DNA sequencing technology development** that utilizes surface-enhanced Raman spectroscopy (SERS) [15];
2. Second a rigorous mathematical development of analytical models to predict experimental SERS signal intensity distributions for biomolecular quantification;
3. Third, a versatile SERS-based quantitative method to monitor the catalyst-free click reaction efficiency for small molecule conjugation;
4. Fourth, theoretical work using molecular dynamics simulations for analyzing the mechanical behavior of a molecular motor involved in DNA processing are described; and
5. Consequently, these research efforts provide a foundation for the novel use and integration of SERS-SBS into microfluidic systems for a wide range of applications, such as high-throughput DNA sequencing and genetic diagnostics, as well as the theoretical framework to investigate DNA polymerase function at the atomic level [15].

However, maybe the most expressive and the most convincing bioengineering aspect to genomics are different levels and manners of **cloning**. There are practically four different types of cloning at different levels (molecular, cellular and whole organism) with different/particular purposes and techniques applied. In essence, the cloning is process equivalent to copying, or replication, since scientists are producing identical copies of molecule, cells, or whole organism [14].

Cloning is achieved on four levels:

1. **Molecular (DNA)**
2. **Cellular (all cells are progeny of the single cell)**
3. **Whole organism (genetically identical to a parent)**
4. **Therapeutic (in order to avoid immunological incompatibility of embryonic cells given to foreign organism)**.

Here, we shall emphasize molecular cloning of DNA molecule known as

(a) either plasmid insertion (cell based protocol) or
(b) polymerase chain reaction (PCR-enzyme based protocol) that will be described in details. Today this reaction is very important not only in research but also is fundamental to forensic medicine in discovering different aspects of criminal events.

Molecular Cloning (DNA)

Cloning is, at molecular level the crux of recombinant DNA technology. **Recombinant DNA technology** refers to a set of techniques that enables scientists to transfer genetic information from one organism to another. It is one of the major technological achievements of the past hundred years, and reflects the desire of biologists to further study genes and their function. In order to study one particular gene it is necessary to produce a high quantity of it. It is possible to separate genes based on size, but how to extract it from the mixture of the thousands of genes that are of the same size?

That's how the idea of cloning was born since cloning, as already mentioned, means making identical copies. To clone the genes, means to make many identical copies of a particular length of DNA. Cloned gene is amplified DNA of certain length, at certain locality within total DNA.

How is gene amplified or cloned? DNA cloning is a technique to reproduce DNA fragments, which can be achieved by two different approaches: (1) cell based (**plasmid insertion and bacteriophage**), and (2) enzyme-based—**polymerase chain reaction** (**PCR**), based. In the cell-based approach, a **vector** is required to carry the DNA fragment of interest into the host cell. **Fig.** 4.2 shows a typical procedure by using plasmids as the cloning vector.

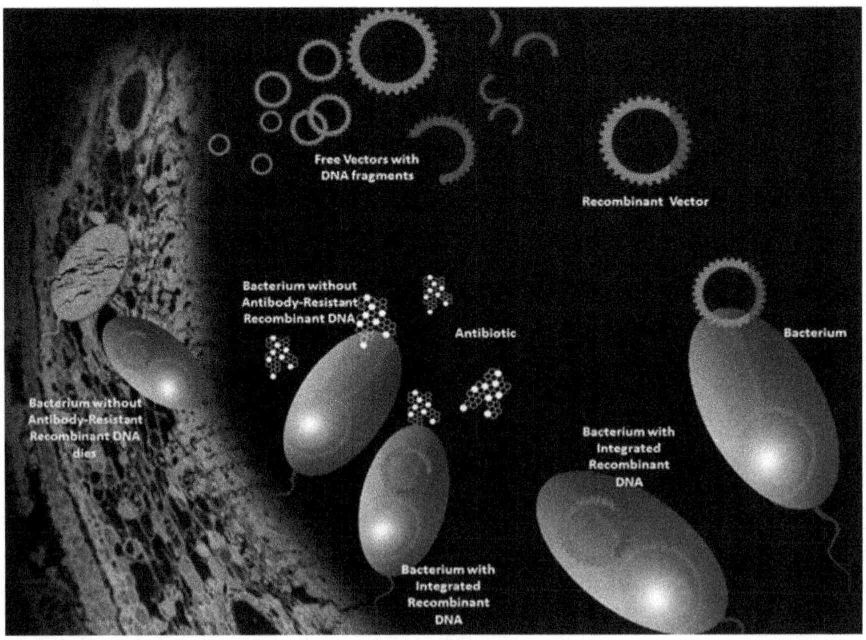

Fig. 4.2 Cloning DNA in the plasmid

The cell based method uses restriction enzymes (endonucleases) which cut DNA at specific sites creating fragments that have unpaired bases or “sticky ends”. These fragments of DNA then can be inserted into a **cloning vector** (such as a **plasmid**) where it pairs with corresponding “sticky ends” in the DNA of the vector.

Here are the essential steps in DNA cloning using plasmids as vectors:

(a) **DNA Recombination**
The DNA fragment to be cloned is inserted into a vector. The recombinant vector must also contain an antibiotic-resistance gene.

(b) **Transformation**
The recombinant DNA enters into the host cell and **proliferates, the process known as “transformation”** because the function of the host cell may be altered. Normal *E. coli* cells are difficult to take up plasmid DNA from the medium. If they are treated with $CaCl_2$, the transformation efficiency can be significantly enhanced. Even so, only one cell in about 10,000 cells may take up a plasmid DNA molecule.

(c) **Selective Amplification**
A specific antibiotic is added to kill *E. coli* without any protection. The transformed *E. coli* is protected by the **antibiotic-resistance gene** whose product can inactivate the specific antibiotic. The numbers of vectors in each *E. coli* cell are not the same, because they may also reproduce independently (see Fig. 4.2).

(d) **Isolation of desired DNA clones**

Kary Mullis is generally credited with inventing PCR in **1983** while working for **Cetus Corporation in Emeryville, California** [1]. Mullis’ role at Cetus was to synthesise oligonucleotides for groups working on, amongst other things, methods to detect point mutations in human genes. Mullis was hatching an idea to detect the point mutations using Sanger-type DNA sequencing, employing DNA polymerase in the presence of an oligonucleotide primer and ddNTPs. The problem was that sequencing a single copy gene within the expanses of the human genome was impossible; the primer would bind in too many places. What he needed was a way to increase the concentration of the specific gene of interest.

PCR (Polymerase Chain Reaction) is another type of DNA cloning or gene amplification. He reasoned that by using two opposed primers, one complementary to the upper strand and the other to the lower, then performing multiple cycles of denaturation, annealing and polymerization he could exponentially amplify the piece of DNA between the primers. This was the key of invention. The idea of PCR was born, but the technique was still very much in its infancy. The *E. coli* DNA polymerase used in the early days was destroyed during the denaturation step so had to be replenished after every cycle. Cetus workers quickly developed the first thermal cycler named “Mr. Cycle”, which automatically added new polymerase after each heating step.

In 1985, Mullis came up with the idea of using polymerase isolated from the extremophile bacterium *Thermophilus aquaticus*. The polymerase, known as **Taq polymerase**, has optimal activity at **72 °C** and can with stand the **94 °C** required for denaturation of the DNA, meaning that many reaction cycles could be performed without being replenishing the enzyme. This breakthrough, together with advances in oligonucleotide synthesis made PCR both cost effective and convenient and it quickly entered mainstream research. As researchers flocked to PCR, improvements to and variations on the process have, and continue to, come quickly and there are now hundreds of PCR-based applications in use in a variety of fields. Stephen Scharf, Mullis' former colleague at Cetus, put it quite nicely: One of PCR's distinctive characteristics is unquestionably its extraordinary versatility. That versatility is more than its "applicability" to many different situations. PCR is a tool that has the power to create new situations for its use and those required to use it. Perhaps the most influential of all techniques enabled by PCR is **massive-scale genomic sequencing**, which itself has transformed the biological and biotechnological research arena.

Mullis received the Nobel Prize for his ground-breaking invention in **1993**. Here are some technical details of PCR.

PCR: Sequence of Events

The purpose of a PCR (Polymerase Chain Reaction) is to make a huge number of copies of a gene. This is necessary to have enough starting template for sequencing.

The Cycling Reactions

There are three major steps in a PCR, which are repeated for 30 or 40 cycles. This is done on an automated cycler, which can heat and cool the tubes with the reaction mixture in a very short time.

1. **Denaturation** at 94 °C: During the denaturation, the double strand melts open to single stranded DNA, all enzymatic reactions stop (for example: the extension from a previous cycle).
2. **Annealing** at 54 °C: The primers are jiggling around, caused by the Brownian motion. Ionic bonds are constantly formed and broken between the single stranded primer and the single stranded template. The more stable bonds last a little bit longer (primers that fit exactly) and on that little piece of double stranded DNA (template and primer); the polymerase can attach and starts copying the template. Once there are a few bases built in, the ionic bond is so strong between the template and the primer, that it does not break anymore.
3. **Extension** at 72 °C: This is the ideal working temperature for the polymerase. The primers, where there are a few bases built in, already have a stronger ionic attraction to the template than the forces breaking these attractions. Primers that are on positions with no exact match get loose again (because of the higher

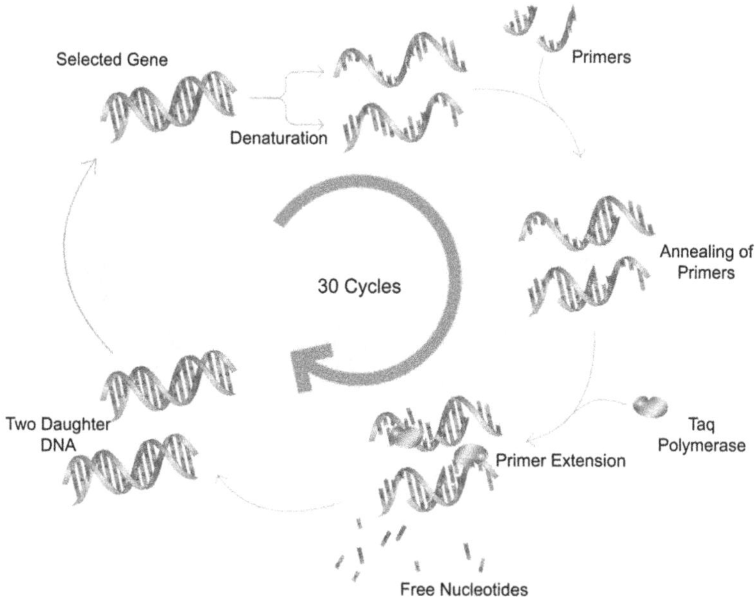

Fig. 4.3 Diagram showing the principle of polymerase chain reaction

temperature) and don't give an extension of the fragment. The bases (complementary to the template) are coupled to the primer on the 3' side (the polymerase adds dent's from 5' to 3', reading the template from 3' to 5' side; bases are added complementary to the template).

Figure 4.3 shows the different steps in PCR. Because both strands are copied during PCR, there is an **exponential** increase of the number of copies of the gene. Suppose there is only one copy of the wanted gene before the cycling starts, after one cycle, there will be 2 copies, after two cycles, there will be 4 copies, and three cycles will result in 8 copies, and so on, reflecting the exponential amplification of the gene in PCR.

Is there a gene copied during PCR and is it the right size?
Before the PCR product is used in further applications, it has to be checked if:

1. There is a **product formed**. Though biochemistry is an exact science, not every PCR is successful. There is for example a possibility that the quality of the DNA is poor, that one of the primers doesn't fit, or that there is too much starting template.
2. The product is of the **right size**. It is possible that there is a product, for example a band of 500 bases, but the expected gene should be 1,800 bases long. In that case, one of the primers probably fits on a part of the gene closer to the other primer. It is also possible that both primers fit on a totally different gene.
3. **Only one band is formed**. As in the description above, it is possible that the primers fit on the desired locations, and also on other locations. In that case, you can have different bands in one lane on a gel (Fig. 4.4).

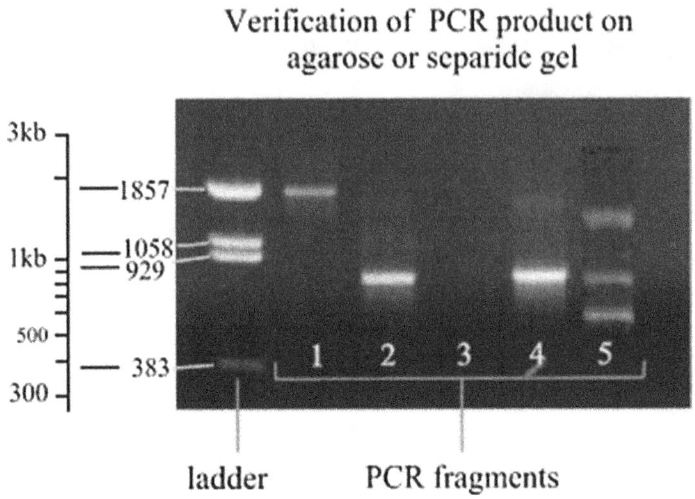

Fig. 4.4 Verification of the PCR product on gel. Credit to: Ref [15]

4. **The ladder is a mixture of fragments with known size** to compare with the PCR fragments. Notice that the distance between the different fragments of the ladder is logarithmic. Lane 1: PCR fragment is approximately 1,850 bases long. Lane 2 and 4: the fragments are approximately 800 bases long. Lane 3: no product is formed, so the PCR failed. Lane 5: multiple bands are formed because one of the primers fits on different places.

Complex organisms (animals, plants) can be modified using recombinant DNA techniques to produce **genetically modified organisms.**

Bioengineering aspect to DNA cloning and significant cellular events

A. **The Role of Bioinformatics**: Control of gene expression: The expression of a particular gene can be controlled at different steps:

   1. **Transcription factors**
   2. **Alternative splicing**
   3. **Affected translocation of the transcript out of the nucleus**
   4. **Degradation of mRNA**
   5. **Altered rate of translation**
   6. **Inactivation of protein due to irregular or blocked signaling mechanism**

B. **PCR technique**, In vitro fertilization and Therapeutic cloning are the most prominent technical/bioengineering approaches to molecular and cellular solutions with application to the biological/medical problems [14, 15].

References

1. https://www.google.com/#q=dna+replication
2. Brown, L., Holme, T.: Chemistry for Engineering Students. Mary Finch, Books/Cole, Cengage Learning, Belmont (2011). ISBN 13:078-1-4390-4791-0
3. Johnson, A.T.: Biology for Engineers, 1st edn. Taylor & Francis, London (2011). ISBN 0-19-856516
4. Andjus, R.K.: General Physiology and Biophysics [in Serbian]. Modules 1–7. Center for Multidisciplinary Studies. University of Belgrade, Belgrade (2002).
5. Guyton, A.C.: Textbook of Medical Physiology. W.B. Saunders, Philadelphia (2000)
6. Saltzman, M.W.: Biomedical Engineering. Cambridge University Press, New York (2009). ISBN 978-0-521-840099-6
7. Pavlovic, M., Balint, B. (eds.): Stem Cells and Tissue Engineering. Springer Briefs in Electrical and Computer Engineering. Springer, New York (2013). ISBN 978-1-4614-5504-2
8. Berger, S.A., Goldsmith, W., Lewis, E.R. (eds.): Introduction to Bioengineering. Oxford University Press, Oxford (1996). ISBN 0-19-856516
9. Domach, M.M.: Introduction to Biomedical Engineering. Prentice Hall, Upper Saddle River (2009). ISBN 10: 0-13-602003-8
10. Palsson, B.P., Bhatia, S.N.: Tissue Engineering. Prentice Hall, Englewood Cliffs (2003). ISBN 0130416967
11. Bronzino, J.D. (ed.): The Biomedical Engineering Handbook, vol. 1 & II, 2nd edn. CRC, Boca Raton (2000). ISBN 0-8493-0461-X
12. Rashidi, H.H., Buehler, L.: Bioinformatics Basics. Applications in Biological Science and Medicine. CRC, Boca Raton (2000). ISBN 0-8493-2375-4
13. Jones, N.C., Pevzner, P.A.: An Introduction to Bioinformatics Algorithms. Massachusetts Institute of Technology, Cambridge (2004). ISBN 0-262-10106-815
14. Savulescu, J.: The ethics of cloning and creating embryonic stem cells as a source of tissue for transplantation: time to change the law in Australia. Aust. N. Z. J. Med. **30**, 492–498 (2000)
15. Palla, M.: Novel Engineering Approaches for DNA Sequencing and Analysis. Doctoral Thesis (2013). http://dx.doi.org/10.7916/D8K0728C

Chapter 5
Proteomics: Enzyme: Structure, Function, Kinetics, and Engineering Aspects

Proteins are the machinery of living tissue that builds the structures and carries out the chemical reactions necessary for the life.

Michael Behert Deloitte (1953–)

Proteins are often defined metaphorically as the workhorses of the cells. They are of extreme importance for many functions of the cells since they can be: enzymes, receptors, carriers, ion channels, contractile elements, antibodies, antigens, hormones, being also major macronutrients in the human diet. The word originates from Greek's "Proteos" which means "the first" or prioritized molecule, and although today we think it is DNA, the link with DNA and incredible spectrum of proteins and their functions including enzymatic, makes this molecule(s) unconditional necessity of the cell and driving force for activity of each cellular compartment.

Proteins: Synthesis, Structure and Function

Proteins, often defined metaphorically as the **workhorses** of the cells, are of extreme importance for many functions of the cells since they can be: enzymes, receptors, carriers, contractile elements, antibodies, antigens, hormones, being also major macronutrients in the human diet [1–3].

M. Pavlovic, *Bioengineering*, DOI 10.1007/978-3-319-10798-1_5

How Are the Proteins Made in the Cell?

The existence of proteins has a root in a genetic code of DNA. Therefore, in the nucleus, dsDNA "unzip" at location of gene, for the protein to be made. From one of the strands at the "unzipped" site until the end of the gene, mRNA complementary to DNA is created (**transcription**). The mRNA leaves the nucleus and protein synthesis occurs in the cytosol at ribosomes with the help of transfer tRNA (**translation**). After attaching at the "start" codon (methionine), amino acid specific tRNA helps to attach subsequent amino acid sequentially (**elongation**), until the stop codon is reached [3]. The protein is then de-attached from the ribosome. This is in essence sequence of events that are presenting the **central dogma of molecular biology**: **mRNA is synthesized on the DNA template**, **while the protein is synthesized on mRNA template**. A single gene can produce more than one protein through the mechanism known as **alternative splicing** [3].

The protein structure can be: **primary** (linear), **secondary** (helicoid due to formation of hydrogen bonds α-helix, δ-sheet), **tertiary**—tridimensional arrangement of these secondary structures with one polypeptide chain and **quaternary** (structural relationships between all of the polypeptides with a complete protein that contains multiple subunits) [3]. **The 3-D structure of the protein determines its function. One of the major goals of bioinformatics is to understand the relationship between amino acid sequence and 3-D structure of proteins.**

Enzymes: Structure, Function and Kinetics of the Reactions

Enzymes are the proteins with a particular role in the cell, the role of catalyzing the thousands reactions that occur in the cell. A **catalyst** is the molecule that speeds up a chemical reaction, but is not consumed or generated in the reaction [1, 3, 7–11]. Enzymes can speed up the reaction by a million or more (106–1,012) time more with respect to unanalyzed reaction. In the presence of right enzymes the reactions that would take thousands of years, can occur within seconds.

Breakthrough: Leonora Maud Menten (1879-1960) and Leonor Michaelis (1875-1949), physician and biochemist who formulated enzyme kinetics in 19th Century and open the door to mathematical modeling and furthered computing, for protein conformation approach to the field of enzyme kinetics.

Enzymes are very specific, acting to speed up only certain reactions. They achieve their specificity by only recognizing their substrates and no other potential substrates. Enzyme stabilizes the transition state of the enzymatic reaction [3]. The transition state has the highest free energy; formation of the transition state is rate-limiting step in the overall sequence of reactions in this equation:

$$E + S \rightleftharpoons ES \rightleftharpoons E + P$$

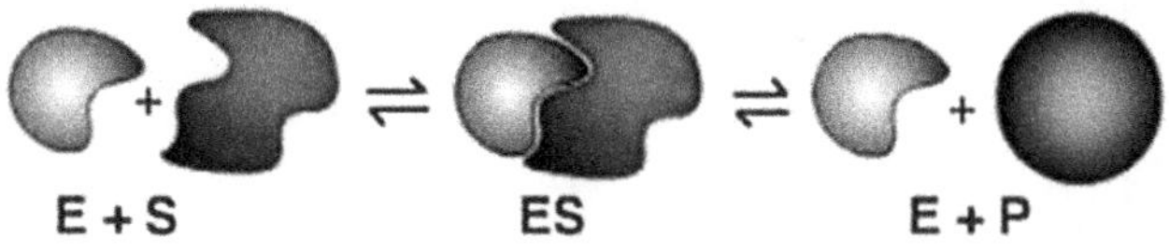

$$\frac{-dS}{dt} = \frac{dP}{dt} = \text{Velocity} = \frac{V_{max} \bullet S}{K_M + S}$$

The change in Gibb's free energy between the transition state and the substrate is called the activation energy ($\Delta G^{\ddagger}$). Enzymes are capable of lowering the activation energy of a reaction and thereby speed up the reaction.

Four essential conditions for enzymatic reaction are:

1. Concentration of the enzyme
2. Concentration of the substrate
3. Concentration of the hydrogen ions (pH)
4. Temperature

There are other conditions that can modify the enzyme function such as the activators or inhibitors and different types of activation or inhibition. They can have different type of mathematical expression [4].

The prototype of enzymatic reaction quantified in a very refined way is Michealis–Menten kinetics. The kinetic properties of many enzymes can be described using Michaelis–Menten equation:

$$v_0 = \frac{V_{max}[S]}{K_M + [S]}$$

Here, V_{max} represents the maximum velocity achieved by the system, at maximum (saturating) substrate concentrations. K_M (the Michaelis constant; sometimes represented as K_S instead) is the substrate concentration at which the reaction velocity is 50 % of the V_{max}. $[S]$ is the concentration of the substrate S. Actually the K_M is showing how strong the affinity of the enzyme for the substrate is, while the V_{max} is telling us how quickly the enzyme is elaborating the substrate.

Figure 5.1 shows a plot of the Michaelis–Menten equation's predicted reaction velocity as a function of substrate concentration, with the significance of the kinetic parameters V_{max} and K_M graphically depicted [5].

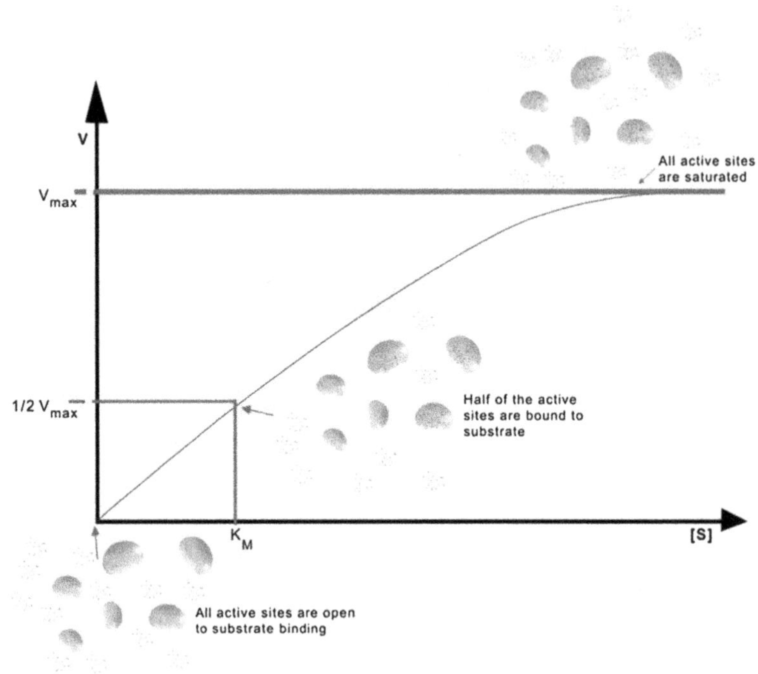

Fig. 5.1 Critical Elements of Michaelis–Menten kinetics (K_m, Vmax)

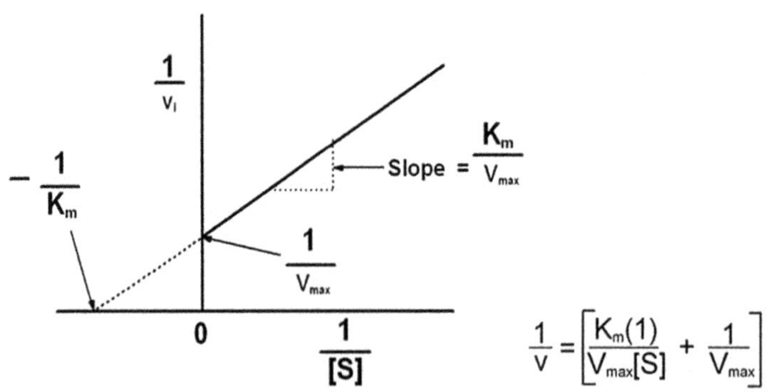

Fig. 5.2 Lineweaver-Burk curve and equation (Saturation kinetics linearized in double reciprocal form)

The **Leineweaver-Burke plot** (or double reciprocal plot) can be used to determine the values of V_{max} and Km from experimental data. They can be determined **graphically** from this double-reciprocal plot. Plots of 1/v versus 1/[S] yield straight lines with a slope of K_m/V_{max} and an intercept on the ordinate at $1/V_{max}$ (Lineweaver-Burk Plot) (Fig. 5.2) [6].

A mathematical model of the kinetics of single substrate-enzyme-catalyzed reaction was first developed by Victor Henry in 1902 and by Leonor Michaelis and

Leonora Maud Menten in 1913. These types of enzymatic reactions are known as *saturation* kinetics. They used mostly that approach, but the significance of their work is that it has **opened the door to mathematical modeling of biochemical and physiological processes in the future**. It is very possible that these scientists appeared too early to be completely understood and appreciated by their co-temporaries, so they were not awarded the Nobel Prize for their fascinating and original work and development of the science. Today, only in the field of enzymes we know many different enzymatic reactions that present differently processed substrates and different enzymatic models, which can be computed and algorhytmized.

Emphasizing Bioengineering Aspects to Proteins and Enzymes

Abzymes are the antibodies with catalytic activity [12–15]. Dual activity of antibodies is predicted by Linus Pauling and broke the **central dogma** of the Immunology that antibodies once secreted from immune cells do not re-circulate into the cell [12]. Alarcon Segovia with his group has shown that anti-DNA autoantibody secreted in autoimmune and lymphoproliferative diseases are re-circulating into the living cells, translocate into nucleus and hydrolyzing DNA of the penetrated cells [12]. This is an amazing field in autoimmunity which involves dual role of anti-DNA antibodies and indicates their role in pathogenesis of the disease.

Figure 5.3a is showing the enzyme Cyclooxygenase 2 (COX-2) fluorescent staining with engineered antibody to the enzyme fluorescently labeled and expression in

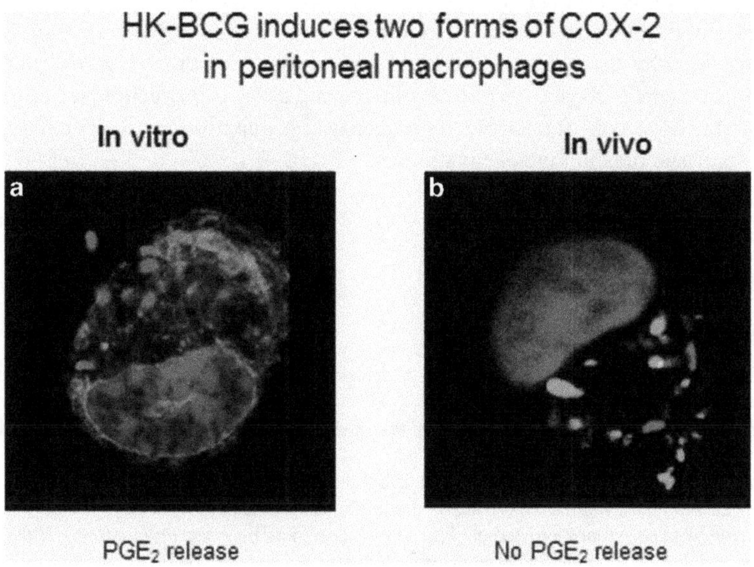

Fig. 5.3 (**a**, **b**) COX-2 (Cyclooxygenase 2) Fluorescent Labeling: activated macrophage (M. Pavlovic, Lab experiment unpublished data, 2006)

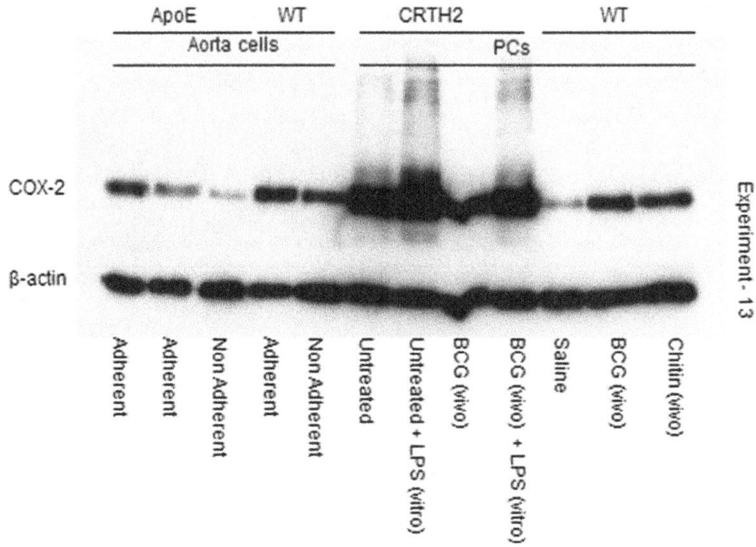

Fig. 5.4 COX-2 Western Blot (Pavlovic M, Lab experiment, unpublished data, 2006)

macrophages after phagocytosis of Bacillus Calmette-Gerene, taken from author's experimental work (unpublished data).

Enzymes are widely used in industry and have significant medical application as diagnostic and curative approaches. Therefore, the **development of biosensors using enzymes as integral components is proceeding rapidly** (M. Schuler). Two examples of immobilized enzyme electrodes are those used in determination of glucose and urea by using glucose oxidase and urease immobilized on the electrode membrane. Scarce enzymes are finding increasing uses, as the techniques of genetic engineering now make it possible to produce usable quantities of such enzymes (for example, tissue plasminogen activator, etc.) (Fig. 5.4).

References

1. https://www.google.com/#q=dna+replication
2. Brown, L., Holme, T.: Chemistry for Engineering Students. Mary Finch, Books/Cole, Cengage Learning, Belmont (2011). ISBN 13:078-1-4390-4791-0
3. Johnson, A.T.: Biology for Engineers. Taylor & Francis, Boca Raton (2011). ISBN 13: 978-1420077636
4. Andjus, R.K.: General Physiology and Biophysics [in Serbian]. Modules 1–7. Center for Multidisciplinary Studies. University of Belgrade, Belgrade, Serbia (2002)
5. Guyton, A.C.: Textbook of Medical Physiology. Sounders Elsevier, Philadelphia (2000)
6. Saltzman, M.W.: Biomedical Engineering. Cambridge University Press, New York (2009). ISBN 978-0-521-840099-6

7. Pavlovic, M., Balint, B. (eds.): Stem Cells and Tissue engineering. Springer, New York (2013). ISBN 978-1-4614-5505-9
8. Berger, S.A., Goldsmith, W., Lewis, E.R. (eds.): Introduction to Bioengineering. Oxford University Press, Oxford (1996). ISBN 0-19-856516
9. Domach, M.M.: Introduction to Biomedical Engineering, 2nd edn. Upper Saddle River, Prentice Hall (2009). ISBN 10: 0-13-602003-8
10. Palsson, B.O., Bhatia, S.N.: Tissue Engineering. Prentice Hall, Englewood Cliffs (2003). ISBN 0130416967
11. Bronzino, J.D. (ed.): The Biomedical Engineering Handbook, vol. 1 & II, 2nd edn. CRC, Boca Raton (2000)
12. Rashidi, H.H., Buehle, L.: Bioinformatics Basics. Applications in Biological Science and Medicine. CRC, New York (2000). ISBN 0-8493-2375-4
13. Jones, N.C., Pevzner, P.A.: An Introduction to Bioinformatics Algorithms. Massachusetts Institute of Technology, Cambridge (2004). ISBN 0-262-10106-815
14. Savulescu, J.: The ethics of cloning and creating embryonic stem cells as a source of tissue for transplantation: time to change the law in Australia. Aust. N. Z. J. Med. **30**, 492–498 (2000)
15. Palla, M.: Novel Engineering Approaches for DNA Sequencing and Analysis. Doctoral Thesis (2013). http://dx.doi.org/10.7916/D8K0728C

Chapter 6
Communication I: Neural System and Regulation of Communication

"I think, therefore I am." (Cogito, ergo sum)

Decartes (1596–1650)

As we already know, the nervous system functions together and in accordance with the endocrine system to maintain homeostasis. Our sensory organs have distinguished receptors sensitive to different signals conveying and sending signals directly to the brain. These sensory organs and their neurons are part of the central nervous system (CNS). They can bring to the brain both external and internal signals. Information from all the senses is integrated in the brain centers located in the brain and spinal cord. There are two types of neurons according to their function: motor and sensory neurons. Motor neurons transmit the signals from central nervous system to the periphery: glands and muscles for example. Sensory neurons are ascendant pathways and they transducer the signal from the periphery to the appropriate centers of the CNS (brain and spinal cord). Sensory and corresponding motoneuron creates the reflex arch. Thus, the nervous system controls the major muscular activities of the body, visceral smooth muscle activity, and secretion of exocrine and endocrine glands.

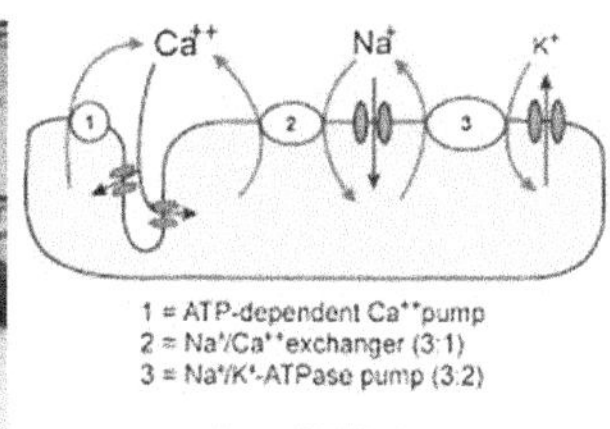

Alan Lloyd Hodgkin (1914-1998) and Andrew Huxley (1917-2012) breakthrough: ("conductance-based model"), mathematical model that describes through the set of non-linear differential equations how action potentials in neurons are initiated and propagated. NP in 1963.	**Luigi Galvani (1737-1798) and Alessandro Volta (1745-1827); first ideas on the bioelectrical phenomenology which will induce development of methods for electrophysiological measurement.**

M. Pavlovic, *Bioengineering*, DOI 10.1007/978-3-319-10798-1_6

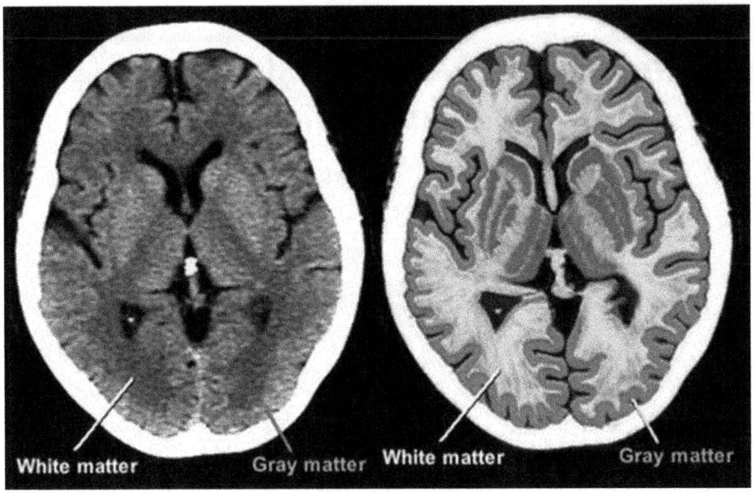

Fig. 6.1 White and gray matter in the brain as detected by MRI imaging

The nervous system is composed of the central NS (CNS) and peripheral NS (PNS) [1–3]. The CNS consists of the brain and spinal cord. The PNS have the cranial and spinal nerves. The CNS has grey and white matter. Grey matter in the brain is composed of neuronal cell bodies, neuropil, glial cells, and capillaries. On the other hand, white matter has no cell bodies and mostly consists of bundles of myelinated axons that act as a bridge between the different grey matter regions of the nervous system. The function of grey matter is to route sensory or motor stimuli to interneurons of the CNS in order to create a response to the stimulus through chemical synapse activity.

As we already know, the nervous system functions together and in accordance with the endocrine system to maintain homeostasis. Our sensory organs have different receptors sensitive to different signals that send signals directly to the brain (Fig. 6.1).

These sensory organs and their neurons are part of the central nervous system (CNS). They can bring to the brain both external and internal signals. Information from all the senses is integrated in the brain and spinal cord (Fig. 6.2).

The primary components of the nervous system include the central and peripheral nervous system. The central nervous system is composed of the brain and spinal cord. The peripheral nervous system is composed of the cranial and spinal nerves. All parts of the nervous system are composed of a common cellular subunit—the neuron.

The parts of the neuron are the cell body, dendrites, axon and ganglia. The cell body contains the nucleus and metabolic machinery of the cell. Dendrites form the extensions into tissues that may synapse to one or many other neurons. The axon is a long cytoplasmic process, also called the nerve fiber. Ganglia, which are groups of neuronal cell bodies that lie peripheral to the central nervous system in vertebrates,

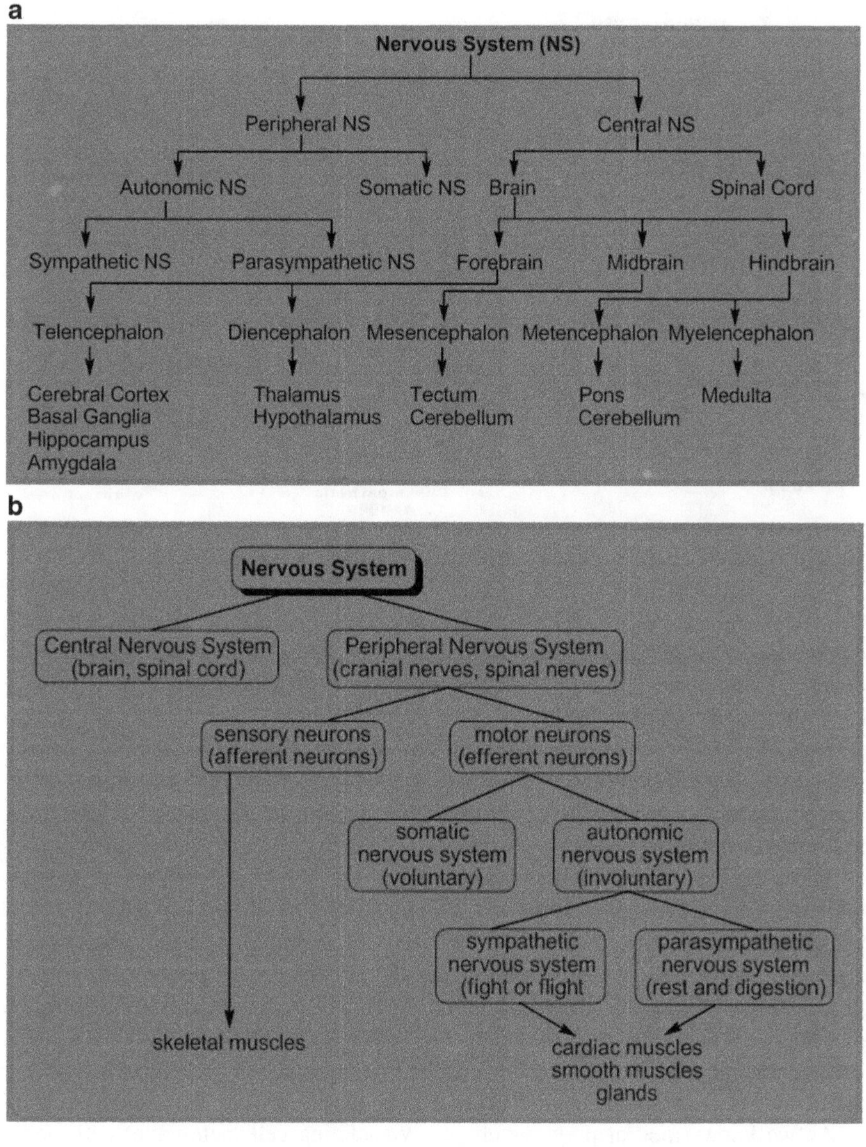

Fig. 6.2 (**a**, **b**) Diagrams showing compartments of nervous system

have long axons converging into trajectories. **Neural cell (neuron)** has different shapes, size and structure, dependent on the location and role in the nervous system, but the global structure is the same. The essential function of the neuron is to convey the information in order to communicate with appropriate members of the **circuit** responsible for creating signals for certain effectors-function.

Thus, dendrites receive the input signals, which are integrated by the neural cell and conducted by the axon to the output terminals. Nodes of Ranvier facilitate fast

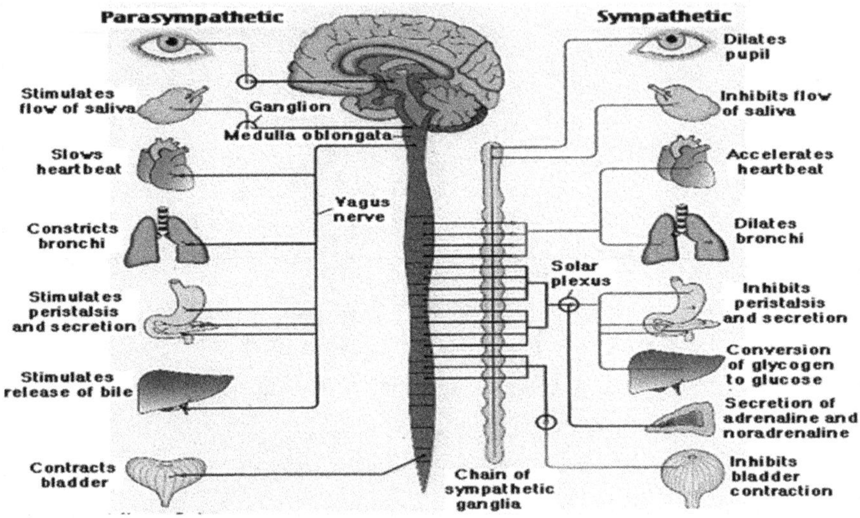

Fig. 6.3 Autonomous nervous system-relationship with CNS and effectors

conduction of the action potential in some neurons. Flow of information is from the dendrite to the axon. The axon conducts the integrated signal, while axon terminals carry the output signal.

Special ligands called neurotransmitters released at the axon terminal interact with the dendrite of the adjacent neuron (dopamine, serotonin, acetylcholine, adrenalin, peptidergic neuronal active peptides, etc.). Signals ate transmitted as **an action potential** or **nerve impulse**.

There are two types of neurons according to their function: **mot**or and **sensory neurons**. Motor neurons transmit the signals from central nervous system to the periphery: glands and muscles, as effectors, for example. Sensory neurons are ascendant pathways and they transduce the signal from the periphery to the appropriate centers of the CNS (brain and spinal cord). Sensory and corresponding motor neuron create the **anatomic structure for reflex arch**. Thus, the nervous system controls the major muscular activities of the body, visceral smooth muscle activity, and secretion of exocrine and endocrine glands (Fig. 6.3).

As we know (look at previous chapter visualizing cell membrane) cell membranes have electrical polarity caused by their permeability to some ions, even in the resting state. The resulting resting potential is approximately **–70 mV**, caused by separation of charge across the plasma-membrane; the inside surface is more negative than the outside of the neuron [1, 2, 12]. The muscle cells and neurons belong to the group of **excitable tissues.** That means that the potential difference across the membrane can be changed by the movement of small number of ions from inside to outside, or vice versa. Due to impermeability of the membrane for the ions (since they are insoluble in the lipid bilayer), they must diffuse through the pores created by ion channels [2]. Some of them are always open, while some of them only if appropriate stimulus occurs (voltage-gated a ligand-gated channels). K^+ leaky ion channels allow passage of K^+ out of the cell (although it is intracellular ion) leaving behind excess negative charge. Thus, the resting membrane potential approaches **the Nernst potential** of K^+. (See the scheme of action potential from the previous chapter 5). So, an action potential results from depolarization of the plasma membrane when voltage-gated Na^+ channels are open. Repolarization back to the normal-resting membrane potential is caused by the opening of voltage-gated K^+ channels [1–3].

The Nernst Potential

How is it possible to separate charge across the membrane? Let us take a simple example. Assume that we have a membrane separating two compartments having channels that are only permeable to potassium and no other ions can permeate. Initially the channels are closed and we add 100 mM of KCl to the lower compartment (say the interior of the cell) and 10 mM of KCl to the upper compartment (the outside). As we have added a neutral salt, there will be the same number of cations than anions in the lower compartment and the same will be true for the upper compartment (even though the total number of ions is ten times lower in the upper compartment). The consequence of the electroneutrality in each side will be zero charge difference across the membrane and consequently the membrane potential difference will be zero. This is because from eq. (3) we can write that **$V = Q/C$**, where V is the potential difference and Q is the excess charge. The permanent thermal motion of the ions will make them move randomly but they will not be able to cross the membrane because they are poorly permeant through the bilayer and the channels are closed.

Suppose that at one point we open the channels. Then, as there **are ten times more K^+ ions** in the bottom compartment than in the top compartment, there will be ten times more chances of an ion crossing up than down. This initial situation is schematically pictured in Fig. 6.4, where a K^+ ion (the blue balls) is crossing the channel in the upward direction leaving a Cl^- ion behind (red balls). This flow, which is proportional to the concentration gradient, increases the top compartment by one positive charge and the lower compartment by one negative charge, producing a charge separation. This charge separation introduces non-random new electrostatic force acting on the ions as electrostatic force with tendency to drive the ions from the top compartment into the bottom compartment and at the same time it tends to break the flow in the opposite direction. The final result is that the charge separation will build up a **voltage across the membrane ($V = Q/C$) that will continue to increase until the flow in both directions becomes equal due to the increased electrostatic force that will tend to balance the flow produced by the concentration gradient.** When that happens, any ion that crosses in one direction will be counterbalanced by another crossing in the opposite direction, maintaining an equilibrium situation. This potential difference is then called the **equilibrium potential** or **Nernst potential** and it is expressed by **Nernst equation.**

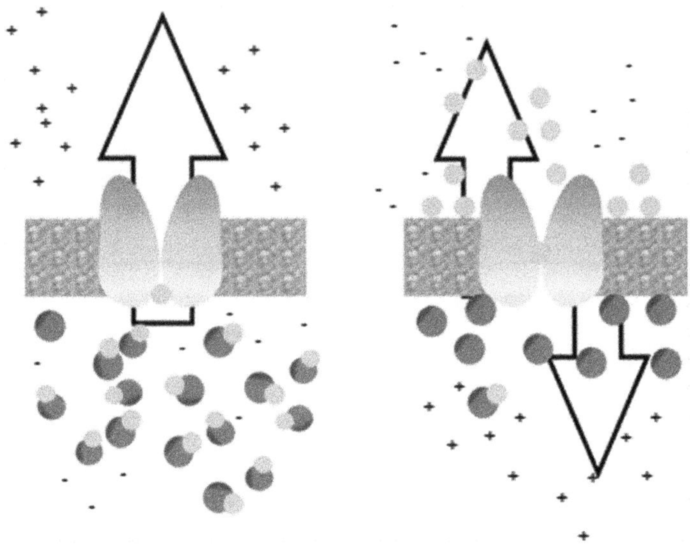

Fig. 6.4 Active transport across the membrane

The above discussion can be put in more quantitative terms by expressing the net ion flow *j* in terms of the chemical and electrical gradients:

$$j = -D\left[\left(\frac{dC}{dx}\right) + C\left(\frac{zF}{RT}\right)\left(\frac{dv}{dx}\right)\right]$$

Where *D* is the diffusion coefficient, *C* is the concentration, *R* is the gas constant, *V* is the voltage, *z* is the valence, *F* is the Faraday constant, and *T* is the temperature. When j = 0 (no net flow), it can be integrated and we get:

$$V = \frac{RT}{xF}\ln\left(\frac{C_0}{C_i}\right)$$

which is the **Nernst equation** that relates the voltage V across the membrane which is in equilibrium with the concentration gradient established by the concentrations Co, outside, and Ci, inside. It is customary to call this voltage the **equilibrium potential** of the ion "N" En, and by calling the external and internal concentrations of the ion "N" No and Ni, respectively we can rewrite eq. (1) as follows:

$$En = \frac{RT}{zF}\ln\left(\frac{N_0}{Nt}\right)$$

Finally, let us only roughly classify neurotransmitter signaling:

Neurotransmitter Signaling

- **The synapse**
  Signaling through gated receptors
- **A ligand-gated ion channel**
  Signaling through indirectly gated receptors
- **The D1-dopamine receptors is an indirectly-gated ion channel (associated to G-protein)**
  Signaling through indirectly-gated ion channels

Emphasizing Bioengineering Aspects to Nervous System

There is tremendous contribution of bioengineering to development of both research and clinical approaches to the understanding of structure, function and pathophysiology of Nervous system. As a very complex system, both central and peripheral nervous systems together with sensorial organs are a target of modern approaches in bioengineering efforts. They include bioinformatics, computer modeling, nanotechnology, biomagnetic, tissue engineering and a spectrum of modern techniques in order to understand normal and prevent or cure pathological evolution of different diseases, for example Multiple Sclerosis (MS). For most of demyelinating diseases it is thought that they are either degenerative or autoimmune. These models are consistent with the new theory and experimental data designed to answer the more than hundred years old question: can the brain or peripheral nerve repair itself? Today it is known that especially in MS as one of the most prominent demyelinating diseases that is possible.

MS is considered autoimmune disease in which the immune system with a broken tolerance cannot recognize its own antigens and therefore is attacking (in this case) the protective covering of axons, e.g. myelin sheaths. This is in part antimicrobial attack without positively known causative agent, within which the immune system misidentifies its own healthy tissue as foreign and attacks it. In MS, the process of demyelination will attenuate the efficient signal transduction and denuded axons will eventually die off and disintegrate. When an immune system attacks the myelin, it produces complex reaction known as inflammation. Recently, it has been discovered that there is a reversible process called remyelination, in both mice and humans [8–11]. Longitudinal human clinical study has discovered that there are four different patterns of demyelination in this disease which can explain the variability of the symptoms between the patients [8, 9]. This can implement the models proposed at the first line. At the second line, the implementation is inevitable and tightly linked to the remyelination phenomenon, which can be strong at the beginning of the diseases but may not translate necessarily into good response in the late stage of MS.

These natural processes strongly support our necessity for development of the spectrum of the models of Action Potential (AP) propagation through the axon during different clinical stages of demyelinating diseases and during different pathological patterns of myelin and RN status. AS for MS, it is known that early stage is characterized by intense inflammation, localized demyelization and limited remyelination. The key to remyelination in MS is a natural human autoantibody labeled as "Number 22" [10]. The antibody is natural to the host, found across species and considered old and primitive in evolutionary terms. It is the body's first and fastest line of defense known as "innate immunity". It is found that it promotes remyelination in about 50 % of lesions in mice in a virus that mimics MS.

So far, the MRI results of the clinical study in patients with 80 % natural remyelination has shown a strong correlation between evidence of remyelination on biopsy and a pattern of "ring enhancement" in MS lesion, indicating that this could greatly enlarge the pool of candidates for clinical trials [11].

The **discovery of the process of remyelination** and its causative factor raises the inevitable questions that could be critically investigated through modifications in our models, such as: is there a critical period for remyelination, how many myelin is needed for success, how many axons need to be repaired to prevent diffuse tissue damage, can axon outgrowth be promoted through remyelination, etc. The work of Hodgkin and Huxley enabled development of other models for axonal behavior, for example if it is injured [5–7] as presented by author [7] in Figs. 6.5 and 6.6.

Although it is known that nerves in the peripheral nervous system (PNS) will regrow; however, where the critical components for the nerves' reconstruction are missing, such as where a gap exists, regeneration and restoration of function is nearly impossible. As in recent CNS work, many scaffolds have been used for the replacement of the lost tissue and designed artificial and natural tubes to help direct PNS nerves to reconnect across the gap [1–3]. In addition, other cell types and materials can be added to the scaffolds that will allow for the ingrowth and outgrowth from the conduits [4].

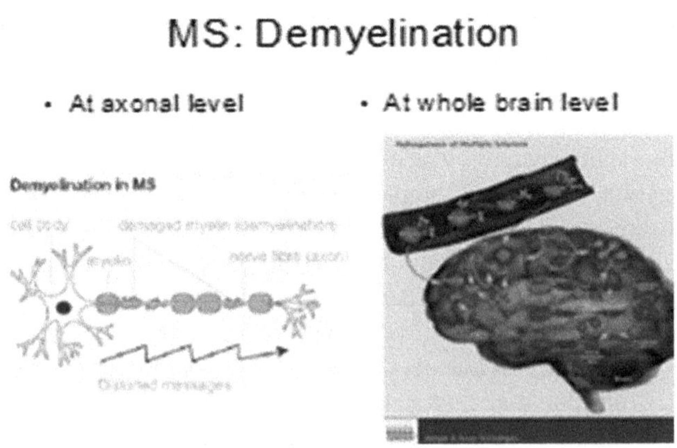

Fig. 6.5 Process of demyelination

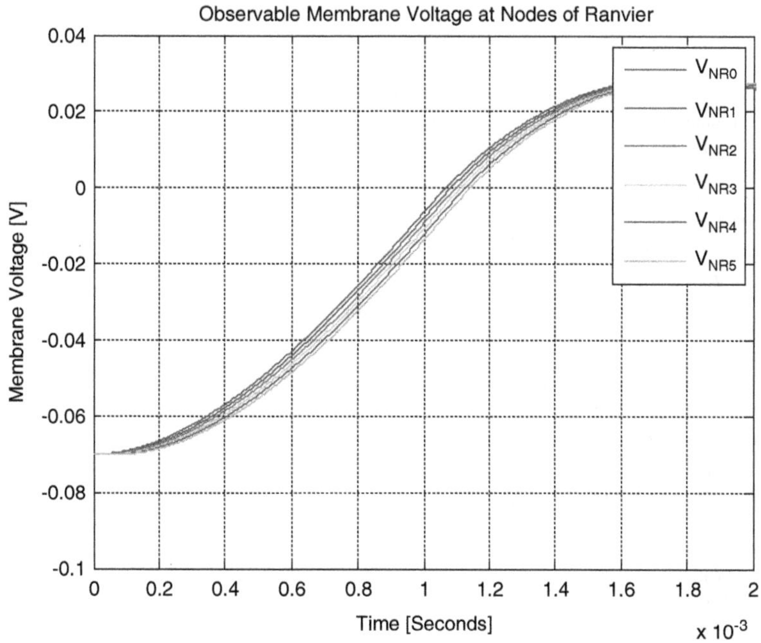

Fig. 6.6 AP rising edge for healthy axon condition (Morales JG. Zhuang H, **Pavlovic M:** A biologically inspired myelinated neuron axon model using a system identification approach MD-Medical Data 2011; 3(2): 119-126)

This is only the indication of the impact that **nanomedicine** will have on treatments delivered to the brain for diseases and illnesses ranging from glioblastoma to restoring the CNS after trauma. As the increased resolution of the nanoscale slowly makes it way to the brain, the problems that were once thought worked out are now being understood better. The manipulations of CNS tissues have many possibilities from restoration to enhancement. With some of the new delivery techniques we will be able to assist those with mental disabilities to either restore them to their original state, or even give a rich life to those who, up until now, had none.

Nanoscience and medicine will not only break through the age-old blood brain barrier to deliver medication, but also break down many of the preconceived notions that performing reconstructive brain surgery was impossible. The seeds have already been planted and so far they have yielded fascinating results, and provide hope for those who suffer from devastating CNS injuries either caused by trauma illness or chemical imbalance [4, 5, 7–11].

References

1. Andjus, R.K.: General Physiology and Biophysics [in Serbian]. Modules 1–7. Center for Multidisciplinary Studies. University of Belgrade, Belgrade, Serbia (2002)
2. Hall, G.E., Guyton, A.C.: Textbook of Medical Physiology. Sounders Elsevier, Philadelphia (2011). ISBN 978-1-4160-4574-8
3. Haltzman, M.W.: Biomedical Engineering. Cambridge University Press, New York (2009). ISBN 978-0-521-840099-6
4. Pavlovic, M., Balint, B. (eds.): Stem cells and Tissue Engineering. Springer, New York (2013). ISBN 978-1-4614-5505-9
5. Zhan, X., Gao, M., Jiang, Y., Zhang, W., Wong, W.M., Yuan, Q., et al.: Nanofiber scaffolds facilitate functional regeneration of peripheral nerve injury. Nanomedicine **9**(3), 305–315 (2013 Apr)
6. Hodgkin, A.L., Huxley, A.F.: The components of membrane conductance in the Giant Axon of Loligo. J. Physiol. **116**, 473–496 (1952)
7. Pavlovic, M., Mayfield, J., Balint, B.: In: Furth, B., Agarwal, A. (eds.). Handbook of Medical Healthcare Technologies. Springer, New York (2013). ISBN:978-1-4614-8494-3
8. Lassmann, H., Bruck, W., Luchinetti, C.: Heterogeneity of multiple sclerosis pathogenesis: implication for diagnosis and therapy. Trends Mol. Med. **7**(3), 115–121 (2001)
9. Riverol, M., Sepulcre, J., Fernandez-Alonso, M., Uccelli, A., Brieva, L., Rubio, M., Rodriguez, A., Fernandez-Dieza, B., Villoslada, P.: Antibodies against Epstein–Barr virus and herpes virus type 6 are associated with the early phases of Multiple Sclerosis. J. Neuroimmunol. **192**(1–2), 184–185 (2007)
10. Wingerchuk, D.M., Lucchinetti, F.C., Noseworthy, H.J.: Multiple sclerosis: current pathophysiological concepts. Lab. Invest. **81**, 263–281 (2001)
11. Erickson, K.I., Voss, M.W., Prakash, R.S., Basak, C., Szabo, A., Chaddock, L., Kim, J.S., Heo, S., Alves, H., White, S.M., Wojcicki, T.R., Mailey, E., Vieira, V.J., Martin, S.A., Pence, B.D., Woods, J.A., McAuley, E., Kramer, A.F.: Exercise training increases size of hippocampus and improves memory. Proc. Natl. Acad. Sci. U. S. A. **108**, 3017–3022 (2011)
12. Morales, J.G., Zhuang, H., Pavlovic, M.: A biologically inspired myelinated neuron axon model using a system identification approach. MD-Medical Data **3**(2), 119–126 (2011)

Chapter 7
Communication II (Endocrine Control)

The endocrine system is also essential to communication. This system utilizes glands located throughout the body, which secrete hormones that regulate a variety of things such as metabolism, digestion, blood pressure and growth. While the endocrine system is not directly linked to the nervous system, the two interact in a number of ways.

Kendra Cherry (Co-temporary psychologist)

This chapter is answering questions such as: how is the operation of trillions of our cells coordinated? What kind of communications exists in human body? Cells communicate with each other either directly, or through signaling molecules to relay signals from outside and inside the cells. The communications via *molecules called ligands can be direct or indirect. In direct cell communication, ligands are bound to the surface of the cell. Soluble, diffusible ligands are used for communication between the cells that are not physically connected or are separated by long distances. The target cells have specialized proteins known as receptors, which are located on the cell surface but anchored in the cell membrane. Ligand can be any molecule, but in the case of endocrine regulation, it is specific hormone.*

Communication II: Signal Transduction Pathways and Endocrine Regulation of Communication

Really, how is the operation of trillions of our cells coordinated? What kind of communications exists in human body? Cells communicate with each other either directly, or indirectly—through signaling molecules to convey signals from outside and inside the cells. The communications via molecules called **ligands** can be direct or indirect. In direct cell communication, ligands are bound to the surface of the cell [1–5].

M. Pavlovic, *Bioengineering*, DOI 10.1007/978-3-319-10798-1_7

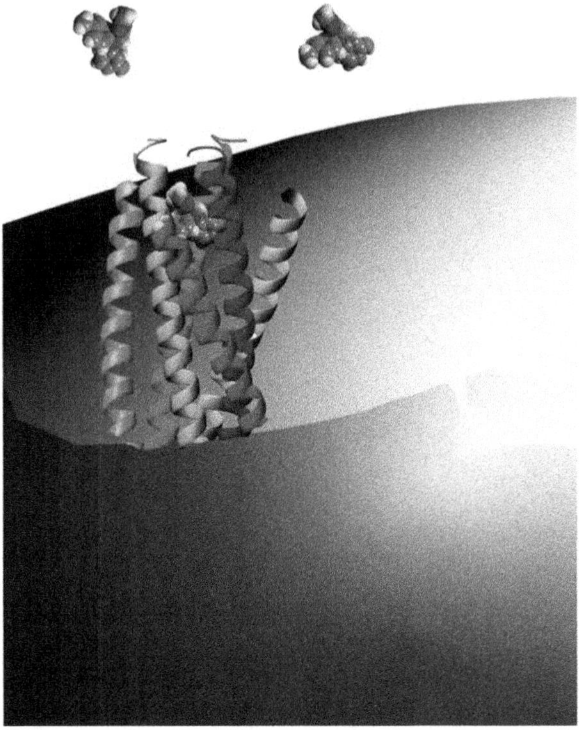

Fig. 7.1 Ligands binding to receptors: model (*membrane – spanning receptor is anchored/fixed to the membrane*)

Soluble, diffusible ligands are used for communication between the cells that are not physically connected or are separated by long distances. The target cells have specialized proteins known as **receptors**, which are located on the surface but anchored in the membrane (Fig. 7.1) [2].

Signal Transduction Pathway is the transmission of molecular signals from a cell's exterior (membrane) to its interior (molecule in cytoplasm or nucleus and nuclear DNA) [1]. *Molecular signals* are transmitted between cells by the secretion of **hormones** and other chemical factors, which are then picked up by different cells. The word hormone is derived from the Greek *hormao* meaning '*I excite or arouse*'. Hormones communicate this effect by their unique chemical structures recognized by specific receptors on their target cells, by their patterns of secretion and their concentrations in the general or localized circulation [1, 2]. *Sensory signals* are also received from the environment, in the form of light, taste, sound, smell, and touch [2]. The ability of an organism to function normally is dependent on all the cells of its different organs communicating effectively with their surroundings. Once a cell picks up a hormonal or sensory signal, it must transmit this information from the surface to the interior parts of the cell—for example, to the nucleus. This occurs via **signal transduction pathways** that are very specific, both in their activation and in their downstream actions. Thus, the various organs in the body respond in an appropriate manner and only to relevant signals.

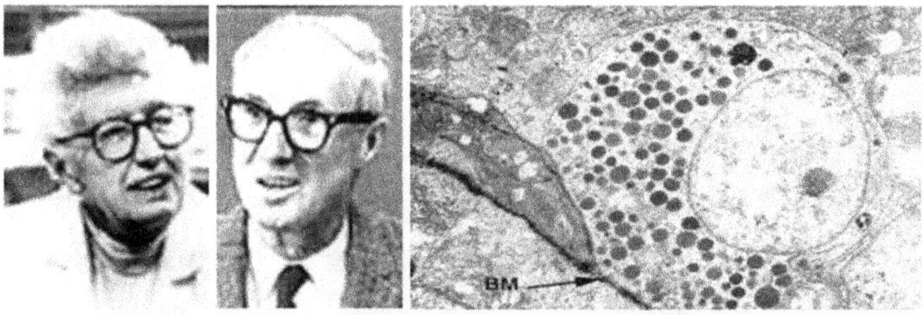

Berta Vogel Sharrer (1906–1995) and Ernst Scharrer (1905–1965) breakthrough: secretory neurons found in the fish brain. Established the concept of neuro-secretion (1928), the basis for DNES concept (D-iffuse N-euro E-ndocrine S-ystem), as an electron micrograph of entero-endocrine cell is showing and was also the subject of this author's Doctoral Thesis in 1984.

All signals received by cells first interact with specialized proteins in the cells called receptors, which are very specific to the signals they receive. These signals can be in various forms. The most common are chemical signals, which include all the hormones and neurotransmitters secreted within the body as well as the sensory (external) signals of taste and smell. The internal hormonal signals include **steroid** and **peptide** hormones, **neurotransmitters**, and **biogenic amines**, all of which are released from specialized cells within the various organs. The external signals of smell, which enter the nasal **compartment** as **gaseous** chemicals, are dissolved in liquid and then picked up by specialized receptors. Other external **stimuli** are first received by specialized receptors (for example, light receptors in the eye and touch receptors in the skin), which then convert the environmental signals into chemical ones, which are then passed on to the brain in the form of electrical impulses through peripheral nerve signaling pathways in order to reach CNS (Fig. 7.2) [1, 2].

Once **a receptor** has received a signal, it must transmit this information effectively into the cell. This is accomplished either by a series of biochemical changes within the cell or by modifying the membrane potential by the movement of ions into or out of the cell. Receptors that initiate biochemical changes can do so either directly via **intrinsic** enzymatic activities within the receptor or by activating **intracellular** messenger molecules) (Fig. 7.3). Receptors may be broadly classified in four groups that differ in their mode of action and in the molecules that activate them.

1. The largest family of receptors is the **G-protein-coupled receptors (GPCRs)**, which depend on **guanosine triphosphate** (GTP) for their function. Many neurotransmitters, hormones, and small molecules bind to and activate **specific G-protein-coupled receptors** [4].
2. A second family of membrane-bound receptors is the **receptor tyrosine kinases (RTKs)**. They function by **phosphorylating** themselves and recruiting downstream signaling components [4]. Receptors that have inherent tyrosine kinase activity bind molecules that have a specific SH2 domain (src homology domain). In turn, another accessory protein may be activated such as SOS (son-of-sevenless). This can activate a monomeric G-protein known as Ras that essentially acts as a signal transduction switch [4]. Its activation can lead to

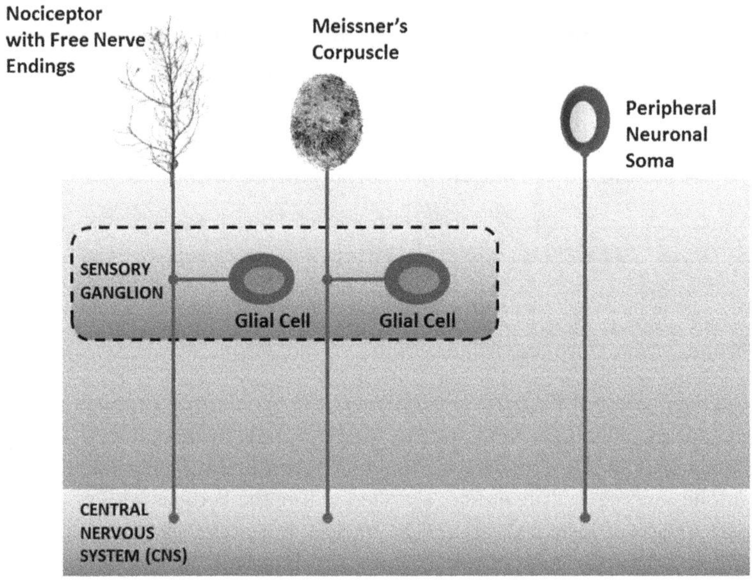

Fig. 7.2 Comparison of peripheral nerve signaling pathways with and without ganglion into the central nervous system

phosphorylation of Raf, MEK and eventually to mitogen activated protein kinase (MAPK) which can initiate transcription (termed the MEK-MAPK pathway).

3. Receptors for GH, prolactin, erythropoietin, insulin and a variety of cytokines and growth factors do not have inherent protein kinase activity but are **associated with a protein that has tyrosine kinase activity**. One of these proteins, known as JAK (just another kinase) may activate downstream effectors that include the STAT proteins—the JAK-STAT pathway. Binding of insulin to its receptor induces phosphorylation of insulin receptor substrate proteins (IRS) which activates further signal transduction pathways including activation of nuclear transcription factor κB (NF-κB). In essence, there is a cascade of protein phosphorylations that ultimately end in the nucleus to induce transcription. The *transcription factor targets for kinases that are activated by protein and peptide hormones* include **c-jun** and **c-fos** which make up the heterodimeric AP-1 complex, the serum response factor (often targeted by the **MAP kinase** dependent pathway), and nuclear **CREB-P** (cAMP response element binding protein) which is phosphorylated by protein kinase A and enhances transcriptional activity of closely positioned promoters [4].
4. **Ion channels** are proteins open upon activation, thereby allowing the passage of ions across the membrane. Ion channels are responsive to either *ligands* or to *voltage* changes across the membrane, depending on the type of channel. The movement of ions changes the membrane potential, which in turn changes cellular function [1].

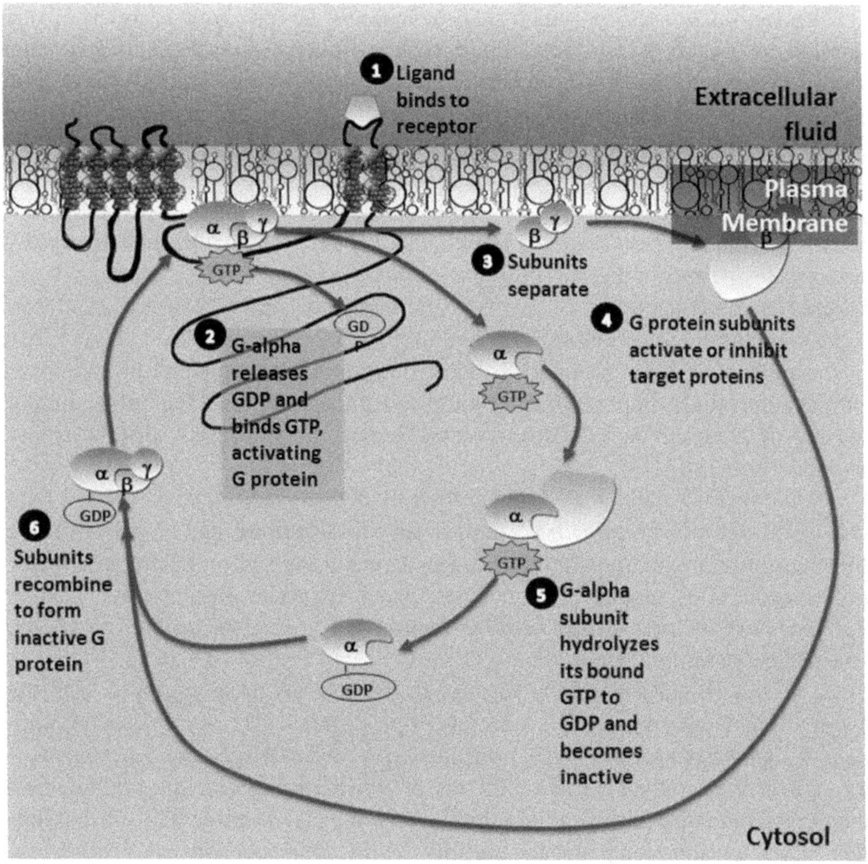

Fig. 7.3 Signal transduction pathways and transcriptional (nuclear) actions of protein and peptide hormones

Chemical Structures of the Three Major Classes of Human Hormones

Hormones are derived from amino acids, from cholesterol, or from phospholipids [1–4]. By far the most numerous are the *protein or peptide hormones*, ranging in size from just 3 to over 200 amino acids. Some hormones, such as insulin, are made up of two sub-units joined by disulfide bonds between two cysteine molecules whilst the glycoprotein hormones of the anterior pituitary gland are not only made up of two protein sub-units but also have complex sugar moieties attached. Other hormones include those derived from tryptophan (serotonin and melatonin) and those derived from fatty acids (eicosanoids).

The *steroid hormones*, which include vitamin D and those secreted by the adrenal cortex and gonads, are derived from cholesterol. All adrenal and gonadal steroids have the same basic ring structure and despite superficial 2-D structural similarity, the side chains and spatial orientation generate specificity. The third group of hormones is those *derived either from tyrosine or from tryptophan.* A single tyrosine molecule yields the catecholamines: epinephrine and norepinephrine, the latter being both a neurotransmitter and a hormone. In the endocrine system, these hormones are secreted by the adrenal medulla and are rapidly broken down once released into the circulation. The thyroid hormones are formed by the conjugation of two tyrosine molecules and resemble steroid hormones in binding to serum proteins and in the mechanism of action. Tryptophan is the precursor of serotonin (5-hydroxytryptamine) and melatonin synthesis. Finally, hormones derived from lipids and phospholipids include the major classes of eicosanoids including prostaglandins, prostacyclins, thromboxanes and leukotrienes [4].

Steroid receptors are located within the cell (cytosol and/or nucleus). They bind cell-permeable molecules such as steroids, **thyroid hormone**, and vitamin D. Once these receptors are activated by ligand, they **translocate** to the nucleus, where they bind specific DNA sequences to modulate gene expression. Receptors for *peptide hormones* **are located on the cellular membrane** (glucagon, insulin, bombesin, enkephalin, endorphin, etc.).

To convey the message from the signal, the cell has to be equipped with the system of molecules which will successfully propagate it. The intracellular component of signal propagation (signal transduction) is receptor-specific. A given receptor will activate only very specific sets of downstream signaling components, thereby maintaining the specificity of the incoming signal inside the cell. In addition, signal transduction pathways **amplify** the incoming signal by a signaling cascade (molecule A activates several molecule B's, which in turn activate several molecule C's) resulting in an appropriate physiological response by the cell (Figs. 7.3 and 7.4).

Several small molecules within the cell act as intracellular messengers. These include **cAMP**, cyclic **guanosine monophosphate (cGMP)**, **nitric oxide (NO)**, and **Ca^{2+} ions**. Increased levels of Ca^{2+} in the cell can trigger several changes, including activation of signaling pathways, changes in cell contraction and **motility**, or secretion of hormones or other factors, depending on the cell type. Increased levels of **nitric oxide** cause relaxation of smooth muscle cells and vasodilation by **increasing cGMP levels** within the cell. Increasing cAMP levels can modulate signaling pathways by activating the enzyme protein kinase A (PKA).

One of the most important functions of cell signaling is to control and maintain normal physiological balance within the body [1–5]. Activation of different signaling pathways leads to diverse physiological responses, such as cell **proliferation**, death, differentiation, and metabolism. Signaling pathways in cells may also interact with each other and serve as signal integrators. For example, negative and positive feedback loops in pathways can modulate signals within a **pathway**; positive

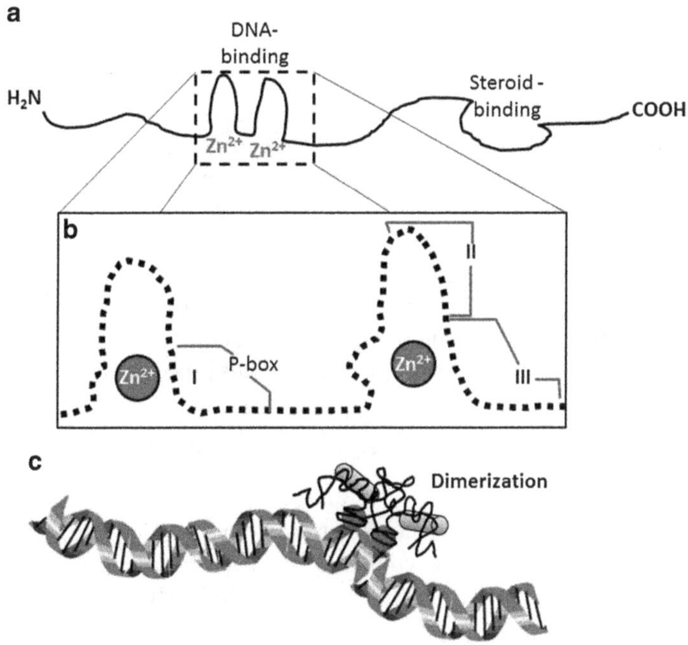

Fig. 7.4 Steroid receptors, zinc fingers and DNA binding

interactions between two signaling pathways can increase duration of signals; and negative interactions between pathways can block signals.

A. *Generalized structure of all steroid hormone receptors showing the different domains, location of the zinc fingers and the regions of the receptor responsible for transcriptional activity (TAF).*
B. *Two-dimensional structure of the zinc fingers of the DNA binding domain (DBD) in a single receptor. I, II and III indicate the helical regions of the DBD. The first helix contains the P box which determines the specificity of the DNA binding. The three amino acids that determine whether the receptor will combine with a glucocorticoid response element (GRE) or an estrogen response element (ERE) on the DNA are indicated. Arrows indicate the different amino acids that convert GRE specificity to ERE specificity. Amino acids shown as solid circles indicate those that are important for dimerization of two receptors.*
C. *Diagram showing dimerization of two receptors and helix I of each receptor slotting into the helix of the DNA. The base sequences of the ERE and GRE are shown plus the palindromic sequence. An example of a direct repeat sequence is also shown.*

The functions of hormones as regulatory, controlling molecules in the body can be broadly grouped into several categories: reproduction and sexual differentiation; development and growth; maintenance of the internal environment; and regulation of metabolism and nutrient supply. A single hormone may affect more than one of these functions and each function may be controlled by several hormones [2, 4, 5]. For example, thyroid hormone is essential in development as well as many aspects of homeostasis and metabolism, whilst glucocorticoids, such as cortisol, are important both in growth and nutrient supply and are also modulators of immune function. The roles several hormones play in one function is exemplified by the control of blood glucose which involves the pancreatic peptide insulin and its counter regulatory hormone, glucagon, as well as cortisol, growth hormone and epinephrine. Hormones act in concert and thus, an abnormality in a controlled variable, such as blood glucose concentration may result from defects in the control of any one of several hormones.

As mentioned, signal transduction occurs when an extracellular signaling molecule activates a cell surface receptor. In turn, this receptor alters intracellular molecules creating a response. There are two stages in this process:

1. A signaling molecule activates a specific receptor protein on the cell membrane.
2. A second messenger transmits the signal into the cell, eliciting a physiological response.

Chemical signal processing is altogether more elusive. However, it can still be conceived of, at the level of an individual cell, as a state-dependent transformation from input signals to output responses within a given cellular context. Schematically:

$$\text{INPUTS} \times \text{STATE} \times \text{CONTEXT} \rightarrow \text{OUTPUTS}$$

Such a formula serves to delineate those aspects of the system that are important and around which this paper will be organized. The INPUT signals are typically under the control of the experimenter. The STATE refers to the internal condition of the particular cell being studied, prior to stimulation by signals, and depends on the type of cell, its age and passage number, growth conditions, etc. The CONTEXT refers to the external environment in which the cell finds itself, including other cells as well as the medium in which the cells are present. The OUTPUTS are typically measures of cellular response over time. The complex molecular machinery within the cell implements the transformation from INPUTS to OUTPUTS. Both the STATE and the CONTEXT can conceal much biological subtlety.

What is the difference between **autocrine**, **paracrine** and **endocrine** types of secretion within endocrine system and/or DNES?

Autocrine signaling is a form of signaling in which a cell secretes a hormone, or chemical messenger (called the autocrine agent) that binds to autocrine receptors on the same cell, leading to changes in the cell. This can be contrasted with paracrine signaling, intracrine signaling, or classical endocrine signaling. Complete endocrine communication is under control of hypothalamo-pituitary orchestration of leading hormone.

An example of an autocrine agent is the cytokine interleukin-1 in monocytes. When this is produced in response to external stimuli, it can bind to cell-surface receptors on the same cell that produced it.

Paracrine signaling is a form of cell signaling in which the target cell is close to ("para" = alongside of or next to, but this strict prefix definition is not meticulously followed here) the signal releasing cell [2, 4]. The signal chemical is called the paracrine agent or paracrine hormone. The distinction is sometimes made between paracrine and autocrine signaling. In both types of signaling, the signal is limited to other cells in the local area. However, paracrine signaling affects cells of a different type than the cell performing the secretion, while autocrine signaling affects cells of the same type. Examples of paracrine signaling agents include growth factor and clotting factors. Growth factor signaling plays an important role in many aspects of development [4]. In mature organisms paracrine signaling functions include responses to allergens, repairs to damaged tissue, formation of scar tissue, and clotting. Overproduction of some paracrine growth factors has been linked to the development of cancer. Other examples of paracrine agents are somatostatin and histamine.

Endocrine signaling: A hormone/chemical released by a specialized group of cells into the circulation and acting on a distant target tissue defines the 'classical' endocrine and neuroendocrine signaling mechanism [4]. The examples are numerous, but let us take that thyroxin is secreted from thyroid gland and has effect on the brain and heart.

Soluble, diffusible ligands are used for communication between the cells that are not physically connected or are separated by long distances. The target cells have specialized proteins known as receptors, which are located on the surface but anchored in the membrane. **Once a receptor has received a signal, it must transmit this information effectively into the cell**. This is accomplished either by a series of biochemical changes within the cell or by modifying the membrane potential by the movement of ions into or out of the cell.

Effector protein elicits a response within the cell: usually activation or inhibition of a system. An effector molecule is usually a small molecule that selectively binds to a protein and regulates its biological activity. In this manner, effector molecules act as ligands that can increase or decrease enzyme activity, gene expression, or cell signaling. Effector molecules can also directly regulate the activity of some mRNA molecules (ribo-switches).

In some cases, proteins can be considered to function as effector molecules, especially in cellular signal transduction cascades. The term effector is used in other fields of biology. For instance, the effector end of a neuron is the terminus where an axon makes contact with the muscle or organ that it stimulates or suppresses.

Example of effectors are: Allosteric effectors, Bacterial effectors, Fungal effectors.

How are the hormones regulated within the body?

The correct (i.e. within an acceptable range) concentration of hormones must be maintained because hormones have powerful effects on the body. Feedback systems are an ideal means of controlling hormone levels because they involve constant monitoring and making adjustments to keep hormone levels stable. That is particularly important in the case of hormone levels because:

- Hormones can affect target organs at low concentrations so even a small quantity can sometimes be too much.
- The length of time during which hormones remain active is limited so more hormones must be secreted as necessary to replace those that are inactivated.

Feedback Mechanism for Regulation of Hormone Secretions

- *Negative feed-back.* One distinctive feature of hormones whose secretion is regulated through the hypothalamus and pituitary is that they regulate their own secretion through negative feedback inhibition [2, 4]. What this means is that a hormone from a peripheral gland, for example, cortisol, binds to its receptor on cells in the hypothalamus and adenohypophysis, and has the effect of inhibition. The secretion of hormones is subject to negative feedback control, and there are several ways by which this is achieved [1, 2, 4]. Feedback loops may involve the hypothalamo-pituitary axis that detects changes in the concentration of hormones secreted by peripheral endocrine glands or a single gland may both sense and respond to changes in a controlled variable. The integration of feedback loops involving several hormones may be complex. Disturbances in feedback loops are clinically important and their significance in diagnosis is pivotal.
- *Positive feedback*: production of a product by the target tissue stimulates additional hormone production through hypothalamo-hypophyseal control system [4].

Emphasizing Bioengineering Aspects to Endocrine Control

The functions of hormones as regulatory, controlling molecules in the body can be broadly grouped into several categories: reproduction and sexual differentiation; development and growth; maintenance of the internal environment; and regulation of metabolism and nutrient supply. A single hormone may affect more than one of these functions and each function may be controlled by several hormones (Fig. 7.5) [2, 4, 5].

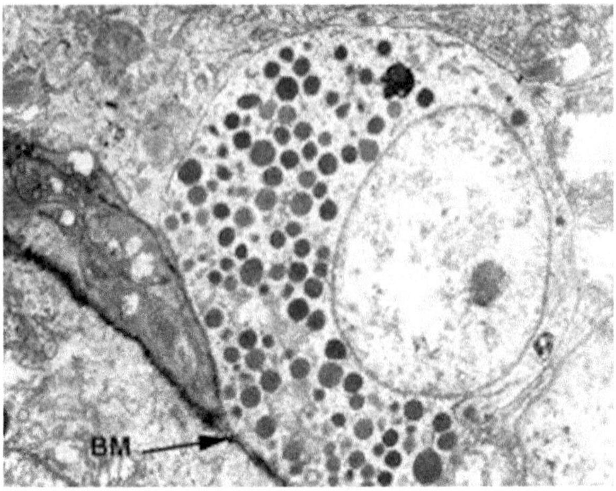

Fig. 7.5 Glucagon-related cells of pancreas with granules filled with hormone. Electron microscope micrograph. Similar can be found in: Mirjana Pavlovic (1984): Protein nutrition and some histological–histochemical properties of rat duodenal mucosa, Belgrade University of Medicine, Belgrade, Serbia—PhD Thesis

Some of Bioengineering Solutions Applied in Hormonal Regulation

There are different and numerous programs all over the world that are supporting research on and development of new technologies for the diagnosis, monitoring, or treatment of diabetes and other endocrine and metabolic diseases [4, 6]. Studies are aimed at developing: technologies and resources to analyze and interpret *protein structure*; large networks of either metabolite pathways (*metabolomics*) or proteins (*proteome*); and resources and tools to address the many roles of cellular proteins in the physiology and pathophysiology of metabolic diseases, such as protein capture reagents, or 3-D protein structure. The program supports research addressing *the development and implementation of glucose sensors* as well as their combination with *insulin delivery systems to form a 'closed-loop' artificial pancreas*. The program also promotes studies that focus on the development of *in vivo* molecular and functional imaging techniques to visualize and monitor physiological or metabolic processes and tissues (e.g., pancreatic beta cell mass, human brown adipose tissue). The work on transformation stem cells into functional; beta-pancreatic island cells, is going on as well [6].

One of the greatest bioengineering achievements in the field of endocrinology is probably artificial pancreas [7]. Artificial pancreas (AP) systems offer an important improvement in regulating blood glucose concentration for patients with type 1 diabetes, compared to current approaches. It consists of sensors, control algorithms and an insulin pump [7]. Different AP control algorithms such as proportional-integral-derivative, model-predictive control, adaptive control, and fuzzy logic control have been investigated in simulation and clinical studies in the past three decades. The variability over time and complexity of the dynamics of blood glucose concentration, unsteady disturbances such as meals, time-varying delays on measurements and insulin infusion, and noisy data from sensors create a challenging system to AP [7] (Fig. 7.6).

Adaptive control is a *powerful control technique* that can deal with such challenges. There is a spectrum of adaptive control techniques for blood glucose regulation with an AP system developed [7]. The investigations and advances in technology produced impressive results, but there is still a need for a reliable AP system that is both commercially viable and appealing to patients with type 1 diabetes [7]. Accurate closed-loop control is essential for developing artificial pancreas (AP) systems that adjust insulin infusion rates from insulin pumps. Glucose concentration information from continuous glucose monitoring (CGM) systems is the most important information for the control system. Additional physiological measurements can provide valuable information that can enhance the accuracy of the control system. The artificial pancreas works in a unique way as it uses a polymeric gel which responds directly to changes in glucose levels by releasing greater or smaller amounts of insulin [7]. The **insulin is delivered into the peritoneum, which allows the insulin to work on reducing blood glucose quicker than insulin that is delivered into the fat under the skin, as is the case with injections and insulin pumps**.

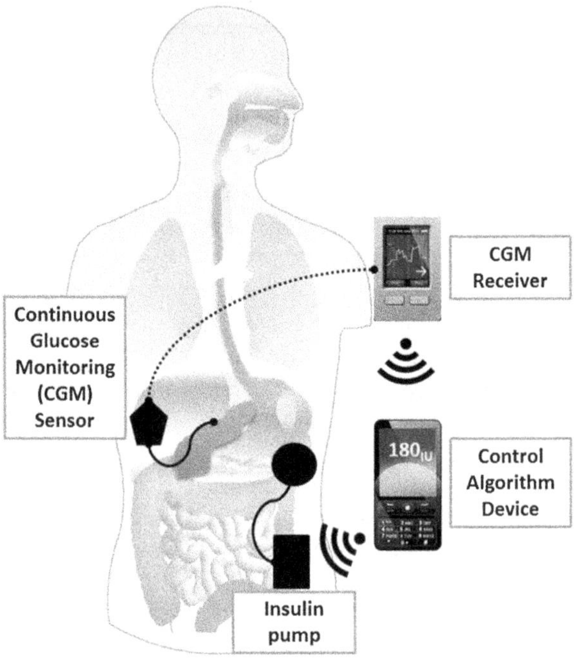

Fig. 7.6 (**a, b**) Artificial pancreas overview

The newest artificial pancreas has been developed by Professor Joan Taylor of De Montfort University with Bruce Renfrew and Mike Phillips of the Renfrew Group. Speaking about the technology, Prof. Taylor said: "This incredible device will not only remove the need to manually inject insulin, but will also ensure that perfect doses are administrated each and every time. By controlling blood glucose so effectively, we should be able to help reduce related health problems." The artificial pancreas is a great piece of British design and engineering and is due to soon start clinical trials to test the product. Studies showed that the artificial pancreas system was able to increase the amount of time study participants spent in the right blood glucose level range by 22 % [7]. The system is still being tested for safety and has yet to be used in non-controlled conditions. Therefore this newest artificial pancreas is not available for public trials yet.

References

1. Andjus, R.K.: General Physiology and Biophysics. Modules 1–7. Center for Multidisciplinary Studies. University of Belgrade, Belgrade, Serbia (2002)
2. Hall, G.E., Guyton, A.C.: Textbook of Medical Physiology. Sounders Elsevier, Philadelphia (2011). ISBN 978-1-4160-4574-8
3. Saltzman, M.W.: Biomedical Engineering. Cambridge University Press, New York (2009). hardback. ISBN 978-0-521-840099-6

4. Nussey, S., Whithead, S.: Endocrinology: An Integrated Approach. BIOS Scientific, Oxford (2001). http://www.ncbi.nlm.nih.gov/books/NBK22/
5. Pavlovic, M.: Protein Nutrition and Histologic—Histochemical Properties of Rat Duodenal Mucosa. Doctoral Thesis, University of Belgrade, Belgrade, Serbia (1984)
6. Pavlovic, M., Balint, B.: Stem Cells and Tissue Engineering. Springer Briefs in Electrical and Computer Engineering. Springer, New York (2013). pp. i–xii, 1–153. ISBN 978-1-4614-5504-2
7. Turksoy, K., Cinar, A.: Adaptive control of artificial pancreas system—a review. J Healthc. Eng. **5**, 1–22 (2014)

Chapter 8
Communication III (Immunological Control)

Once you die it only takes a few weeks for these organisms to completely dismantle your body and carry it away, until all that's left is a skeleton. Obviously your immune system is doing something amazing to keep all of that dismantling from happening when you are alive.

Marshall Brain (1951–)

One of controlling systems in the body which the body also uses for communication (externally and internally) is the Immune system, based upon existence of MHC molecules fundamental for recognition, and other molecules responsible for antigen-presentation and immediate or postponed reaction to it. The fundamental unique feature of immune system cells is the capability of distinguishing "self" from "non-self" cells and proteins. Communication between different cell types of the immune system is critical in the recognition of self, surveillance, defense, and clearance of foreign invaders. These signaling mechanisms involve direct cell–cell signaling as well as autocrine and paracrine signaling. The essential feature of particular cells of immunological system is memory and although still known at the level of phenomenology, presents the basis for vaccines.

Breakthroughs in Immunology and background for vaccines (small pox and rabies) and later, Rational Vaccine Design (RVD): Edward Jenner (1749–1823), Louis Pasteur (1822-1895), Elie Metchnikoff (1845–1816), NP 1908 for discovery of phagocytosis.

M. Pavlovic, *Bioengineering*, DOI 10.1007/978-3-319-10798-1_8

Communication III: Immune System and Regulation of Communication

The Adaptive Immune System: Signaling Mechanism

The unique feature of immune system cells is their capability to distinguish "self" from "non-self" (cells and proteins). Communication between different cell types of the immune system is critical in the recognition of self, surveillance, defense, and clearance of foreign invaders [1, 2]. These signaling mechanisms involve direct cell–cell signaling as well as autocrine and paracrine signaling. Direct cell to cell signaling is the best presented through antigen presentation of antigen presenting cells [macrophages, dendritic cells (DC) and B-cells] to T-naïve cells which will process the information on antigen epitopal features through T-Cell Receptor (TCR) and become educated, memory T-cells. This principle is used in rational designed vaccines (RVD) against bacteria and viruses, especially. B-cells also have a memory after acquiring antigen, but that process in the B-cells is less understood, although interesting novel discoveries are emerging [2–5]. The molecular base of immune system memory cells is still elusive and although we are **using the term**, **we still do not understand** the fundamental processes leading to that memory. Yet, with existing knowledge in mind we can design successful vaccines sometimes. There is still a lot to be understood about this very subtitle type of communication.

The communication between antigen-presenting cells and T-cells is based on the existence of **MHC** molecules (Major Histocompatibility Class) I and II that could be found on APC, reflecting two different types of molecules that are differently processed through different APC and then presented to T-cells. Almost all cells in the organism can be infected and they possess MHC molecules, but only immune system, Antigen Presenting Cells can efficiently communicate to T-cells through their MHC molecule and TCR of T-cells in order to prime them and teach them about antigen epitope features. In response, T-cells (T CD4+ and/or TCD8+) will either produce spectrum of cytokines in response in order to induce the production of specific antibodies from B-cells, or express cell-mediated cytotoxicity and kill infected cells, thus eliminating antigen [6].

T-Cell Receptor Signaling

During antigen recognition, cell–cell communication is mediated by a ligand on the surface of APC (it is MHC complex), binding to a receptor on the T cell (TCR complex). The body contains an inventory or repertoire of T cells bearing a large variety of a very different TCRs; each version of TCR is capable of recognizing a single antigen, so that population contains T-cells, that will recognize virtually any antigen. Each T-cell expresses only one type of TCR, specific for a particular antigen. Thus, millions of T-cells each with an antigen specific receptor continually sample the surface of Apes to determine whether the presented peptide matches the binding site for the receptor they carry. This scanning process takes place in the lymph draining nodes where circulating APCs and T cells most meet. T cell briefly binds to MHC-peptide complex on the surface of APC. It is not a simple matching process but requires multiple signals: the involvement of co-stimulatory molecules, for full T-cell activation (B7 has to match to CD28). When a match occurs, signal transduction pathways are activated. This, multiple receptor-legend interactions must simultaneously occur before the specialized signal of antigen recognition can be transduced into T-cell. After recognition, T-cells will be activated and undergo clonal expansion caused by autocrine signaling of a cytokines called interleukin 2 (IL-2) which is a growth factor. Thus, many copies of the antigen-specific T-cells are produced.

Cytotoxic T-cells (Tc), activated through its binding to peptide-MHC Class I ligand, will kill the target cells by releasing granules with degrading enzymes for target, effector cells (granzymes, perforins, etc.) and destroy virus-infected cell, freeing other host cells from further infection.

Second category of T-cells, T-helper–Th cells with matching peptide MHC class II will employ paracrine signaling with secretion of cytokines which bind either macrophages or B–0 cells and activate them. So, activated macrophages will engulf and kill antigen, while B-cells will secrete antibodies and neutralize antigen (Fig. 8.1).

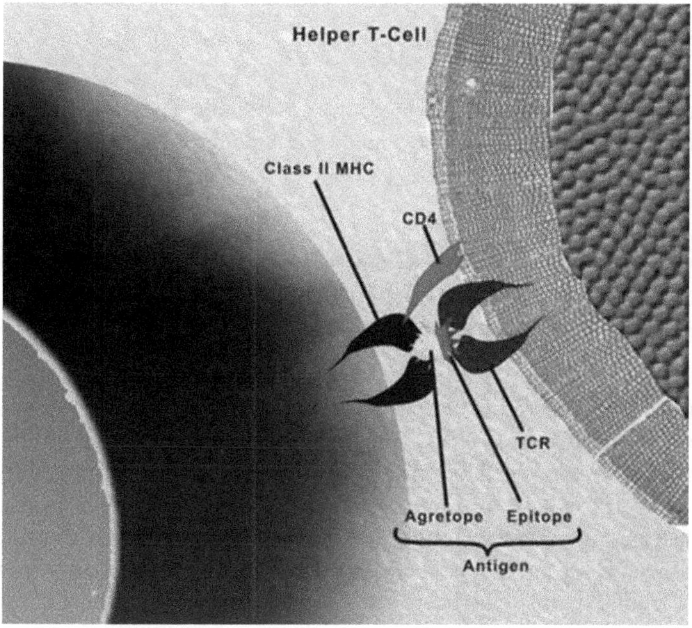

Fig. 8.1 Antigen presentation

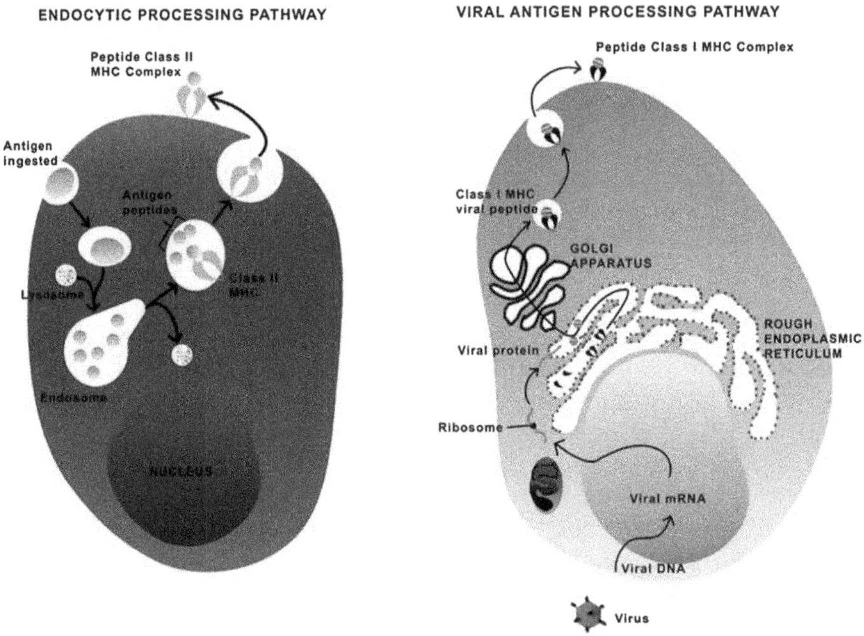

Fig. 8.2 Two different types of antigen processing in APC

Cytokine Signaling

Cytokines play an important role in adaptive immunity. They act using autocrine, paracrine and endocrine mechanism and they are numerous (Fig. 8.2).

Emphasizing Bioengineering Aspect to Immunological Control and Communication: Engineering Vaccines and Rational Vaccine Design (RVD)

A vaccine is any preparation used as a preventive inoculation to confer immunity against a specific disease, usually employing an innocuous form of the disease agent, as killed or weakened bacteria or viruses, to stimulate antibody production. The word vaccine is derived from *vacca*, (Latin for cow) [7]. The science of vaccination began with **Edward Jenner** in 1796 and his observation that milk maids who contracted cowpox due to their exposure to farm cows became immune to small pox from the pus in the blisters formed by cowpox [8]. Jenner subsequently tested his hypothesis on an 8-year old boy and was successful [8]. He inoculated the boy with cowpox blisters initially from which he developed only a mild illness.

Later the same boy was inoculated with small pox or variola particles. The child showed immunity developing no sign of disease [8]. Thus, a new field of preventive medicine was born and vaccine development has been a major biomedical concern ever since.

Despite the overall success of vaccination efforts in this modern era, there is still a great need for new and improved vaccines which cannot be met easily [9]. Even though vaccination is probably the most beneficial therapy that a physician can provide a patient, there are still significant roadblocks to the development and licensing of new vaccines [9]. The greatest roadblock is the lack of a complete understanding of how the human immune system "works". Early vaccines were developed using technology from the 19th and 20th centuries: inactivation by heat, chemicals, and irradiation to produce a killed vaccine, vaccination with a serologically related virus a' la Jenner, and attenuation by tissue culture passage to produce live vaccines with substantially reduced virulence [9]. These methodologies have failed to usher in vaccines against new and emerging diseases. Unmet targets for vaccine development include some of the more difficult infectious agents, such as *human immunodeficiency virus* (HIV), *Ebola*, cytomegalovirus, *Dengue* virus, *Human Parvovirus B19* and *severe acute respiratory syndrome coronavirus*; bacteria, such as *Pseudomonas aeruginosa*, *Neisseria gonorrhea*, *or Mycobacterium tuberculosis*; and parasitic diseases, such as malaria or hookworm disease [9, 10]. In the upcoming years vaccines towards diseases of this caliber will be developed by improvements on the basic techniques mentioned above and through the use of new technologies based on the expanding understanding of the immune response [9, 10].

In today's world, vaccine design is not limited towards elimination and prevention of infectious diseases. For example, Bioterrorism has brought renewed interest to new and large-scale vaccine development [9]. Furthermore in the developed world chronic illnesses are of greater concern than infectious diseases. Thus, there are the trials for the vaccines that will also be developed as therapies against disease for autoimmune diseases (lupus—SLE), cancer, hypertension, Alzheimer's dementia, contraception, and to promote the cessation of bad habits, such as smoking [9].

What all the aforementioned diseases have in common, particularly those of an infectious nature, is the involvement of the immune system. Understanding the immune system requires the work of not only natural scientists—biologists, chemists, physicists etc.—but also the involvement of applied scientist's as well; namely engineering and computational experts. Such an interdisciplinary approach is promising as research endeavors are now a part of the post genomic era, a common acceptance that all diseases have a genetic component. In microbiology the pathogen's genome is equally important to the host's genome in the establishment of a disease state. Experts in human and microbial genomic exploitation and information extrapolation, both of which rely on the domain expertise of the applied scientist must form a bridge with the biologist in an effort to develop better and more effective vaccines. While it is no secret that disciplines such as bioengineering, bioinformatics, and artificial intelligence cannot replace traditional wet-lab biology, it cannot be disputed that these disciplines and their associated tools have accelerated biomedical research at astounding rates. Undeniably, the development of a successful vaccine towards one ailment may very well serve as a gateway to the elucidation of novel immune system mechanisms as well as a vaccine model towards other disease targets.

Rational Vaccine Design

The idea behind ration or cellular vaccine design as it pertains to viruses is that viral properties (proteins of their "body") can be exploited for sensible vaccine design. This is similar to the design of subunit vaccines which are made from microorganism fragments such as viral surface proteins. Vaccines towards the Human Papilloma Virus (HPV) and Hepatitis B Virus (HBV) are both subunit vaccines. The difference here is the design of "super vaccines" which rely on epitopes-antigenic determinants, usually made of protein, that are recognized by the immune system. Super-vaccines are therefore compact forms of "pseudo-virus" that cover the diversity of the virus being studied.

Rational vaccine design seeks to manipulate the immune system to "work harder". This might be possible if the number of responding immune cells targeted by a vaccine is increased upon vaccination and later on during an immune challenge. Thus it is practical to explore improved vaccine design that is based on the cellular arm of the immune system while focusing on a specific pathogen—Parvovirus B19, Dengue virus, Ebola virus, etc. In the post genomic era all potential antigens, which are coming into consideration for inclusion into a vaccine formulation, are well known [9, 10]. This knowledge has been exploited in the context of reverse vaccinology–driven approaches, which in combination with comparative genomics enable us to select the most highly conserved and promising antigens for vaccine design [9, 10]. Therefore the issue isn't identification of the best epitopes. Rather the roadblocks to rational vaccine design (RVD) are as follows.

Roadblocks Toward RVD [11]

1. Knowledge on the effector mechanisms responsible for the clearance of these pathogens is by and large fragmentary.
2. The availability of tools enabling the stimulation of predictable immune responses of the adequate quality following vaccination. In fact, highly purified antigens are often less immunogenic than more complex preparations, rendering essential their co-administration with potent adjuvants (chemical agents that stimulate the immune system).
3. The need to bridge the translational gap, as well as current stringent regulations for vaccine testing

There are 3 roadblocks listed but 1 holds more importance than the remaining 2, the first. There are still unanswered questions as to how exactly the immune system responds to pathogens. This is why both computer simulation studies and clinical studies of infected individuals are necessary. Yet the latter can be enhanced by the former. Mathematics can help us understand some of the complex cellular and molecular processes that make up the immune system [7]. Thus modeling the activity of a vaccine's impact on the immune system can provide insight which can ultimately lead to the development—roadblock #2—and clinical testing—roadblock #3—of novel vaccines. Hence overcoming the first roadblock must happen first, and modeling and simulation studies may help that to occur sooner while enhancing our knowledge base of how exactly the immune system processes pathogens. This greater understanding can be achieved both by analyzing models that formalize the biological ideas and by using mathematical methods to extract information from experiments that may not be accessible to a more intuitive biological approach [12].

The Adaptive Immune System

The human immune system has two primary components: **the innate system** and the **adaptive system**. *The innate, nonspecific system* is present at birth and treats all infectious agents equally, meaning it does not distinguish between different species or types of viruses etc. [13]. Consequently vaccination efforts are not targeted towards stimulation of the innate line of defense. On the contrary, the *adaptive, specific, immune system* refers to defenses that involve specific recognition of a microbe once it has breached the innate immunity defense [13]. It is this system that is the target of vaccine development as it possesses the ability to confer memory or "immunity" to the individual or host.

The adaptive immune system is very complex and is based on the activity of white blood cells (WBCs) called lymphocytes. There are two major types of lymphocytes: B cells and T cells, each of which is involved in a specific branch of adaptive immunity. B cells contribute to the fluid or humoral response (by secreting antibodies into the blood), while T cells regulate and integrate the cell mediated response [13]. Both responses collectively represent the adaptive immune system.

The high degree of microbe specificity seen in adaptive immunity is a testimony to the complex molecular interactions that take place between the cells involved. These lymphocytes have to be activated by other white blood cells (WBCs). It is this comingling of cells that allows the human host to develop memory against pathogens it has seen before, and in terms of vaccination to create effective memory against the pathogen.

The Humoral Arm of Immunity

The humoral arm of immunity involves two major players: B cells and antibodies. B cells are a type of lymphocyte capable of secreting antibodies, and antibodies are proteins capable of binding antigens [13, 14]. An antigen is simply any chemical or particle that triggers an immune response. Antigens are typically recognized as being foreign to the host by the immune system. Humoral refers to the fact that antibodies are generally found in body fluids or humor. Thus the humoral response is most effective against pathogens such as viruses and bacteria that are circulating freely where the antibodies can contact them [13, 14].

Antibodies

Antibodies are a type of globular proteins that are very soluble [13]. They are often referred to as immunoglobulins due to their structure and function. Antibodies are generally produced in response to an antigen on a pathogen that has invaded the human host, and are capable of recognizing this antigen [13]. A single pathogen typically possesses several antigens which trigger the activation of several different antibodies at the same time [13].

The role of antibodies in immunity is of great significance. Antibodies bind the epitopes of a pathogen with both specificity and affinity [13]. The closer the fit is between the antibody and its epitope—a fragment of an antigen—the higher the affinity or binding energy between the pair [8]. Regarding specificity, antibodies are capable of discerning between structural isomers as well as minor differences in the amino acid sequence of a protein [13]. Though the antibody does nothing to the epitope, it marks the pathogen for elimination from the system by specific immune mechanisms carried out by other immune components. Ultimately this can lead to clearance of the pathogen.

B Lymphocytes

B Lymphocytes or B cells are primarily involved in antibody production. They are produced in the bone marrow where they undergo maturation before entering circulation [13]. Each B cell bears fixed immunoglobulins or antibodies on its surface which serve as a B cell receptors (BCR) capable of recognizing the same antigen or epitope [13, 15]. This accounts for the pronounced specificity of B cells towards specific pathogens. B cells must be activated by a specific epitope in order to trigger an immune response.

Prior to activation B cells are considered naïve. When a B cell's immunoglobulins bind to the epitope for which they become specific, the B cell is activated [13]. Once activated the B cell undergoes a process referred to as proliferation or clonal expansion [13, 15]. During clonal expansion the activated B cell proliferates into two classes of cells: plasma or effector cells and memory cells [8]. Effector B cells or plasma cells secrete antibodies while memory cells are long-lived and responsible for the enhanced secondary response to an antigen [13]. B cell proliferation serves two purposes. The first is to produce additional cells that can search the body for the pathogen bearing the epitope and the second is to confer immunity to the human host. This is achieved by the effector cells and memory cells respectively. Without B cells there are no antibodies, the two immune system components that mediate the humoral response. Thus vaccination towards any pathogen must seek to activate B cells.

The Cell Mediated Arm of Immunity

It can be argued that the cell mediated arm of immunity or cell mediated response picks up where the humoral arm leaves off. Intracellular antigens, such as a virus within an infected cell are not exposed to circulating antibodies and are therefore inaccessible to the humoral response [8]. T cells, the major players in the cell mediated response, probably evolved in response to this aspect of pathogenicity—the need to combat intracellular parasites [13].

Like B lymphocytes, T lymphocytes are produced in the bone marrow; however they migrate to the thymus, and organ in the upper chest, to undergo maturation [13, 15]. Also like B cells T cells undergo clonal expansion to form effector and memory cells [15]. The primary difference between the two lymphocytes is that unlike B cells, T cells never interact with native antigen or epitopes. On the contrary T cells are stimulated for activity *via* a process known as antigen presentation.

Antigen Presenting Cells

T cells do not interact directly with native antigens. Rather processed antigens are "presented" to T cells on their surfaces by a special groups of cells commonly referred to as antigen presenting cells (APCs) [13]. APCs take up antigen, process them *via* chemical reactions, and then place an antigenic fragment/epitope into a specialized receptor on their surface [13]. The processed, surface bound epitope is now fit for interaction with T cells. APCs then migrate to specialized regions of the immune system where T cells are abundant [13]. One primary location is the lymph node through which lymphatic fluid flows [13].

There are three major types of APCs all of which are WBCs: dendritic cells, macrophage, and B cells [13]. Macrophages and dendritic cells are found predominantly in lymphatic tissue and fluids and very important in their role as APCs [13]. Both ingest pathogens *via* phagocytosis—cell eating—and present pathogen specific antigens on their surfaces [13]. B cells are a component of the humoral response and their presentation of antigens is slightly different. B cells are not phagocytes; rather they take in antigens *via* endocytosis—cell invagination—after the antigen binds their antibody receptor [13]. All three types of APCs enhance the immune response by their interaction with T cells following antigen ingestion.

The Major Histocompatibility Complex

APCs have the ability to interact with T cells. This is accomplished only after these cells have loaded processed antigens into specific receptors on their cells. These specific receptors are referred to as the Major Histocompatibility Complex (MHC) or Human Leukocyte Antigen (HLA) system [13].

MHC proteins are protein receptors on cells in the human body. They function as cell recognition agents in the immune system. All non-immune system cells contain MHC type 1 receptors (MHC-1) and cells of the immune system contain type 2 MHC receptors (MHC-2). The type 3 MHC receptors are less understood but they are found on a variety of cells in the body [8]. MHC receptors vary among members of the human population. The name is derived from the necessity of a match or similarity between these receptors in organ donors and recipients; hence the histocompatibility—histo meaning tissue [13].

The concept is clear. When a cell, for example an epithelial cell, is infected by a virus, it digests the virus and loads a small epitope into its type 1 MHC receptor. Cells of the immune system that are involved in the cell-mediated response, such as T cells, can then interact with the infected cell to elicit an immune response. The T cell interacts with the MHC receptor *via* its own receptor—T cell receptor (TCR). The communication between the two cells is enough to trigger an immune response.

The existence of different subgroups of MHC molecules is not due to mere coincidence. There are different types of T cells that can interact with infected cells, different APCs, and different MHC receptors [8]. Therefore, this produces an opportunity to develop a vaccine that can target multiple immune systems cells that are involved in cellular immunity. Of primary importance are the cytotoxic T cells and the helper T cells.

Cytotoxic T Cells

Different classes of T cells are recognized by their surface markers or cluster differentiation [CD] protein receptors. Cytotoxic T cells are a group of lymphocytes that are recognizable by their CD8 markers [13]. CD8—cluster differentiation 8—is a transmembrane glycoprotein and also a co-receptor [14]. It has a preference for interacting with the MHC-1 receptors and antigens [13]. Upon activation, these CD8+ positive T cells undergo clonal expansion into memory cells and effector cells and it is the effectors cells that are capable of cytolytic activity [13]. Their ability to kill infected body cells renders these T lymphocytes most effective at eliminating viruses and other intracellular parasites from the human host. Dendritic cells are the APCs that are most effective at activating CD8+ T cells [14, 15].

Helper T Cells

While cytotoxic T cells interact heavily with dendritic cells, another type of T cell favors interaction with macrophages and B cells. This class is referred to as the CD4+ positive group of T cells. CD4 is a surface glycoprotein found on several cell types such as T-helper cells, macrophage, dendritic cells, and monocytes [13]. The latter of the three are phagocytes, which are capable of engulfing pathogens that have entered the body, and later presenting them to the helper T cells [13, 14]. The CD4 functions as a co-receptor and helps to activate the helper T cell. It also interacts with the MHC-2 antigens on the surface of the phagocytic cell [13].

Helper T cells are involved in recruiting or activating other cells of the immune system, namely B cells. B cells are critical as they are mainly involved in the humoral response system which produces antibodies. When activated by helper T cells, B cells also undergo clonal expansion to form memory cells and antibody producing plasma cells [13].

Example from Author's Collaborative Work: RVD for Ebola Virus

The Ebola virus (EBOV) is extremely lethal with mortality rates ranging from 23 to 90 %. Rational vaccine design toward the Ebola vaccine seeks to treat the immune system as a decoder as it is responsible for the processing of incoming "information". Enhancing the immune system's output by controlling its various components could ultimately lead to the discovery of novel vaccine development strategies and deeper understanding of the humoral and cell mediated immune responses of immunity.

No licensed Ebola vaccine exists and classical protocols for vaccine design do not comply. One solution, rational vaccine design (RVD) is based on two parameters:

1. Identification of epitopes, antigenic peptides that mediate the cellular immune system and
2. Exploitation of the immune system's ability to recognize and remember vaccines.

The Ebola virus not only poses a safety threat for bioterrorism, but serves as an excellent model to study for all viruses that cause human disease. In the post genomic era vaccine design will take on new techniques as well as reinvent some of the older means of producing vaccines. This type of progress will require the involvement of not only the microbiologists, but engineers, and informaticians as well to name a few. Current vaccines against the Ebola have yet to be tested at a time of crisis and there is much doubt surrounding their expected rate of efficacy.

The human immune system though very complex can be studied using the Ebola model with the hope of not only saving lives, but setting an example for rational vaccine design—exploitation of the virus' antigens and the immune systems natural ability to recognize, remove, and repair. Thus a super-vaccine geared towards Ebola should at least contain viral epitopes towards both MHC-2:CD4+ and MHC-1:CD8+ cell interactions. Activation of the former will result in B cell activation and expansion and subsequently antibody production towards the Ebola virus. Activation of the latter will supply the immune system with a fleet of cytotoxic effector cells capable of eliminating virus on contact. Additional epitopes could be used to involve other components of the immune system such as Natural Killer (NK) cells, macrophage, complement and other non-specific responses.

To assess RVD feasibility, EBOV proteins were computationally analyzed for epitope identification. To evaluate vaccine efficacy, mathematical models for virus dynamics were simulated using MATLAB. Models relied on data from EBOV cultivation in cell-cultures, and were extended with novel equations to consider memory B- and T-cell production.

First, RVD towards the EBOV is feasible. Computer-based protein analysis identifiednovelEBOVpeptidesforvaccinedesign.Akeyepitope–**EAIVNAQPKCNPN… MHNQDG**– was extracted from a three-dimensional structure of an EBOV protein bound to human antibody **KZ52**. Secondly, **vaccine efficacy can be assessed using mathematical models**. Multiple simulations of the *models revealed generally unknown parameters such as the virus' birth and cellular infection rates*. The models also quantified the cellular immune response necessary for vaccine efficacy in an individual; the specifications of what the vaccine must accomplish.

These results show that computer-aided (CADE) RVD is feasible and that mathematical models can establish RVD guidelines for the development of an EBOV vaccine, and not only that one (Fig. 8.3).

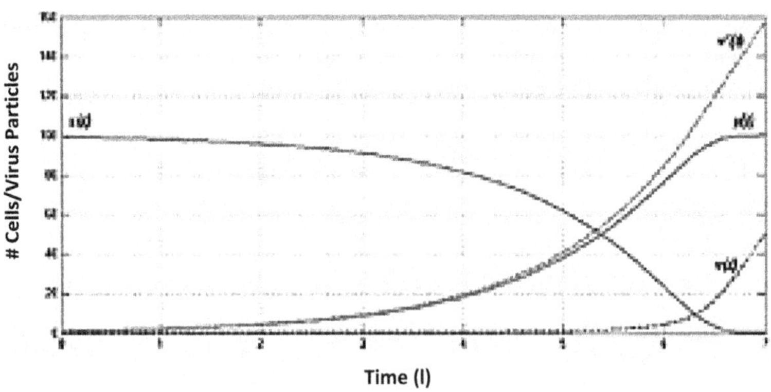

Fig 8.3 Ebola dynamics in unvaccinated system. Sophia Banton, Zvi Roth and Mirjana Pavlovic. Mathematical Modeling of Ebola virus Dynamics as a Step towards Rational Vaccine design. In: KEHerold, We Bentley, and J. Vossoughi (Eds): SBEC 2010 IFMBE Proceedings 32.:196-200
*This chapter and details of theoretical and practical approach are initiated and performed mostly by Sophie Banton, graduate student at FAU at that time, and Dr. Zvi Roth to whom the author is giving the full credit for. My participation as an immunologist was also useful

References

1. Pavlovic, M., Balint, B.: Stem Cells and Tissue Engineering. Springer Briefs in Electrical and Computer Engineering. Springer, New York (2013). pp. i–xii, 1–153. ISBN 978-1-4614-5504-2
2. Pavlovic, M., Cavallo, M., Kats, A., Kotlarchyk, A., Zhuang, H., Shoenfeld, Y.: From Pauling's abzyme concept to the new era of hydrolytic anti-DNA autoantibodies: a link to rational vaccine design? – A review. Int. J. Bioinform. Res. Appl. **7**(3), 220–238 (2011)
3. Pavlovic, M., Kats, A., Cavallo, M., Shoenfeld, Y.: Clinical and molecular evidence for association of SLE with parvovirus B19. Lupus **19**(7), 783–792 (2010)
4. Pavlovic, M., Kats, A., Cavallo, M., Chen, R., Hartmann, J.X., Shoenfeld, Y.: Pathogenic and epiphenomenal anti-DNA antibodies in SLE. Autoimmune Dis. 18 pages (2010). Article ID 462841
5. Cavallo, M.F., Kats, A.M., Chen, R., Hartmann, J.X., Pavlovic, M.: a novel method for real-time, continuous, fluorescence-based analysis of anti-DNA abzyme activity in systemic lupus. Autoimmune Dis. (2012). 814048
6. Banton, S., Roth, Z., Pavlovic, M.: A bioengineering approach for rational vaccine design towards the Ebola Virus. BMC Bioinformatics. **11**(S-10), O12 (2010)
7. Banton, S., Roth, Z., Pavlovic, M.: Mathematical modeling of Ebola virus dynamics as a step towards rational vaccine design. In: Herold, K.E., Bentley, W.E., Vossoughi, J. (eds.): SBEC 2010 IFMBE Proceedings, 32, 196–200
8. Tan, S.Y.: Edward Jenner (1749–1823): conqueror of smallpox. Singapore Med. J. **45**(11), 507–508 (2004). PMID: 15510320
9. Rosenthal, K.S., Zimmerman, D.H.: Vaccines: all things considered. Clin. Vaccine Immunol. **13**(8), 821–829 (2006)
10. Jones, S.M., Feldmann, H., Ströher, U., et al.: Live attenuated recombinant vaccine protects nonhuman primates against Ebola and Marburg viruses. Nat. Med. **11**(7), 786–790 (2005). doi:10.1038/nm1258. PMID 15937495
11. Martin, J.E., Sullivan, N.J., Enama, M.E., Gordon, I.J., Roederer, M., Koup, R.A., Bailer, R.T., Chakrabarti, B.K., Bailey, M.A., Gomez, P.L., Andrews, C.A., Moodie, Z., Gu, L., Stein, J.A., Nabel, G.J., Graham, B.S.: A DNA vaccine for Ebola virus is safe and immunogenic in a phase I clinical trial. Clin. Vaccine Immunol. **13**(11), 1267–1277 (2006)
12. Callard, R., Hodgkin, P.: Modeling T- and B-cell growth and differentiation. Immunol. Rev. **216**(1), 119–129 (2007)
13. Guzman, C.A.: Optimizing rational vaccine design. Microb. Biotechnol **2**(2), 136–138 (2009)
14. Tortura, G.J., Funke, B.R., Case, C.L.: Microbiology: An Introduction, 9th edn, pp. 514–516. Pearson Education, San Francisco (2007)
15. Hamilton, S.E., Harty, J.T.: Quantitation of CD8+ T cell expansion, memory, and protective immunity after immunization with peptide-coated dendritic cells. J. Immunol. **169**, 4936–4944 (2002)

Chapter 9
Stem Cells in Regenerative Therapy

Mankind is searching for a key to longevity and there is no doubt that stem cells could be an important answer to this problem.

Ratajczak M. (–present)

This chapter will teach you about origin, classification, features of stem cells and fundaments of stem cell therapy as the segment of cellular–based therapy. Generally, the Stem Cell (SC)—compartment is divided into embryonic and tissue specific or adult SCs. Paul Niehans, M.D., (1882–1971), the originator of cell therapy, wrote: "Cellular therapy is a method of treating the whole organism on a biological basis, capable of revitalizing the human organism with its trillions of cells by bringing to it those embryonic or young cells which it needs. Cells from all organs are at our disposal; the doctor's art is to choose the right cells. Selective cellular therapy offers new life to the ailing or diseased organism".

The first use of stem cells in humans was done by physicians who were tempted to use them in trying to treat hematological disorders. Stem cell transplantation was pioneered using bone-marrow-derived stem cells by a team at the Fred Hutchinson Cancer Research Center from the 1950s through the 1970s led by **Edward Donnall Thomas**, whose work was later recognized with a Nobel Prize in Physiology or Medicine [1]. Thomas' work showed that bone marrow cells infused intravenously could repopulate the bone marrow and produce new blood cells. His work also reduced the likelihood of developing a life-threatening complication called graft-versus-host disease. The first physician to perform a successful human bone marrow transplant was **Robert A. Good** at the University of Minnesota in 1968 [1]. With the availability of the stem cell growth factors (GM–CSF and G–CSF), most hematopoietic stem cell transplantation procedures are now performed using stem cells collected from the peripheral blood, rather than from the bone marrow. Collecting peripheral blood stem cells provides a bigger graft, does not require the donor to be

M. Pavlovic, *Bioengineering*, DOI 10.1007/978-3-319-10798-1_9

subjected to general anesthesia in order to collect the graft, results in a shorter time to engraftment, and may provide for a lower long-term relapse rate.

Breakthroughs: Edward Donall Thomas (1920–2012). Robert A. Good (1922–2003). pioneers of stem cells transplantation. Paul Niehans (1882–1971) pioneer of stem cell therapy. Dr. Edward Donall Thomas is an American physician and a Nobel Laureate in Physiology or Medicine 1990. He was awarded the Nobel Prize for his work on the development of cell and organ transplantation. Dr. Thomas shared the award with Joseph Murray.

The first recorded attempt at cellular therapy occurred in 1912 when German physicians attempted to treat hypothyroid children with thyroid cells. Cellular therapy, as practiced today, was developed in the early 1930s by **Paul Niehans**, MD (1882–1971), a Swiss physician who became known as "the father of cell therapy." It soon became popular with celebrities as a means of rejuvenation. A 1990 article in *In Health* magazine described Niehans as a "public relations genius" and stated that the Clinique La Prairie, which he had founded in Clarens-Montreaux, Switzerland, had attracted 65,000 patients. Its 1999 one-week "revitalization program" coasted about $8,000 [1].

Generally, the Stem Cell (SC)—compartment is divided into embryonic and tissue specific or adult SCs [1]. Embryonic SCs (ES or ESC) are by definition the "master cells" with the largest spectrum of differentiation potential, e.g. capable of differentiating into every type of cells either *in vitro* or *in vivo*. Thanks to the presence of embryonic body, these cells have ability to develop into three primary layers: endoderm, ectoderm and mesoderm [1]. The discovery of SCs inside cell mass of embryos and in adult tissue has revolutionized the medical field by introducing new therapeutic dimensions into previously untreatable diseases and injuries. Several experimental or preclinical studies have suggested that application of embryonic SC could be promising in the treatment of various diseases [2–6]. However, recognition of appropriate ethical aspects, regulatory acts and standardization in embryonic SC mediated regenerative medicine is needed as it is still the matter of controversy. Besides, permanent, persistent and accurate updating of the facts regarding their phenotypic, functional, and immunologic characteristics is an essential requirement for safe clinical application of SCs. Some authors stand that the initial theory that embryonic SCs are ignored by immunocompetent hosts was overlooked. On the contrary, they think that it is even more evident that embryonic SCs could protect themselves actively by several immunomodulatory mechanisms against T lymphocytes and natural killer cells of host, and actively participate in immune-mediated events. Recent isolation of fetal SCs from several sources either at the early stages of development or during the later trimesters of gestation, sharing similar growth kinetics and expressing markers of pluripotency, provides strong support to the statement that these cells may be biologically closer to embryonic SCs . In fact, they represent intermediates between embryonic and adult mesenchymal SCs with regards to proliferation rates and plasticity features, thus being able to confer an advantage over postnatal mesenchymal SCs derived from conventional adult sources.

Historically, bone marrow was the primary source of SCs for transplant [1]. However, peripheral blood and umbilical (cord) blood are also currently used as sources. SCs derived from these sources may have therapeutic potential (without severe adverse effects) only when given to the individual from whom they were derived (autologous transplants) or from an immunologically matched donor (allogeneic transplants) [1].

Despite the fact that the ideal type and source of cells have not yet been defined, immature SCs are capable of colonizing different tissues due to ability of homing and trans–differentiation or lineage–plasticity, in the settings of regenerative medicine. Furthermore, there are several facts suggesting that adult SCs and even differentiated somatic cells, under appropriate microenvironmental cues or signals, are able to be "reprogrammed" and contribute to a much wider spectrum of differentiated progeny than previously anticipated. This has been demonstrated by using tissue–specific SCs—which like embryonic SCs—do not express CD45 as an exclusive hematopoietic marker. Consequently, adult mesenchymal SCs and endothelial precursors seem to be clinically applicable for cell–mediated, regenerative therapy of patients with myocardial, brain, vascular, liver, pancreas and some other tissue damages.

It is widely accepted that allogeneic transplants are still the most efficient treatment for patients with liver failure and Chronic Myelogenes Leukemia (CML) [1]. However, there is a lack of donors and some alternative therapeutic approaches are therefore, needed. Transplantation of mature hepatocytes has been evaluated, but the long–term efficacy remains unclear and the paucity of donor cells makes this strategy quite limited. The use of SC–therapy transplantation is perhaps a more promising alternative approach.

The intensification of myeloablative radio–chemotherapy enlarged the use of SC transplantats, as well as the introduction of cell–mediated therapeutic approaches in regenerative medicine resulting in increased needs for both specific blood–derived progenitor/cells, and practical operating procedures inducing minimized cellular damages during their collection or processing and storage in frozen state. Therefore, successful performance of SC transplants or cell–mediated therapy requires efficient collection, processing, and (cryo) preservation procedures for obtaining an acceptable cell yield and post–thawing recovery, as well as advantageous clinical outcome. For wound healing in the skin, epidermal stem cells and bone-marrow progenitor cells both contribute. Thus, it is likely that organ-specific progenitors and hematopoietic stem cells are involved in repair, even for other organ repair. In summary, stem cells could be described as:

- Foundation cells for every organ, tissue and cell in the body
- A "blank microchip" that can ultimately be programmed to perform any number of specialized tasks
- Undifferentiated "blank" cells that do not yet have a specific function

- Self-sustaining and capable of replicating themselves for long periods of time
- Under proper conditions, begin to develop into specialized tissues and organs [1]

These unique characteristics make stem cells very promising potential for supplying cells and tissues instead of organs in a spectrum of devastating diseases from diabetes type1 to stroke, spinal cord injuries, and myocardial infarction [1–7]. In the situation when the number of people needing organ and tissue transplants exceed the number of donated organs and tissues, this is the promise and hope, which deserves a deep and serious consideration. However, despite rapidly growing knowledge on adult stem cell sources, features and use, there are still some fundamental remaining questions regarding them that include: Does only **one common type of stem cell migrate to different organs** and repair tissue or are **there multiple types of stem cells**? Does every organ have stem cells (some of which have not yet been discovered)? Are the stem cells programmed to divide a finite number of times or do they have unlimited cell proliferation capacity?

According to their **functionality**, stem cells can be divided in two categories: normal and cancer stem cells [1].

1. *Normal stem cells* are immature cells that can replicate, or renew them, and are able to differentiate, or mature into all the cells that an organism or particular organ system needs. In other words, they possess a kind of immortality marked as self-renewal because these cells can divide indefinitely to produce more copies of them. Each stem cell is unspecialized, but it can produce progeny that mature into the various cell types of, say, the brain or the immune system. Once this maturation occurs, these adult stem cell heirs may divide rapidly but only a limited number of times [1–7]. The primary purpose of **adult stem cells** is healing. Finding out how adult stem cells store information and transform themselves into other cells with different properties is a fascinating topic for exploration [12–17]. Stem cells are so named because cells are derived from a main stem or mother set of cells. This is similar to a tree trunk that provides the stem from which other cells grow and branch out into other types of cells.
2. *Cancer stem cells*. Finding cancers' stem cells is a rapidly growing area of research [5, 7–11]. These cancer-causing cells, which make up a tiny fraction of cells within tumors, have properties similar to those of stem cells [5]. Cancer stem cells make up only a tiny number of the total cancer cells in a leukemia patient, which makes the cells next to impossible to find. Therefore, it seems that promise of this line of research can only be realized, by studying adult stem cells as well as embryonic stem cells (ES). The latter are still ethical problem and therefore substantially controversial because an early embryo is destroyed when researchers remove stem cells from it. An alternative is to take the stem cells from embryos that carry a genetic defect for specific diseases. Are cancer cells transformed normal stem cells? Researchers have traditionally thought of cancer as a collection of cells, all growing exponentially. According to the new research, conventional cancer therapies do an effective job killing the majority of cells within the tumor, but they may miss cancer stem cells. As a result, cancers often reoccur. Even hematologic and some non-hematologic malignancies treated by autologous stem cell transplant and high dose chemotherapy, have shown that regardless of survival

rate of some cancers, the final outcome is death, due to recurrence of cancer. The reason is (among others) in the fact that clinicians are injecting also cancer cells with healthy stem cells during re-infusion after apheresis collection, which accumulate and renew with a time to the critical level causing relapse or death. Ontogeny (development of an organism) and oncology (cancer development) share many common features. From the 1870s the connection between development and cancer has been reported for various types of cancers [1]. Existence of "cancer stem cells" with aberrant cell division has also been reported more recently [5]. The connection between cancer and development is clearly evident in teratocarcinomas. As early as 1862, Virchow discovered that the germ cell tumor teratocarcinoma is made up of embryonic cells [1]. In 1970, Stevens derived embryonic carcinoma cells from teratocarcinoma (a spontaneous tumor of germ cells that resembles development gone awry) [1]. This tumor may contain several types of epithelia: areas of bone, cartilage, muscle, fat, hair, yolk sac, and placenta. These specialized tissues are often adjacent to an area of rapidly dividing unspecialized cells. The terato-carcinomas are able to differentiate into normal mature cells when transplanted into another animal. This alternation between developmental and tumor cells status demonstrates how closely development and cancer are related. The present-day challenge is to decode the common molecular mechanism and genes involved in self-renewal for cancer cells and stem cells.

The very new concept in the field of cancerogenesis is the cancer stem cell (CSC) [5]. Cancer stem cells share many characteristics with normal stem cells, including self-renewal and differentiation. CSC is defined as "a cell within a tumor that possesses the capacity to self-renew and to cause the heterogeneous lineages of cancer cells that comprise the tumor". These cells have functionality allowing them the capability of causing an invasive group of cells (tumorigenic) that create metastasis [7–11]. There are two theories with respect to CSC entity [11]:

A. Stochastic/Clonal evolution Model	B. Hyerarchic/Cancer Stem Cell Model
This model states that all cancer cells hold tumorigenic potential and they are the product of clonal evolution by the acquisition of genetic mutations and epigenetic changes	Tumors show hierarchy, with a subpopulation of cancer cells having a tumorigenic potential much greater than that of other cancer cells

There are two well defined yet different models of Cancer Stem Cell within scientific community none of which completely can describe the features of Cancer Stem Cell:

(a) **Stochastic (clonal evolution) model**: This model states that all cancer cells hold tumorigenic potential and they are the product of clonal evolution by the acquisition of genetic mutations and epigenetic changes.
(b) **Hierarchical (cancer stem cell) model**: Tumors show hierarchy, with a subpopulation of cancer cells having a tumorigenic potential much greater than that of other cancer cells. Tumor contains hierarchical organization consisting of stem cells at top, which are cells within a tumor with the capacity to self-renew and generate heterogeneous lineages of cancer cells, progenitors, and differentiated cells which are no longer able to produce tumors (Fig. 9.1).

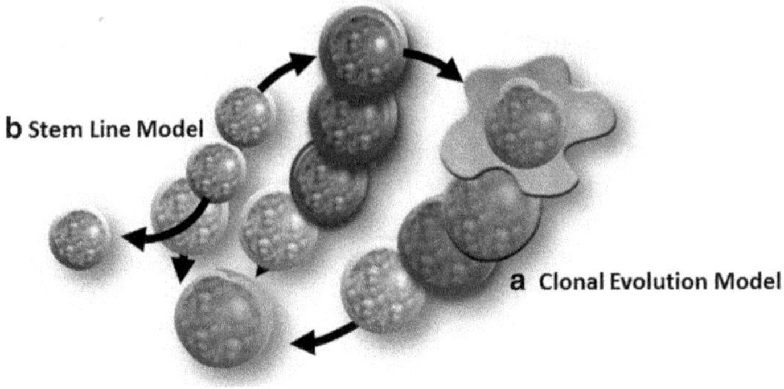

Fig. 9.1 (**a**) and (**b**) Two leading theories and cancer stem cell models: (**a**) Clonal evolution and (**b**) Cancer stem cell/Hierarchical model

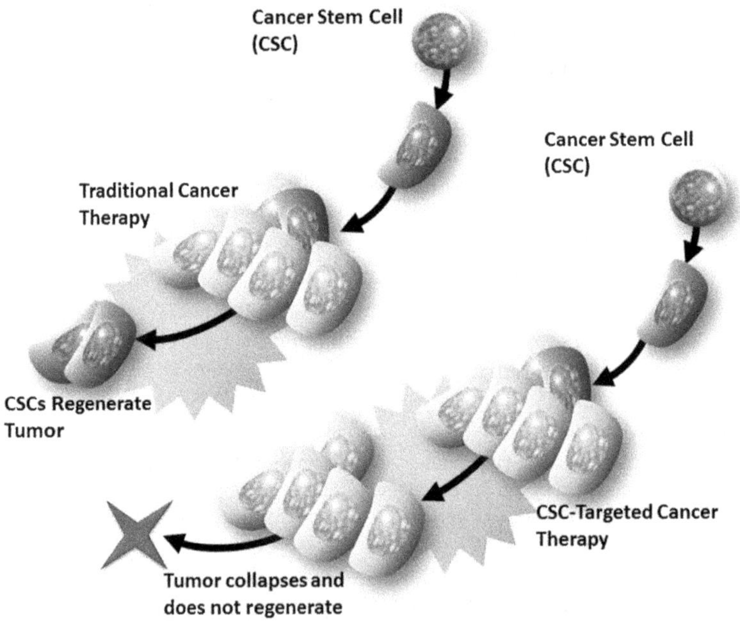

Fig. 9.2 Concept of cancer stem cell targeted therapy

Unlike cancer cells in a tumor, CSCs are capable of establishing new tumors when xenotransplanted into NOD/SCID animal models [5]. Although it has been shown that cancer cells are able to proliferate at a faster rate than CSCs, they have slight tumor initiating potential.

Therapeutic strategies that focus on targeting CSC markers will help address the ineffectiveness of traditional cancer therapies, which would otherwise result in therapy resistance and relapse (Fig. 9.2; Table 9.1).

Table 9.1 Cancer stem cell markers

Cancer	Cancer Stem Cell Markers																			Reference
	CD24	CD44	ALDH1	CD90	CD29	CD117	CD133	α2β integrin	α6 integrin	CD166	Nanog	ABCG2	CD96	CD34	NCAM1	CD271	CD105	POU5F1	CD38	
Breast	■	■	■																	Hajj et al, 2003 ; Ferro de Beça, 2012
Prostate		■					■	■	■											Maeda, 2009 ; Hoogland, 2013
Colon	■	■			■		■			■										L. Vermeulen, 2008
Brain							■													Zeppernick, 2008
Lung							■													Bertolini, 2009
Pancreatic	■	■																		Li et al, 2007 ; Zhu, 2012
Hepatic		■		■			■					■								Zhang, 2013 ; F. Yan, 2008 ; Shengyong, 2007
Ovarian						■	■				■									Luo, 2011 ; Zhang et al, 2011 ; Siu, 2013
Acute Myeloid Leukemia													■	■					■	Hosen, 2007 ; Horton, 2012 ; D Bonnet, 1997
Wilm's Tumor			■												■					Shukrun, 2013 ; Pode-Shakked, 2012
Melanoma			■													■				Luo Y, 2012 ; Civenni, 2011
Gastric	■	■																		Takaishi, 2009 ; Zhang, 2011
Renal																	■			Sandlund, 2006 ; Bussolati, 2008 ; Azzi, 2011
Thyroid																		■		Soon-Hyun Ahn, 2013

[a]Upgraded by: Mirjana Pavlovic, Jennifer Tarakmi, John Mayfield and Shimon Knutsen

Organogenesis from Adult Stem Cells and Problems with Different Tissues

How do a small number of stem cells give rise to a complex three dimensional tissue with different types of mature cells in different locations? This is the most fundamental question in organogenesis. The hematopoietic and nervous systems employ very different strategies for generating diversity from stem cells. The hematopoietic system assiduously avoids regional specialization by stem cells. Hematopoietic stem cells are distributed in different hematopoietic compartments throughout the body during fetal and adult life, and yet these spatially distinct stem cells do not exhibit intrinsic differences in the types of cells they generate. This contrasts with the nervous system, where even small differences in position are associated with the acquisition of different fates by stem cells.

While local environmental differences play an important role in this generation of "neural diversity," we must accept that intrinsic differences between stem cells are also critical. Part of the reason why different types of cells are generated in different regions of the nervous system is that intrinsically different types of stem cells are present in different regions of the nervous system. To understand the molecular basis for the regional patterning of neural stem cell function, we are now studying how these differences are encoded.

Therapeutic Implications for TCSCs as a New Concept

To prove the stem-cells derived from bone marrow (BM) and peripheral blood, including hematopoietic stem cells, are indeed transformed into solid-organ specific cells, several conditions must be met:

1. The origin of the exogenous cell integrated into solid-organ time must be documented by cell marking, preferably at the single-cell level.
2. Cell should be processed with a minimum of "ex vivo" manipulation (e.g. culturing) which may make them more susceptible to crossing lineages.
3. The exogenous cells must be shown to have become an integral morphologic part of the newly acquired tissue.
4. Transformed cells must be shown to have acquired the function of the particular organ into which it has been integrated both by expressing organ-specific proteins and by showing specific organ function.

Organ/Tissue specific niche (like in BM, liver, etc.)—exists as a deposit (storage) of the adult stem cells in a specific location (Fig. 9.3). These cells are circulating in a very low number in the blood [18]. Accumulating evidence suggests that stem cells may also actively migrate/circulate in the postnatal period of life. Stem cell trafficking/circulation may be one of the crucial mechanisms that maintains the pool of stem cells dispersed in stem cell niches of the same tissue, that are spread throughout different anatomical areas of the body. This phenomenon is very well

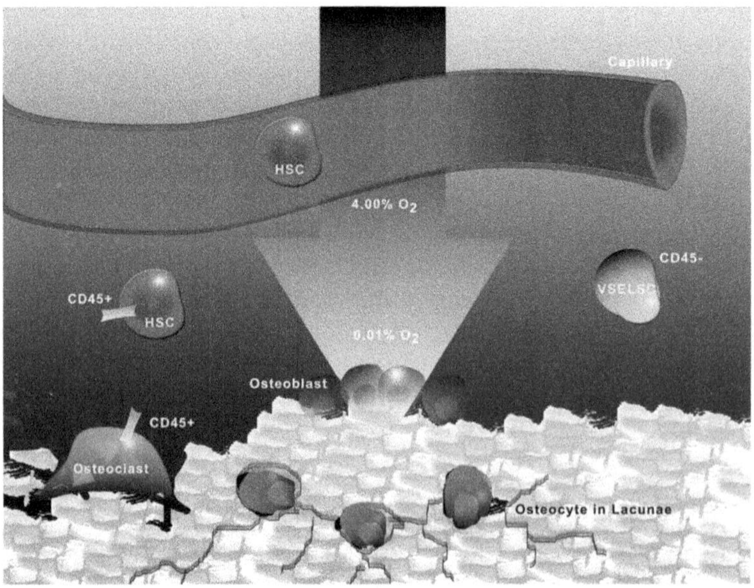

Fig. 9.3 Differences in phenotypes (external and internal markers) between HSC and VSEL from mouse bone marrow

described for HSC, but other, already tissue committed stem cells (TCSC) (for example, endothelial, skeletal muscle, skeletal or neural stem cells) are probably circulating as well [18].

BM is the home of migrating stem cells with not only hematopoietic stem cells within their niches, but also a small number of TCSC, which might be the reason why many authors think that HSC may transdifferentiate, although we do not have a direct proof for that. They might have plasticity, but not necessarily the "transdifferentional" potential [18]. What is differentiated in the tissue of injection might be TCSC characteristic for that tissue. It has been shown that number of these cells is decreased with ageing (long living and short living mice and humans). It would be interesting to identify genes that are responsible for tissue distribution/ expansion of TCSC. These genes could be involved in controlling the life span of the mammals. Therefore, BM stem cells are a heterogeneous population of cells with HSC and TCSC, the morphological and functional characteristics of which are different from HSC. Their number among BM MNC is very low (1 cell per 1000–10000 BM MNC) within young mammals and might play a role in small injuries [1]. In severe injuries like hart infarct or stroke they have no possibility to reveal their full therapeutic potential. The allocation of these cells to the damaged areas depends on homing signals that maybe inefficient in the presence of some other cytokines or proteolytic enzymes that are released from damaged tissue-associated leukocytes and macrophages [17]. We can envision, for example that metalloproteinases released from inflammatory cells may degrade SDF-1 locally, and thus perturb homing of CXCR4+TCSC. There is possibility that these cells while "trapped" in BM are still in: "dormant" stage-not fully functional, and need the appropriate activation signals by unknown factors [18]. These cells also, at least in some cases could be attracted to the inflammatory areas, and if not properly incorporated into the damaged tissue they may transform and initiate tumor growth. In summary, between the pools of tissue committed stem cells, there are probably those already committed to transdifferentiate into neural cells, or cells of tissues and organs other than neural, but we still do not have the control over their tracking, homing and finally regenerative capacity in the given tissue, which is a fundamental prerequisite for successful regenerative therapy [12–17].

The Concept of VSEL

In a discovery that has the potential to change the face of stem cell research, a University of Louisville scientist has identified cells in the adult body that seem to behave like embryonic stem cells [18–45]. The cells, drawn from adult bone marrow, look like embryonic stem cells and appear to mimic their ability to multiply and develop into other kinds of cells. The finding, presented the first time at the 47th Annual Meeting of the American Society of Hematology (ASH) in Atlanta, was announced December 12 at the society's news conference. A study by Ratajczak's team published in 2005 year in the journal "Leukemia" was the first to identify a

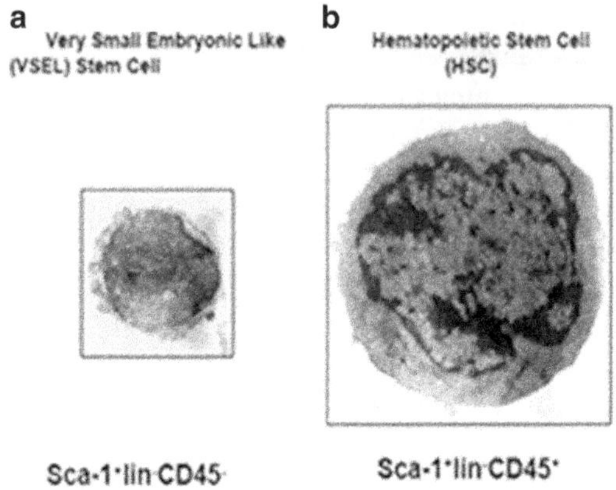

Fig. 9.4 TEM comparing morphological features of VSEL and HSC. With kindness of Dr. Ratajczak

type of stem cell in adult bone marrow that acts differently than other marrow stem cells [18]. The newly-identified cells, called "very small embryonic-like" (VSEL) stem cells, have the same ultra structure and protein markers as embryonic stem cells [18–40]. Ratajczak and several other researchers from University of Louisville in the presentation at the ASH meeting showed that VSEL stem cells mobilize into the bloodstream **to help repair damaged tissue following a stroke** [30]. In further research advance, Ratajczak's team also has grown VSEL cells in a lab and has stimulated them to change into nerve, heart and pancreas cells [30]. The difference in markers between HSC and VSELs in mouse are shown in (Fig 9.3), while the differences in ultrastructure are shown in (Fig. 9.4).

Along with this new concept, there is a premise that in regenerative therapy done before, with hematopoietic stem cells (considered to have plasticity and multipotency) the VSELs were "contaminats" that actually contributed to positive regenerative clinical outcome, since they have those capabilities [18]. This is an interesting concept which should be seriously considered in humans.

Thus, since VSELs have been found in human cord blood and bone marrow, they seem to be of a critical importance for consideration of stem cell transplant choice based upon the phenotype and number of stem cells aimed to be transplanted within a given clinical scenario. Despite conflicting data about this population [43–45], they are getting more confirmation in scientific community [31–40]. These cells have a great potential and like induced stem cells, can potentially eliminate the need for embryonic stem cells given that in adult organism they have all necessary components (parameters) that embryonic cells have, with a highest potency for lineage differentiation [41, 42].

1. Morphological studies have discovered that VSELs are unusually small (3–4 μm) eukaryotic cells which do possess several features of embryonic cells. Thus, the strategy based on FACS sorting of these cells should consider whether other adult tissues have those primitive little cells bigger than thrombocytes but smaller than erythrocytes [40–46] (Fig. 9.5).

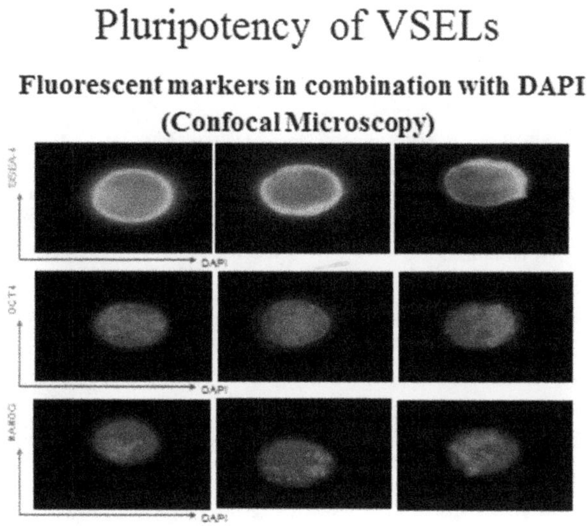

Fig. 9.5 Fluorescent label of Hallmarks of pluripotency in human VSELs intranucelar region. With kindness of Dr. Ratajczak

2. These cells also express high nucleo/cytoplazmatic ratio and smaller cytoplazmatic region compared to HSCs and mature granulocytes. Beside the fact that it has confirmed the features such as: size, confocal microscopy has also confirmed that VSELs express Oct-4, a hallmark of pluripotency of embryonic stem cells. In sum, morphological studies have discovered that VSELs are unusually small eucariotic cells with several fundamental features of embryonic stem cells except tumotigenicity (pluripotency, sphere formation, embryonic bodies and small size) [28].
3. These cells in a suitable medium perpetuate self-renewal longer, “without jumping” into differentiation, while on the other side they are capable of differentiating into bigger number of cell types in a suitable/conditioned medium into most of the cell (pancreatic cells, neural cells, cells of heart muscle and liver) which makes them suitable for expansion and reparative and regenerative purposes [28–30].
4. VSEL cells are, accordingly, a unique and distinguished entity rather than state with the features of plasticity, that questions plasticity of HSCs, suggesting strongly that that particular feature of BM stem cells could be in essence artifact caused by contamination of VSELs. Finally, the discovery of VSELs in the CB, PB and BM of humans indicate their significance with respect to other features. Some other researchers before Ratajczak have not succeeded to completely isolate this fraction [46–48], probably due to bad technique of isolation and timing. More extensive and deeper studies in the future will show what is true and possible.
5. Key advantages associated with VSELs, seem to be that they avoid the ethical or moral dilemmas associated with the use of embryonic and fetal cells, the potential negative biological effects associated with ESCLs such as their propensity for tumor formation, and the use of autologous stem cells to avoid immune rejection.

Table 9.2 VSELs: pros and cons with respect to different findings

Parameters of VSELs *necessary to detect in order to be able to consider their pluripotent function*	Authors: Cons *Dulak, May 2013* *Weissman, August, 2013*	Authors: Pros *Kassmer et al, December, 2013* *Bhartya et al, 2011–2013* *Wang J, X.Guo et al, 2013* *W. Wojakowski, 2013* *R. Taichman, 2013, Chang et al, 2013, Havens et al, 2013*
DNA amount	Little	Abundant
Formation of spheres	No	Yes
Octapeptide-4 expression	No	Yes
Differentiation into other lineages/ blood cells	No	Into epithelial cells and cardiac cells, multipotent tissue progenitors in vitro and in vivo

Note: The other two authors who commented existence of VSELs in negative sense were Alison Abbot (Nature, 2013) who gave a short reviews on the matter but not her own results, and Paul Koepfler who initiated negative comments on his blog (www.ipscell.com/tag/russel–taichman/)

6. The studies on mouse model suggest necessity for the human studies on VSELs since it would be of great interest to check if these intriguing population of stem cells are also involved in caloric intake, longevity and regenerative features of this distinctive stem cell entity (48). While this paper was prepared for print a recent report from Ratajczak's group in the form of Editorial, explained many aspects of conflicting data in VSELs history in a very professional way strongly suggesting that VSELs are rather detectible entity than the state of stem cell [46–50] (Table 9.2).

The Concept of Mesenchymal Stem-Cell (with Dental Pulp Cells as an Example)

Many human tissues are the source of stem cells responsible for tissue development and regeneration. Beside BM (*bone marrow stromal stem cells*, BMSCs), currently it is considered that dental pulp is practically the most approachable and the most important source of adult mesenchymal stem cells [49–54]. Within the last eight years, several populations of stem cells from dental pulp were isolated and characterized: (1) (*dental pulp stem cells*-DPSCs), (2) (*stem cells from human exfoliated decidual teeth*, SHEDs) and (3) (*immature dental pulp cells*, IDPCs) [51–54]. These cells are of the ectomesenchymal origin, located in perivascular niche, highly proliferative, clonogenic, multipotent and similar to BMSCs.

In in vitro conditions, they can differentiate with certain intercellular differences toward odontoblasts, hondrocytes, osteoblasts, adipocytes, neurons/glial cells, smooth and sceletal muscle cells. In in vivo, conditions, after implantation, they show different potential for dentine formation, as well as osteogenesis; after transplantation in mouse with compromised immune system, they make good grafts in different tissues and are capable of migrating into the brain, where they survave a certain time while reaching neurogenic phenotype. DPSCs have immunomodulatory effect, as they can be involved into immune response during infection of dental pulpe by NF-kB activation, and by inhibiting T-lymphocyte proliferation, suggesting their immunosupressive effect [51–54]. The future research should give us the complex data on the molecular and functional characteristics of dental pulp stem cells, as well as differences between different populations of these cells. Such research would fundamentally contribute to the better knowledge on the dental pulp stem cells, which is necessary due to their potential clinical application in *in vivo* cell transplantatioin, tissue engineering, and gene therapy (in vivo and ex vivo). Actually, by the isolation of IDPCs, which are the most primitive, but also the most plastic, (similar to embryonic stem cells), they are opening the new perspectives in a potential therapeutic application of these cells not only in regeneration of dentine, but also the regeneration of periodontal tissue and bone-junctional tissue of craniofacial region, as well as in the therapy of neurotrauma, myocardial infarction and connective tissue damage (Table 9.3).

However, the shift of the logic and turning of the sense, entitling phenotypically defined populations as stem cells, (although only some of them within that "cluster" are stem cells indeed) have introduced so much confusion into this discipline, that it

Table 9.3 Expression of protein or gene profiles in some dental stem cells in in vitro cultivation and relationship toward BMSC

Antigen	DPSC[a]	SHED	PDLSC	BMSC
CD14	–	–	–	–
CD34	–	–	–	–
CD44	++	++	++	++
CD45	–	–	–	–
CD106	+	+/–	+/–	++
CD146	++/+/–	++/+/–	++/+/–	++/+/–
3G5	+/–	+/–	+/–	+/–
Stro-1	++/+/–	++/+/–	++/+/–	++/+/–
α-smooth muscle actin	++/–	++/–	++/–	++/+/–
Colagen type-I	++	++	++	++
Colagen type-III	++/+	++/+/–	++/+/–	++/+
Alkaline phosphatase	++/+/–	++/+/–	++/+/	++/+/–
Osteocalcin	++/+	++/+/–	++/–	+/–
Osteonectin	++/+	++/+	++/+	++/+
Osteopontin	+/–	+/–	+/–	+/–
Sialoprotein of the bone	–	–	–	–
Skleraksis	+	+	++	+
Sialophosphoprotein of the dentine	–	–	–	–

[a]DPSC-Dental Pulp Stem Cells; SHED-Stem Cells From Human **Exfoliated Decidual Teeth; PDLSC-Periodontal Ligament Stem Cells; (++) strong expression; (+) weak expression; (–) negative; (/) subpopulation**

is very difficult to perform corrections nowadays. It is at the same time the reason why many discoveries that enable stem cell therapy on the rodents, do not work on humans. One has to be very critical with respect to stem cell markers and its functional properties in order not to make a mistake in stem cell therapy (Figs. 9.6 and 9.7).

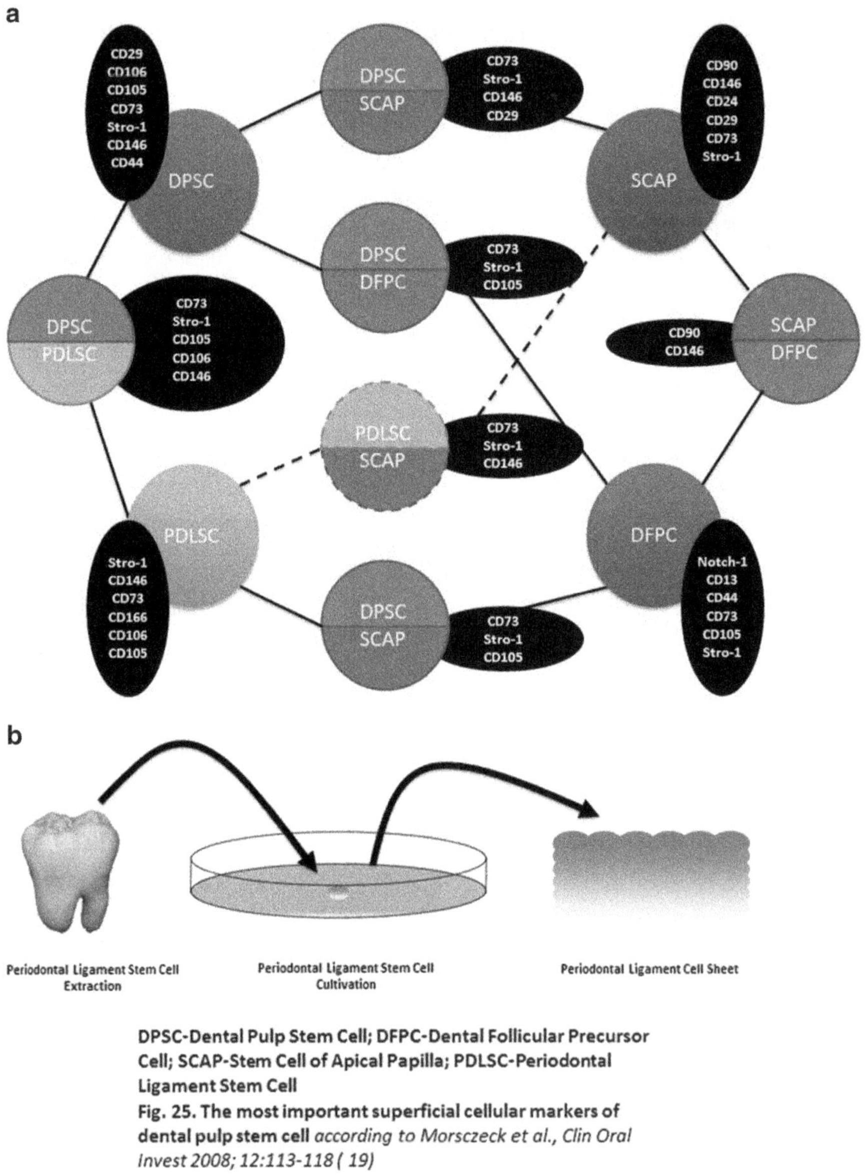

Fig. 9.6 The most important superficial cellular markers of dental pulp stem cell according to DPSC-Dental Pulp Stem Cell; DFPC-Dental Follicular Precursor Cell; SCAP-Stem Cell of Apical Papilla; PDLSC-Periodontal Ligament Stem Cell, Morsczeck et al., Clin Oral Invest 2008; 12:113-118 [19]

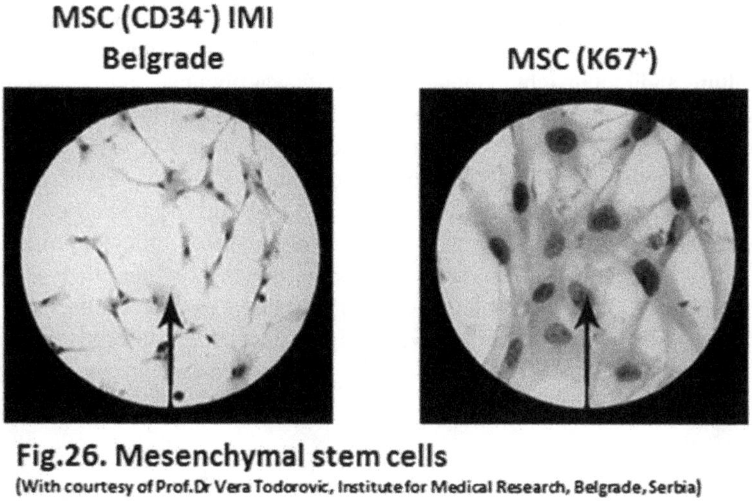

Fig. 9.7 Mesenchymal stem cells

Mobilization as a New Non-invasive Therapeutic Concept

The classification of patients into "good" or "poor" mobilizers is based on CD34+ cell count in their peripheral blood (PB) after granulocyte–colony-stimulating factor (G-CSF) injection. CD34+ cells mobilized into peripheral blood (PB) are considered a more convenient source of hematopoietic stem and progenitor cells than their bone marrow (BM) counterparts, in autologous transplantation protocols. Besides going through a less invasive collection procedure than BM aspiration, leukapheresed CD34+ cell collections ensure a rapid hematologic recovery as a function of transplanted dose of these cells, and their cell cycle status. Patients unable to mobilize sufficient number of CD34+ cells for efficient transplantation procedure are designated as poor mobilizers. Whereas numerous studies were dedicated to defining predictive factors for successful mobilization, 3 only a few characterized the phenotype of mobilized CD34+ in good versus poor mobilizers 4,5 and none explored the functional and metabolic properties of mobilized cells in these two groups of patients. Thus, Ivanovic et al., (2009) hypothesized that, apart from their mobilization from marrow to the blood, the response to G-CSF of CD34+ cells also includes activation of proliferation, metabolic activity, and proliferative capacity. In this study, mobilized PB CD34+ cells purified from samples obtained by cytapheresis of multiple myeloma or non-Hodgkin's lymphoma patients of both good (>50 CD34+ cells/mL) and poor (50 CD34+ cells/mL) mobilizers, were studied [55]. The initial cell cycle state of CD34+ cells after selection and their kinetics of activation (exit from G0 phase) during ex vivo culture were analyzed.

Their proliferative capacity was estimated on the basis of ex vivo generation of total cells, CD34+ cells, and colony-forming cells (CFCs), in a standardized expansion culture. Indirect insight in metabolic activity was obtained on the basis of their survival (viability and apoptosis follow-up) during the 7-day-long conservation in hypothermia (4 °C) in the air or in atmosphere containing 3 % O_2/6 % CO_2. The results have shown that CD34+ cells obtained from good mobilizers were in lower proportion in the G0 phase, their activation in a cytokine-stimulated culture was accelerated, and they exhibited a lower *ex vivo* expansion efficiency than those from poor mobilizers. The resistance to hypothermia of good mobilizers' CD34+ cells is impaired. The inevitable conclusion was that a good response to G-CSF mobilization treatment is associated with a higher degree of proliferative and metabolic activation of mobilized CD34+cells with a decrease in their expansion capacity [56].

Emphasizing Bioengineering Aspects to Stem Cell Engineering

New Concepts in Adult Stem Cell Research with Development of New Strategies: Personal Experience in the Light of Significance of Growing Information

Background and significance. Edward Thomas developed bone marrow transplantation as a treatment for leukemia. Initially the process was successful only if the donor was an identical twin of the patient. With the development of immunosuppressant drugs to counter organ rejection now many patients are treated for leukemia, a plastic anemia, sickle cell anemia, hurlers syndrome, severe combined immunodeficiency (SCID) and *Wiskott-Aldrich syndrome* as a result of his development in bone marrow transplantation. Dr. Edward Thomas was also awarded the *National Medal of Science 1990*. The primary role of adult stem cells in a living organism is to maintain and repair the tissue in which they reside. As an adult, stem cell is an undifferentiated cell found among differentiated cells in a tissue or organ. It can renew itself, and differentiate to yield the major specialized cell types of the tissue or organ. Within past ten years tremendous piece of work has been done with regard to development of the concepts of "*stemness*", primitive stem cell patterns used in regenerative purposes, and concept of *cancer stem cells*, with significant impact on the development of *new strategies* for their detection and targeted intervention. Despite deep skepticism and arguments these three concepts have their basis in scientific approaches and facts, researched and detected in order to support them. Results obtained are already empowering them to "step" into clinical arena [57–69].

Directions and Relevant Studies: We and Others

What is "stemness" [1, 70–74]? **Stemness** has so far been defined as both phenotypically and functionally recognizable cell pattern capable of self-renewal, proliferation and trans-differentiation through the phenomenon of plasticity [1, 74].

One has to be aware of the fact that stem cell category, as an elementary term is assuming the particular functionality. As the entity, or the state, it rationally presents the cell which of its all possible functions possesses at the moment of stemness only those that allow it to survive and sometimes divide: all other functions of this cell are at the potential level. When those possible functions really come up into scenario, that cell is not stem cell anymore. That is why the collections and clusters of different antigens expressed all over the cells in different developmental stages of different tissues (such as kit-receptor, CD117) cannot be the stem cell markers.

The "**stemness**" is the status in which only the oldest, the most primitive part of the genome is activated "with the only purpose to save what is stored in the nucleus of stem cell: genetic information", e.g. potential [1]. The purpose of this event is to save the cell of death and (if it comes to the stimulus for differentiation by asymmetric division) from self-renewal [1]. In that way we are becoming aware that the nature does create the standards that we should rather understand, instead of forcing the nature into our simplified concepts, some of which are very superficial. Tremendous advance which has enabled enrichment of stem cells based upon selection using phenotype as a standard could be appreciated as the advance in this discipline. It has also enabled more direct approach to investigation of stem cells.

However, there are other explanations for this status and one of them was defined by Dov Zipori [70–72]. According to him, this feature is not stem-cell specific, given the fact that it is unaquired. Most importantly, according to Zipori, 'stemness' is a *transient trait* and cannot be predicted on the basis of momentary gene-expression patterns (Fig. 9.8) [73].

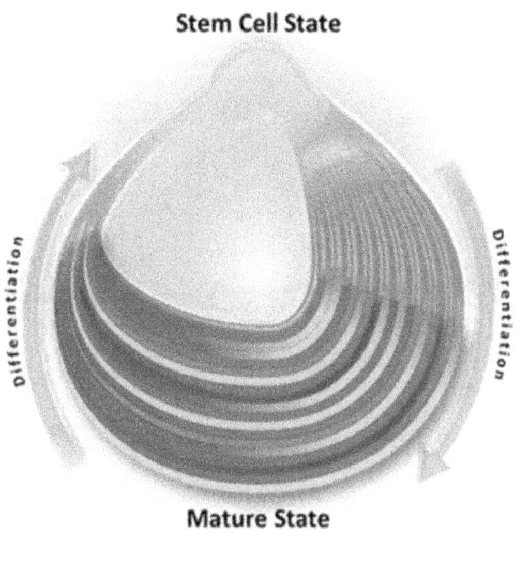

Fig. 9.8 Stem cell maturation according to D. Zipori. Med Sci (Paris) 2011; 27: 303–301

We have started optimization of the primitive stem cell pool in the case of acute myocardial infarction with intention to discriminate possible contamination with very small embryonic—like cells (VSELs) within Hematopoietic Stem Cell (HSCs) pool and determine which subpopulation is the best for regenerative purposes.

Optimization of primitive stem-cell patterns for regeneration and repair has today at least three strong candidates:

- HSCs (hematopoietic stem cells)
- VSELs (very small embryonic-like stem cells)
- MSCs (Mesenchymal stem cells)

The concept of plasticity have been revised by Ratajczak's group which has recently developed and together with us supported the concept of Very Small Embryonic Like Cells (VSELs), shown to be stem cells in bone marrow and other organs in non-hematopoietic compartment, committed to differentiate into some other tissues. These cells can be detected in mobilized bone marrow cells of mice and humans using cell sorter. However, we have also shown that not all the patients must be good mobilizers which require alternative approach [56]. Therefore, exploring the possibility of using adult stem cells for cell-based therapies has become a very expanding area of investigation (Fig. 9.9).

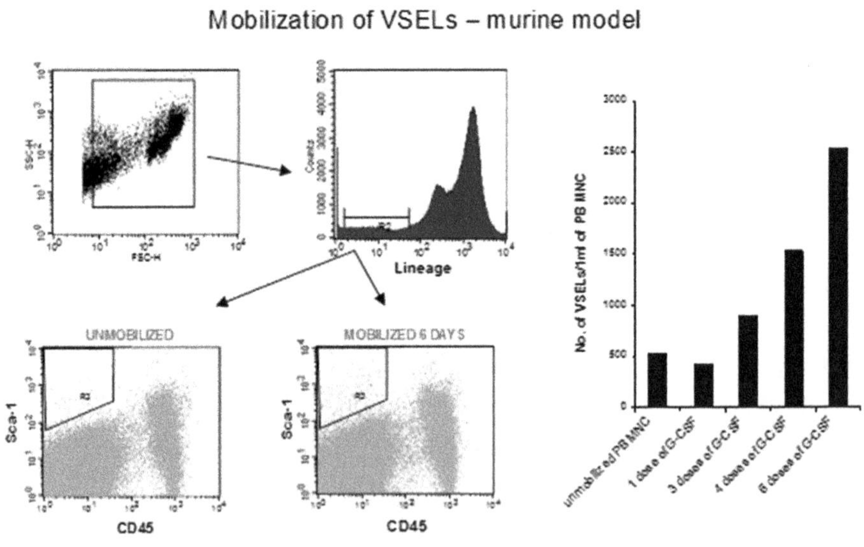

Fig. 9.9 Mobilization of VSELs in murine model by use of G-CSF (Neupogen) in mice and expression of critical markers in mobilized and unmobilized animals. Obtained by kindness of Dr. M. Ratajczak

Breakthrough in stem cell research: John Gurdon (UK) (1933–) and Shinya Yamanaka (Japan) (1963) have won the NP in 2012 for the discovery that mature cells can be reprogrammed to become pluripotent

Breakthrough: Induced stem cells (iPSC)

The Yamanaka lab identified four factors that, when co-transfected and expressed in mouse adult fibroblast cells, caused those fibroblasts to revert back to a pluripotent like state. One year later, the same four factors were used to successfully reprogram human adult fibroblast cells into induced pluripotent stem cells [75, 76]. These four factors are:

Octamer-4 (Oct-4) encoded by the gene POU5F1 is a transcription factor that is highly expressed in undifferentiated embryonic stem cells compared to other somatic cells. Oct-4 expression in embryonic stem cells is critical to maintain them in an undifferentiated, pluripotent state. In fact, if Oct-4 expression is experimentally knocked out, this causes embryonic stem cells to spontaneously differentiate.

SOX2 is a transcription factor critical for the maintenance of pluripotency in embryonic stem cells. SOX2 and Oct-4 work in parallel to co-regulate expression of target genes involved in the maintenance of pluripotency.

c-Myc is a well known proto-oncogene. The c-Myc gene codes for a transcription factor that regulates the expression of many genes involved in the control of cell proliferation, growth, differentiation and apoptosis. Abberant expression of c-Myc on the other hand is associated with tumor formation and cancer. Recent studies have demonstrated that c-Myc is a dispensable reprogramming factor; however, the transcription factor has been shown to greatly improve reprogramming efficiency.

Kruppel-like factor 4 (Klf-4) is a transcription factor that is highly expressed in undifferentiated ES cells and is also expressed elsewhere in the adult organism including the gut, testis and lungs and functions to regulate proliferation, differentiation and cell survival.

Reprogramming as a Therapeutic Event

Recent data have shown the use of reprogramming technologies to cause cancer cells to lose tumorigenicity in chronic myeloid leukemia cells, melanoma cells, and gastrointestinal cancer cells [75–77]. These results suggest that nuclear reprogramming may be a therapeutic strategy for the treatment of cancer [77]. However, these experiments have also revealed that reprogramming technology is not very efficient. Experiments suggest that cancer cells are resistant to reprogramming and this resistance might be related to the role of epigenetic regulations during reprogramming. The fact that transformation of iPSCs is accomplished by erasing the epigenetic modification similar to those found in early embryos demonstrates the significance of epigenetic changes for successful reprogramming, and thus, its role in carcinogenesis [76].

As Dr. Gordana Vunjak-Novakovic published lately, the recent availability of human cardiomyocytes derived from induced pluripotent stem (iPS) cells opens new opportunities to build in vitro **models of cardiac disease**, screening for new drugs, and patient-specific cardiac therapy. **Notably, the use of iPS cells enables studies in the wide pool of genotypes and phenotypes**. The progress in reprogramming of induced pluripotent stem (iPS) cells towards the cardiac lineage/differentiation is going on. The focus is on challenges of cardiac disease modeling using iPS cells and their potential to produce safe, effective and affordable therapies/applications with the emphasis on cardiac tissue engineering. The paper has emphasized implications of human iPS cells to biological research and some of the future needs [76].

Stem cell cancer concept had reached its spike this year [5]. Cancer Stem Cells (CSCs) share properties similar to those described for tissue stem cells: self-renewal and asymmetric division resulting in the generation of daughter cells destined to differentiate, enabling the regeneration of a tissue. They provide tumors with unrestricted dividing potential that resemble embryonic stem cells. This process of dedifferentiation involves mis-regulation of JAK/STAT, Wnt, and Hedgehog signaling [5]. The theory of cellular dedifferentiation may hold key therapeutic potential in the cancer research field. Proliferation in hypoxic conditions is a common feature of cancer stem cells.

The directions in stem cell research development require further work on:

- More firm and precise definition of stemness
- Optimization of best candidate for different tissue engineering manipulations in clinical arena and optimization of scaffolds for particular tissue engineered patterns
- Expanded work on cancer stem cells in order to discriminate origin, underlying causes and mechanisms of that sort of malignancy and particular, selective, targeted therapeutic approach

It is clear nowadays that the presence of multipotent stem cells in the adult might open up new therapeutic opportunities on the basis of tissue and organ replacement. Therefore, the exact definition of stem cells and the ability to isolate them are matters of supreme importance. However, despite the efforts of many investigators who strive to determine their nature, a definitive stem-cell 'portrait' is lacking. Yet, quite recently, two independent studies claimed to have identified a stem-cell-specific group of genes that form a 'stem-cell signature'. In fact, these studies have defined two different and unrelated groups of genes; the conclusion that these signatures characterize stem cells is therefore premature. Experimental and/or technical reasons might explain the disparity of the results from these independent studies, and alternative approaches that might lead to identification of the 'correct' gene-expression profile of stem cells were suggested. But should one expect to find a stem-cell-specific signature using an approach based on the analysis of gene expression? Zipori, argues that **renewal ability is an aquired property and, as such, is not stem cell specific** [72]. **Most importantly, according to him, 'stemness' is a transient trait and cannot be predicted on the basis of momentary gene-expression patterns**. Due to the complexity of the problem, the solution to determining the molecular configurations that dictate a stem-cell state should, therefore, come from an overall genomic and proteomic analysis, coupled with mathematical modeling.

Hematopoietic stem cell transplantation remains a risky procedure with many possible complications. It has traditionally been reserved for patients with life-threatening diseases, such as malignancies. While occasionally used experimentally in nonmalignant and nonhematologic indications such as severe disabling autoimmune and cardiovascular diseases, the risk of fatal complications appears too high to gain wider acceptance. Yet, this is the most-well known and the most developed stem-cell regenerative approach, given that if successfully engrafted, it repopulates and later on recruits the new, healthy bone marrow cells in circulation.

Embryonic stem-cell research is still the matter if controversies at a very stratified levels, although many researchers agree that it might be the source of stem cells with the highest differentiation potential.

The experimental and clinical trials have shown both in animal models and humans the neovascularization and myocardial tissue repair through trans-differentiation into myo-cardiocytes, or some other mechanism. Repair of damaged organ/tissue (myocardial, neuronal, liver, cartilage, bone, etc.) is shown mostly in animal models, although very good data are coming from the Belgrade group in treatment of AMI [78–83]. Maybe the most illustrative of all is the bunch of experimental data suggesting the great potential for stem cell differentiation and homing into damaged tissues either when mobilized or injected into the tissue of interest after apheresis or BM puncture, with or without cryopreservation [1]. Although the adult stem cell regenerative therapy after BM aspiration and apheresis injection into coronary arteries is becoming more and more successful, the most evident success of mesenchymal stem cell treatment at regenerative therapy level in clinical arena is seen so far in children with *osteogenesis imperfecta* where the results with diseased children dramatically visible and easily reproducible. Yet, due to the obstacles already mentioned above, this is not the case with nervous system regenerative treatment, especially in humans.

Apparently, basic adult stem cell research is still evolving, and is the matter of everchangeing issues. Due to our extensive studies, but yet limited knowledge on their behavior and potentials, it is not yet easy to determine how to act in clinical arena. It is obvious that each approach to any particular disease or damage has to be optimized within team work and by bridgeing the gap between fundamental and clinical studies. Knowing molecular level in depth, will help clinicians to orcestrate the team work and overcome critical obstacles in each particular scenario. There is no doubt that adult stem cell therapy (and probably embryonic as well) belong to the future, but we have to act as that we shall belong to the future, as well. Continuous efforts in both molecular and clinical directions will lead to the unique and optimal plan for each particular regenerative treatment. How far away we are from that goal it will be innevitabely shown up in a near future.

References

1. Pavlovic, M., Balint, B.: Stem Cells and Tissue Engineering. Springer, New York (2013). ISBN: 978-1-4614-5505-9 (eBook)
2. Saltzman, M.W.: Biomedical Engineering. Cambridge University, New York (2009). ISBN: 978-0-521-840099-6(hardback)
3. Ratajczak, M., Ratajczak, J., Shin, D.M., Wan, W., Liu, R., Masternak, M.M., Piotrowska, K., et al.: Higher number of stem cells in the bone marrow of circulating low Igf-1 level LaronDwarf novel view on Igf-1, stem cells and aging. Leukemia **25**, 29–733 (2011)
4. Balint, B., Todorović, M., Jevtić, M., Ostojić, G., Ristanović, E., Vojvodić, D., et al.: The use of stem cells for marrow repopulation and in the field of regenerative medicine. Mak. Med. Pregl. 63(Suppl 75), str. 12 (2009)
5. Mayfield, J., Pavlovic, M.: Current modalities and the implications of cancer stem cell engineering in oncological treatment. ART **14**, (1–2), 49–59 (2014)
6. Pavlovic, M., Balint, B.: The use of stem cells to repair the cardiac tissue. Anest. Reanim. Transfuziol. **34**, 129–150 (2006)
7. Li, C., et al.: Identification of pancreatic cancer stem cells. Cancer Res. **67**(3), 1030–1037 (2007)
8. Neve, R.M., et al.: A collection of breast cancer cell lines for the study of functionally distinct cancer subtypes. Cancer Cell **10**(6), 515–527 (2006)
9. Al-Hajj, M., et al.: Prospective identification of tumorigenic breast cancer cells. Proc. Natl. Acad. Sci. U. S. A. **100**(7), 3983–3988 (2003)
10. Reya, T., et al.: Stem cells, cancer, and cancer stem cells. Nature **414**(6859), 105–111 (2001)
11. Singh, S.K., et al.: Identification of a cancer stem cell in human brain tumors. Cancer Res. **63**(18), 5821–5828 (2003)
12. Orlic, D., Kajstura, J., Chimenti, S., et al.: Bone marrow cells regenerate infarcted myocardium. Nature **410**(6829), 701–705 (2001)
13. Zhang, M., Methot, D., Poppa, V., Fujio, Y., Walsh, K., Murry, C.: Cardiomyocyte grafting for cardiac repair: graft cell death and anti-death strategies. J. Mol. Cell. Cardiol. **33**(5), 907–921 (2001)
14. Dow, J., Simkhovich, B.Z., Kedes, L., Kloner, R.A.: Washout of transplanted cells from the heart: a potential new hurdle for cell transplantation therapy. Cardiovasc. Res. **67**(2), 301–307 (2005)
15. http://www.ncbi.nlm.nih.gov/pubmed/17645413
16. http://www.intechopen.com/books/apoptosis-and-medicine/apoptosis-as-a-therapeutic-target-in-cancer-and-cancer-stem-cells-novel-strategies-and-futures-persp
17. http://stemcells.nih.gov/info/Regenerative_Medicine/pages/2006chapter9.aspx
18. Kucia, M., Wojakowski, W., Reca, R., Machalinski, B., Gozdzik, J., Majka, M., Baran, J., Ratajczak, J., Ratajczak, M.: The migration of bone marrow-derived non-hematopoietic tissue commited stem cells is regulated in an SDF-1, HGF—LIF-dependent manner. Arch. Immunol. Ther. Exp. (Warsz). March 28: 1642290 (2006)
19. Pavlović, M.: VSELs concept: review. MNE Medica **1**, 16–17 (2008). ref. p. 43

20. Pavlovic, M.: Very small embryonic like cells (VSELs): review and perspectives in the light of critical data and controversies. (2014). ART **41**, (1–2), 33–47 (2014)
21. Ratajzcak, M.Z., Kucia, M., Reca, R., Majka, M., et al.: Stem cell plasticity revised: CXR4 positive cells expressing mRNA for early muscle, liver and neural cells "hide out" in the bone marrow. Leukemia **19**(1), 29–40 (2004)
22. Kucia, M., Reca, R., Campbell, F.R., Zuba-Surma, E., Majka, M., Ratajczak, J., Ratajczak, M.Z.: A population of very small embryonic-like (VSEL) CXCR4þSSEA-1þOct-4þ stem cells identified in adult bone marrow. Leukemia **20**, 857–869 (2006)
23. Kucia, M., Halasa, M., Wysoczynski, M., Baskiewicz-Masiuk, M., Moldenhawer, S., Zuba-Surma, E., Czajka, R., Wojakowski, W., Machalinski, B., Ratajczak, M.Z.: Morphological and molecular characterization of novel population of CXCR4þ SSEA-4þOct-4þ very small embryonic-like cells purified from human cord blood—preliminary report. Leukemia **21**, 297–303 (2007)
24. Zuba-Surma, E., Kucia, M., Abdel-Latif, A., Dawn, B., Hall, B., Singh, R., Lillard, J.W., Ratajczak, M.Z.: Morphological characterization of very small embryonic- like stem cells (VSELs) by image stream system analysis. J. Cell. Mol. Med. Nov 20: 18031297 (2007)
25. Zuba-Surma, E.K., Wu, W., Ratajczak, J., Kucia, M., Ratajczak, M.Z.: Very small embryonic-like stem cells in adult tissues-potential implications for ageing. Mech. Ageing Dev. **130**, 58–66 (2009)
26. Ratajczak, M., Zuba-Surma, E., Machslinski, B., Ratajczak, J., Kucia, M.: Very small embryonic-like (VSEL) stem cells: purification from adult organs, characterization, and biological significance. Stem Cell Rev. **4**, 89–99 (2008)
27. Kucia, M., Wysoczynski, M., Ratajczak, J., Ratajczak, M.Z.: Identification of very small embryonic like (VSEL) stem cells in bone marrow. Cell Tiss. Res. **331**, 125–134 (2008)
28. Suszynska, M., Zuba-Surma, E.K., Maj, M., Mierzejewska, K., Ratajczak, J., Kucia, M., Ratajczak, M.Z.: The proper criteria for identification and sorting of very small embryonic-like stem cells, and some nomenclature issues. Stem Cells Dev. **23**, 702–713 (2014)
29. Ratajczak, M., et al.: Very small embryonic/epiblast like stem cells (VSELs) and their potential role in aging and organ rejuvenation-an update and comparison to other primitive small stem cells isolated from adult tissues. AGIGMG **4**(4), 235–246 (2012)
30. Denis Rodgerson, D.O., Harris, A.G.: A comparison of stem cells for therapeutic use. Stem Cell Reviews and Reports. Published online: 2 March 2011. http://www.neostem.com/assets/Where do stem cells come from.Pdf
31. Kassmer, S.H., Bruscia, E.M., Zhang, P.X., Krause, D.S.: Nonhematopoietic cells are the primary source of bone marrow-derived lung epithelial cells. Stem Cells **30**(3), 491–499 (2012). doi:10.1002/stem.1003
32. Kassmer, S.H., Krause, D.S.: Very small embryonic-like cells: biology and function of these potential endogenous pluripotent stem cells in adult tissues. Mol. Reprod. Develop. **80**, 677–690 (2013)
33. Kassmer, S.H., Jin, H., Zhang, P.-X., Bruscia, E.M., Heydari, K., Lee, J.H., Kim, C.F., Kassmer, S.H.: Very small embryonic-like stem cells from the murine bone marrow differentiate into epithelial cells of the lung. Stem Cells **31**(12), 2759–2766 (2013)
34. Bhartiya, D.: Are mesenchymal cells indeed pluripotent stem cells or just stromal cells? OCT-4 and VSELs biology has led to better understanding. Stem Cells Int **2013**, 547501 (2013). Epub 2013 Sep 25. Review
35. Anand, S., Bhartiya, D., Sriraman, K., Patel, H., Manjramkar, D., Bakshi, G., Dhamankar, V., Kurkure, P.: Quiescent very small embryonic-like stem cells resist oncotherapy and can restore spermatogenesis in germ cell depleted mammalian testis. Stem Cells Dev. **23**, 1428 (2013)
36. Bhartiya, D., Sriraman, K., Parte, S., Patel, H.J.: Ovarian stem cells: absence of evidence is not evidence of absence. Ovarian Res. **6**(1), 65 (2013). doi:10.1186/1757-2215-6-65.13. Sep 30. [Epub ahead of print]
37. Bhartiya, D., Unni, S., Parte, S., Anand, S.: Very small embryonic-like stem cells: implications in reproductive biology. Biomed Res Int. **2013**, 682326 (2013). doi:10.1155/2013/682326. Epub 2013 Feb 13. Review

38. Bhartiya, D., Kasiviswananthan, S., Shaikh, A.: Cellular origin of testis-derived pluripotent stem cells: a case for very small embryonic-like stem cells. Stem Cells Dev. **21**(5), 670–674 (2012). doi:10.1089/scd.2011.0554. Epub 2012 Feb 8
39. Bhartiya, D., Shaikh, A., Nagvenkar, P., Kasiviswanathan, S., Pethe, P., Pawani, H., Mohanty, S., Rao, S.G., Zaveri, K., Hinduja, I.: Very small embryonic-like stem cells with maximum regenerative potential get discarded during cord blood banking and bone marrow processing for autologous stem cell therapy. Stem Cells Dev. **21**(1), 1–6 (2012). doi:10.1089/scd.2011.0311. Epub 2011 Aug 29
40. Wang, J., Guo, X., Lui, M., Chu, P.J., Yoo, J., Chang, M., Yen, Y.: Identification of a distinct small cell population from human bone marrow reveals its multipotency in vivo and in vitro. PLoS One **9**(1), e85112 (2014). doi:10.1371/journal.pone.0085112. eCollection 2014 Jan 17
41. Wojakowski, W., Tendera, M., Kucia, M., Zuba-Surma, E., Paczkowska, E., Ciosek, J., Halasa, M., Krol, M., Kazmierski, M., et al.: Mobilization of bone marrow-derived Oct-4 + SSEA-4 + very small embryonic-like stem cells in patients with acute myocardial infarction. J. Am. Coll. Cardiol. **53**, 1–9 (2009)
42. Chang, Y.J., Tien, K.E., Wen, C.H., Hsieh, T.B., Hwang, S.M.: Recovery of CD45-/Lin-/SSEA-4 + very smallembryonic-like stem cells by cord blood bank standard operating procedures. Cytotherapy **16**, 560–565 (2013). doi:10.1016/j.jcyt.2013.10.00
43. Knoepfler, P.: Stem cell blog. Taichman are VSELs the sasquatch of the stem cell field? Posted 2: http://www.ipscell.com/tag/russell-taichman/
44. Szade, K., Bukowska-Strakova, K., Nowak, W.N., Szade, A., Kachamakova-Trojanowska, N., Zukowska, M., Jozkowicz, A., Dulak, J.: Murine bone marrow Lin−Sca-1+CD45−very small embryonic-like (VSEL) cells are heterogeneous population lacking Oct-4A expression. PLoS One **8**(5), e63329 (2013). Published online 2013 May 17. doi:10.1371/journal.pone.0063329
45. Miyanishi, M., Mori, Y., Seita, J., Chen, J.Y., Karten, S., Chan, C.K.F., Nakauchi, H., Weissman, I.L.: Do pluripotent stem cells exist in adult mice as very small embryonic stem cells? Logo Stemcr Stem Cell Reports **1**(2), 198–208 (2013)
46. D' Ipolito, G., Diabira, S., Howard, G.A., Menei, P., Roos, B.A., Pc, S.: Marrow-isolated adult, ultilineage inducible (MIAMI) cells, a unique populationof postnatal and old human cells with extensive expamsion and differentiation potential. J. Cell Sci. **117**, 2971–2981 (2004)
47. Jiang, Y., Jahagirdar, B.N., Reinhardt, R.L., Schwartz, R.E., Keene, C.D., Ortiz-Gonzalez, X.R., Reyes, M., Lenvik, T., Lund, T., Blackstad, M., Du, J., Aldrich, S., Lisberg, A., Low, W.C., Largaespada, D.A., Verfaillie, C.M.: Pluripotency of mesenchymal stem cells derived from adult marrow. Nature **418**(6893), 41–49 (2002)
48. Yoon, Y.S., Wecker, A., Heyd, L., Park, J.S., Tkebuchava, T., Kusano, K., Hanley, A., Scadova, H., Asahara, T., Losordo, D.W.: A clonally expanded novel population of multipotent stem cells derived from human bone marrow regenerates myocardium after myocardial infarction. J. Clin. Invest. **115**, 326–338 (2005). PMCID: 546424
49. Erices, A., Conget, P., Minguell, J.J.: Mesenchymal progeniitor cells in human umbilical cord blood. Br. J. Haematol. **109**(3), 209–218 (2000)
50. Huss, R., et al.: Perspectives on the morphology and biology of CD34-negative stem cells. J. Hematother. Stem Cell Res. **9**(6), 783–793 (2000)
51. Zvaifler, N.J., Marinova-Mutafchieva, L., Adams, G., Edwards, C.J., Moss, J., Burger, J.A., Maini, R.N.: Mesenchymal precursor cells in the blood of normal individuals. Arthritis Res. **2**(6), 477–488 (2000). Epub 2000 Aug 31
52. Young, H.E., Steele, T.A., Bray, R.A., Hudson, J., Floyd, J.A., Hawkins, K., Thomas, K., Austin, T., Edwards, C., Cuzzourt, J., Duenzl, M., Lucas, P.A., Black Jr., A.C.: Human reserve pluripotent mesenchymal stem cells are present in the connective tissues of skeletal muscle and dermis derived from fetal, adult, and geriatric donors. Anat. Rec. **264**(1), 51–62 (2001)
53. Ratajczak, J., Kucia, M., Reca, R., Zhang, J., Machalinski, B., Ratajczak, M.Z.: Quiescent CD34+ early erythroid progenitors are resistant to several erythropoietic 'inhibitory' cytokines; role of FLIP. Br. J. Haematol. **123**(1), 160–169 (2003)

54. Zuk, P.A., Zhu, M., Mizuno, H., Huang, J., Futrell, J.W., Katz, A.J., Benhaim, P., Lorenz, H.P., Hedrick, M.H.: Multilineage cells from human adipose tissue: implications for cell-based therapies. Tissue Eng. **7**(2), 211–228 (2001)
55. Ivanovic, Z., Kovacevic-Filipovic, M., Jeanne, M., Ardilouze, L., Bertot, A., Szyporta, M., Hermitte, F., Lafarge, X., Duchez, P., Vlaski, M., Milpied, N., Pavlovic, M., Praloran, V., Boiron, J.M.: CD34+ cells obtained from "good mobilizers" are more activated and exhibit lower *ex vivo* expansion efficiency than their counterparts from "poor mobilizers". Transfusion **50**(1), 120–127 (2010). doi:10.1111/j.1537-2995.2009.02436.x. Epub 2009 Oct 12
56. Kucia, M.J., Wysoczynski, M., Wu, W.: Evidence that very small embryonic-like stem cells are mobilized into peripheral blood. Stem Cells **26**, 2083–2092 (2008)
57. Korbling, M., Katz, R.L., Khanna, A., Ruifrok, A.C., Rondon, G., Albitar, M., Champlin, R.E., Estrov, Z.: Hepatocytes and epithelial cells of donor origin in recipients of peripheral-blood stem cells. New Engl. J. Med. **346**, 738–746 (2002)
58. Kucia, M., Masternak, M., Liu, R., Shin, D.M., Ratajczak, J., Mierzejewska, K., et al.: The negative effect of prolonged somatotrophic/insulin signaling on an adult bone marrow-residing population of pluripotent very small embryonic-like stem cells (VSELs). Age (Dordr.) **35**, 315–330 (2013)
59. Kucia, M., Shin, D.M., Liu, R., Ratajczak, J., Bryndza, E., Masternak, M.M., et al.: Reduced number of VSELs in the bone marrow of growth hormone transgenic mice indicates that chronically elevated Igf1 level accelerates age-dependent exhaustion of pluripotent stem cell pool: a novel view on aging. Leukemia **25**, 1370–1374 (2011)
60. Ratajczak, M.Z., Shin, D.M., Kucia, M.: Very small embryonic/epiblast-like stem cells: a missing link to support the germ line hypothesis of cancer development? Am. J. Pathol. **174**(6), 1985–1992 (2009). doi:10.2353/ajpath.2009.081143. Epub 2009 Apr 30. Review
61. Shin, D.M., Liu, R., Klich, I., Ratajczak, J., Kucia, M., Ratajczak, M.Z.: Molecular characterization of isolated from murine adult tissues very small embryonic/epiblast like stem cells (VSELs). Mol. Cells **29**(6), 533–538 (2010). doi:10.1007/s10059-010-0081-4. Epub 2010 Jun 4
62. Ratajczak, M.Z., Liu, R., Ratajczak, J., Kucia, M., Shin, D.M.: The role of pluripotent embryonic-like stem cells residing in adult tissues in regeneration and longevity. Differentiation **81**(3), 153–161 (2011). doi:10.1016/j.diff.2011.01.006. Epub 2011 Feb 19
63. Ratajczak, M.Z., Liu, R., Marlicz, W., Blogowski, W., Starzynska, T., Wojakowski, W., Zuba-Surma, E.: Identification of very small embryonic/epiblast-like stem cells (VSELs) circulating in peripheral blood during organ/tissue injuries. Methods Cell Biol. **103**, 31–54 (2011). doi:10.1016/B978-0-12-385493-3.00003-6
64. Ratajczak, M.Z., Zuba-Surma, E.K., Shin, D.M., Ratajczak, J., Kucia, M.: Very small embryonic-like (VSEL) stem cells in adult organs and their potential role in rejuvenation of tissues and longevity. Exp. Gerontol. **43**(11), 1009–1017 (2008). doi:10.1016/j.exger.2008.06.002. Epub 2008 Jun 14
65. Ratajczak, M.Z., Liu, R., Marlicz, W., Blogowski, T.S., Wojakowski, W., Zuba-Surma, E.: Identification of very small embryonic/epiblast-like stem cells (VSELs) circulating in peripheral blood during organ/tissue injuries methods in cell biology. Method Cell Biol **103**, 31–54 (2011)
66. Ratajczak, M.Z., Liu, R., Ratajczak, J., Kucia, M., Shin, D.M.: The role of pluripotent embryonic-like stem cells residing in adult tissues in regeneration and longevity. Differentiation **81**(3), 153–161 (2011). doi:10.1016/j.diff.2011.01.006. Epub 2011 Feb 19. Review
67. Ratajczak, M.Z., Zuba-Surma, E., Wojakowski, W., Synska, M., Mierzejewska, K., Liu, R., Ratajczak, J., Shin, M., Kucia, M.: Very small embryonic-like stem cells (VSELs) represent a real challenge in stem cell biology: recent pros and cons in the midst of a lively debate. Leukemia **28**, 473–484 (2014)
68. Suszynska, M., Zuba-Surma, E.K., Maj, M., Mierzejewska, K., Ratajczak, J., Kucia, M., Ratajczak, M.Z.: The proper criteria for identification and sorting of very small embryonic-like stem cells, and some nomenclature issues. Stem Cells Development **23**(5), 1–13 (2014)
69. Havens, A.H., Shiozawa, S.Y., Jung, Y., Wang, J., Mishra, A., Jiang, Y., O'Neill, D.W., Krebsbach, P.H., et al.: Human and murine very small embryonic-like (VSEL) cells represent

multipotent tissue progenitors, in vitro and in vivo. Stem Cells Dev. **23**, 689–701 (2014). doi:10.1089/scd.2013.0362

70. Zipori, D.: The nature of stem cells: state rather than entity. Nat. Rev. Genet. **5**(11), 873–878 (2004). doi:10.1038/nrg1475
71. Zipori, D.: Biology of Stem Cells and the Molecular Basis of the Stem State. Humana, New York (2009). ISBN:160761295
72. Zipori, D.: À la recherche d'une définition moléculaire plus que descriptive pour les cellules souches towards a molecular rather than a descriptive definition of stemness. Med Sci (Paris) **27**, 303–301 (2011)
73. Ramalho-Santos, M., et al.: Stemness: transcriptional profiling of embryonic and adult stem cells. Science **298**(5593), 597–600 (2002)
74. http://www.nature.com/nature/focus/cancerstemcells/
75. Martins, A.M., Vunjak-Novakovic, G., Reis, R.L.: The current status of iPS cells in cardiac research and their potential for tissue engineering and regenerative medicine. Stem Cell Rev Rep **10**, 177–190 (2014)
76. Yamanaka, S.: Patient-specific pluripotent stemcells become even more accessible. Cell Stem Cell **7**(1), 11–14 (2010)
77. Novakovic-Vunjak, G., Scaden, D.T.: Biomimetic platforms for human stem cell research. Cell Stem Cell **8**(3), 252–261 (2011)
78. Balint, B., Stamatović, D., Todorović, M., Jevtić, M., Ostojić, G., Pavlović, M., Lojpur, Z., Jocić, M.: Stem cells in the arrangement of bone marrow repopulation and regenerative medicine. Vojnosanit. Pregl. **64**(7), 481–484 (2007)
79. Obradovic, S., Rusovic, S., Dincic, D., Gligic, B., Baskot, B., Balint, B., et al.: Autologous pluripotent progenitor cells in the treatment of ischemic heart disease. Vojnosanit. Pregl. **60**(6), 725–731 (2003). Review. Serbian. No abstract available
80. Venkatraman, A.X., He, C., Thorvaldsen, J.L., Sugimura, R., Perry, J.M., Tao, F., Zhao, M., Christenson, M.K., Sanchez, R., et al.: Maternal imprinting at the H19-Igf2 locus maintains adult haematopoietic stem cell quiescence. Nature **500**, 345–349 (2013)
81. Balint, B., Ivanović, Z., Petakov, M., Taseski, J., Jovčić, G., Stojanović, N., Milenković, P.: The cryopreservation protocol optimal for progenitor recovery is not optimal for preservation of marrow repopulating ability. Bone Marrow Transplant. **23**(6), 613–619 (1999)
82. Balint, B., Jovicic-Gojkov, D., Todorovic-Balint, M., Subota, V., Pavlovic, M., Goodrich, R.: Plasma constituent integrity in pre-storage vs post-storage riboflavin and UV-light treatment—a comparative study. Transfus. Apher. Sci. **49**(3), 434–439 (2013). doi:10.1016/j.transci.2013.05.035. Epub 2013 Jun 29
83. Balint, B., Pavlovic, M., Todorovic, M., Jevtic, M., Ristanovic, E., Ignjatovic, L.: The use of original *ex vivo* immunoadsorption and "multi-manner" apheresis in ABO/H-mismatched kidney transplants—a phase II clinical study. Transfus. Apher. Sci. **43**(2), 141–148 (2010). doi:10.1016/j.transci.2010.07.002. Epub 2010 Jul 27

Chapter 10
Concept of Drug Delivery

The important thing in science is not so much to obtain new facts as to discover new ways of thinking about them.

Sir William Lawrence Bragg (1890–1971)

The past two decades have hosted a revolution in material science. In many cases it is now possible to manipulate atoms, and molecules within materials one at a time, and therefore, to construct materials with nano-scale (i.e. the size of individual molecules) precision. The potential intersection between nanotechnology and biological sciences is vast. The drug delivery technology landscape is representative field, highly competitive and rapidly evolving. The market involves both numerous startups and major players in the medical device, pharmaceutical and biotechnology industries. This is a market with intensive intellectual property protection. Products that have been brought to market or that are in clinical trials often involve combinations of technologies from multiple players, with complex licensing and strategic partnering relationships. Drug delivery could be thought of an integration of biomaterials, nanotechnology and targeted system evolution, meaning that essential goal is recognition of the targeted tissue.

Introduction

New classes of pharmaceuticals and biologics (peptides, proteins and DNA-based therapeutics) are fueling the rapid evolution of drug delivery technology [1–5]. The drug delivery concept involves targeted therapy by using the systems that will reach the specific target with high precision and deliver drug in a dosage necessary for improvement or destruction of that tissues without "touching" other cells and/or tissues [1]. This is a market with intensive intellectual property protection. Products that have been brought to market or that are in clinical trials often involve

M. Pavlovic, *Bioengineering*, DOI 10.1007/978-3-319-10798-1_10

combinations of technologies from multiple players, with complex licensing and strategic partnering relationships. These new drugs typically cannot be effectively delivered by conventional means.

Breakthrough in drug delivery: Robert Langer, who revolutionized biomedical technology with drug delivery and remote system, to control drug delivery.

Additionally, it has been determined that, for many conventional pharmaceutical therapies, the efficacy may be improved and the side effects reduced if the therapy is administered continuously (although potentially variable rate), rather than through conventional burst release techniques (oral ingestion, injection, etc.) [6]. The benefits from targeted, localized delivery of certain therapeutic agents are another driving force in this market. Additional drivers include the desire to eliminate or minimize the danger of needle stick injuries (and blood-borne pathogens) to healthcare workers, increase patient compliance by simplified or reduced stigma delivery methods, reduced healthcare worker involvement and reduced health care costs. Increasingly, delivery devices and drugs will be more tightly coupled. In some cases, device development is beginning as early as the discovery phase of the pharmaceutical development process.

Development of Nano-Biotechnologies

The past two decades have hosted a revolution in material science [7–12]. In many cases it is now possible to manipulate atoms, and molecules within materials one at a time, and therefore, to construct materials with nano-scale (*i.e.* the size of individual molecules) precision [1–5]. The potential intersection between nanotechnology and biological sciences is vast. Biological function depends heavily on units that have nano-scale dimensions, such as:

- Viruses
- Ribosomes
- Molecular motors
- Components of ECM
- Nanobacteria and Magnetobacteria

The use of biologics, polymers, silicon materials, carbon materials, and metals has been proposed for the preparation of innovative drug delivery devices. One of the most promising materials in this field are the **carbon-nanotubes composites** and **hybrid materials** coupling the advantages of polymers (biocompatibility and biodegradability) with those of carbon nanotubes (cellular uptake, stability, electromagnetic, and magnetic behavior). The applicability of polymer-carbon nanotubes composites in drug delivery, with particular attention to the controlled release by composites hydrogel, is being extensively investigated in the present days [1].

Engineered devices at the nanometer scale are small enough to interact directly with subcellular compartments and to probe intracellular events. The ability to assemble and study materials with nano-scale precision leads to opportunities in both the basic biology (e.g. testing of biological hypotheses that require nano-scale manipulation) and development of new biological technologies (e.g. drug delivery systems, imaging probes, or nanodevices).

Millimeter–scale and micrometer–scale controlled release systems have been well studied, and some systems have been approved for clinical use. As we already know. one of the major advances in recent years has been **further reduction in the size of these systems**: it is now possible to make **polymer delivery systems** that are nanometer in scale, can be easily injected or inhaled, and are much smaller than-and capable of being internalized by-many types of human cells [10, 12–15].

- **Examples of drug-delivery nanoparticles are:**Liposomal systems (Vesicles with targeting poly-ethylene-glycol (PEG groups))
- Solid biodegradable nanoparticles (polymer emulsions with targeting or PEG groups on the surface)
- Dendrimer-polymer conjugates (5 nm)
- Polymer nanoparticles [17]

ECM that surrounds the cells and tissues in the body is composed of fibers that are typically 10 nm to 100 nm in diameter. Although there are a variety of ways of achieving nano-scale delivery systems, including self-assembling systems based on **liposomes or micelles**, the most stable and versatile systems **are miniaturized versions of the synthetic materials** that have already been used in drug-delivery applications. Construction of such a system is usually accomplished by degradable polymers such as **pegylated granules** (PLGA) [15]. These particles can be injected in circulation, or used to release drugs, locally. The encapsulated drugs can be complex, if appropriate methods of fabrication are used to assemble the nanoparticles. For example, it is now possible to make 300 nm particles that have functional DNA within the solid matrix.

Polymer materials have many potential uses in drug delivery, serving as vehicles for drug distribution and release. One usually obtains a complex mixture of particles of different sizes and shapes, however, the methods of fabrication are still imperfect. **Matching method of particle formation with drugs has been one of the major challenges in this area**. Many different ways to make small particles, especially with Nanotechnology have now been described (Fig. 10.1). **Unfortunately, few of these methods are compatible with most drugs**. Finding better ways to

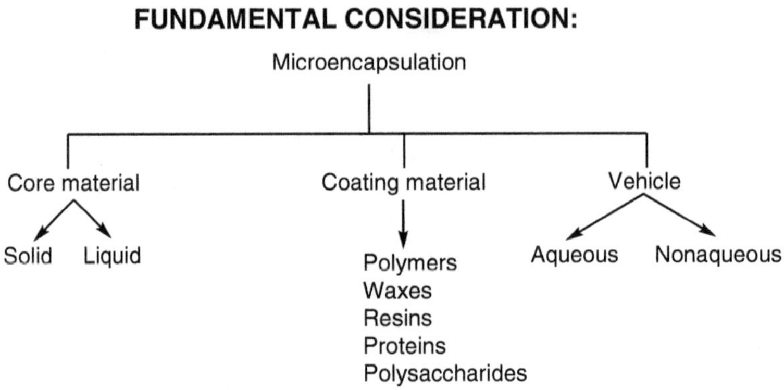

Fig. 10.1 Microencapsulation as the method for Targeted Drug delivery System engineering (From: Ravindra Kumar Gupta: Microencapsulation Techniques, PP)

make controlled particles that are compatible with drug incorporation is a challenge for the future. It is of essential interest that drug delivery system can **recognize** the targeted tissue and due to that entire spectrum of challenging ideas/techniques have been developed by different authors [18, 19].

Challenges

Historically, drug delivery has taken the form of injection, infusion, ingestion, and inhalation, with additional variations of each category. For example, ingestion may be in tablet, capsule or liquid form; inhalation may be via use of a dry powder inhaler, an MDI, or a nebulizer. The challenge for both drug and drug delivery companies is to deliver both existing and emerging drug technologies in a manner that improves the benefits to the patients, healthcare workers and the healthcare system. Areas that are being targeted for improvements through device development include: improved efficacy, reduced side effects, continuous dosing (sustained release), reduced pain from administration, increased ease of use, increased use compliance, improved mobility, decreased involvement of healthcare workers, improved safety for healthcare workers, reduced environmental impact (elimination of CFC's) (Fig. 10.2).

To provide these benefits, a number of approaches are being (or in some cases have been) developed. The common thread running through the approaches is the concept of self-administered, targeted, sustained release with increased bioavailability. **Determining which of the emerging approaches best meets stakeholder needs is a complex, multifaceted problem.** Although **ingestion** is probably the most widely accepted form of delivery it presents difficulties for a number of important classes of drugs. Many drug delivery scientists view **oral delivery** as the ideal drug delivery method. In the case of proteins and peptides, historical oral

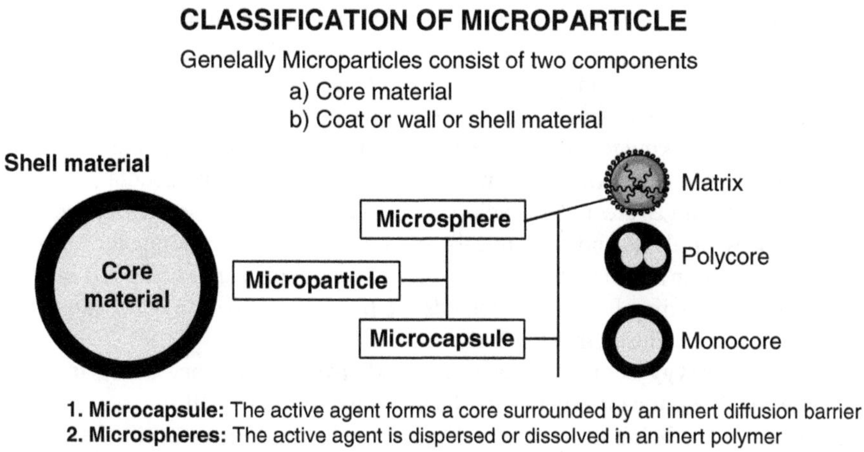

Fig. 10.2 Microparticles and Microspheres for drug delivery (Inspired by the same author)

delivery mechanisms can only delivery bioavailabilities of a few percent. In some cases, dose limiting toxicity levels are caused by lack of selectivity. Although oral delivery meets the need for self-administered drugs, targeted, sustained release and increased bioavailability present the areas of difficulty in meeting the emerging value proposition.

Technologies

To address this difficulty, companies are developing **micro-fabricated drug delivery systems.** Technologies such as **nano-pore membranes** and **micro-particles** enable the drug to survive stomach acids and be released at specific targeted areas of the gastrointestinal tract (GIT). These technologies are being developed to provide more efficient drug absorption and enhanced bioavailability. **Pulmonary delivery** provides a number of benefits particularly with regard to absorption area and avoidance of first pass metabolism in the liver. **However, meeting the sustained-release goal is somewhat problematic. The lungs tend to expel materials that are introduced and it is therefore difficult to keep the drug in the lung long enough for the sustained release to be effective.** Additional challenges revolve around elimination of excipient (enabling delivery of a neat drug), elimination of CFC propellants (in the case of MDI), reduction of the stigma associated with inhalers, and ease of use.

Transdermal patches have been used for a number of years. To improve their effectiveness for a broader range of drugs, devices are being developed that disrupt the skin barrier to allow drug transfer to the interstitial fluid.

Technologies are being developed that range from **ultrasonic disruption of the skin, to micro-projections, to using electro-transport to drive molecules though**

the skin barrier. These technologies are being developed individually and in various combinations. Although from a patient standpoint the elimination of injections is ideal, indications are that injection will remain a necessary means of drug delivery. To minimize the pain, biohazard, cost and inconvenience associated with injections, companies are working to reduce the negative aspects of this delivery method.

Along these lines, **advances in needle-free injection, micro needle injection and MEMS syringes** are under development. To minimize the number of injections required, new implants and time release approaches are in development. Meeting the need for better bioavailability and reduced side effects is not being left only to the mechanical delivery systems; methods are being developed to better target the drug once it is introduced into the body.

Different energy sources (ultrasound, infra-red, laser) are being used to activate the drug once it reaches the targeted location [16–19].

Receptors are being used to target specific cells, and in the event that a targeted cell does not have a required receptor, methods for adding receptors for a specific drug are being developed. Antibodies to targeted molecules on targeted cells can be developed and attached to the surface of drug delivery system.

It is also possible to make **nanostructured material from minerals, and ceramics nanominerals** are formed under mild conditions, in which **DNA molecules can be embedded**. The mineral composition of this particles and **the loading of DNA can be controlled. When cells are cultured on these composite surfaces, DNA is taken up into the cell. The overall composition of the surface (and hence its nanostructure) influences the amount of DNA that is taken up and expressed in the cell.** Other systems have also been used to show that nanoparticles of controlled size and density can be used to facilitate uptake and expression by concentrating DNA at the cell surface.

Some cells will internalize nano-scale particles. If these particles are loaded with drugs, such as **chemotherapy drugs**, then the nanoparticles can be used to deliver high drug doses into the cell interior. Polymeric nanoparticles can also be conjugated with **cell ligands** for targeting specific cell populations that have **surface molecules** which will recognize and bind the ligands, so that the drug content can reach the cell. In this way, it may be possible to make drug carriers that are much smaller than a cell, but capable of delivering large doses of drug directly to the cell's internal machinery.

New research suggests that these particles can be made from materials that respond to **mild external signals (such as light, ultrasound, or magnetic fields)** so that the movement of the particles can be **directed from outside the body**, or the particles could be activated at particular sites. In this way, nanotechnology is providing new methods for using materials in the body: the very small size of the materials makes them suitable for many biological functions. Because of their combination of properties—including subcellular size and controlled release capability and susceptibility to external activation-devices produced by nanotechnology will enable new applications in biological and medical science. Thus, the controlled drug delivery can **extend** and **optimize** drug use.

The most obvious use for controlled drug delivery involves miniaturization of the familiar infusion system. Mechanical pumps, either totally implantable, or requiring catheters, have been used to deliver insulin, anticoagulants, analgesics, and cancer chemotherapy. Although the technology is mature and accepted by many patients, this approach has certain disadvantages. Drugs are maintained in a liquid reservoir prior to delivery, so only agents that are stable in solution and body temperature can be used. The devices are bulky, expensive, and only suitable for highly motivated patients who can visit their physicians, regularly. Because of the opportunity for precisely controlled delivery, however, pumps may find additional important applications, particularly when coupled with implanted biosensors for feedback control.

The disadvantages of pumps can be avoided by using controlled release polymers, as drug-delivery vehicles. They are used to modify the:

- Rate
- Pattern
- Duration of drug release

Polymer drug-delivery systems and mechanisms of controlled drug delivery by polymers:

A. Non-biodegradable

- Polymer membranes that control drug release
- Uniform distribution of drugs through the polymer matrix
- Water–soluble polymers which if cross-linked are dry and if exposed to water swell; drug release is activated by swelling action of the water determining the rate of drug release

B. Biodegradable

- Drug released by degradation or dissolution of polymer matrix
- Cleavage of covalent bonds that links the drug within the polymer matrix
- The rate of release determined by polymer erosion
- The rate of release determined by the kinetics of bond degradation

Controlled delivery of proteins and other molecules:
Recombinant proteins and polypeptides that serve as a drugs are generally less stable than conventional drugs and more difficult to deliver. Protein drugs are difficult to use in humans since they are either:

- Eliminated very quickly when introduced into the body or
- Toxic when delivered systematically at the doses required to achieve a local effect

Controlled–release polymers may overcome these problems by:

- Slowly releasing the protein into the blood over a long period, or
- Releasing protein into a local-tissue site, thus sparing systemic exposure
- Many of these systems have been developed and you can Google for them

Genetically Engineered Cells for Controlled Drug Delivery

Drug delivery strategies can some days be tailored to individuals. **Cells harvested from a patient could be genetically modified to increase production of a protein of interest: genes enhancing the cell's ability to control protein expression could also be added. Transplantation of these cells to the patient may restore normal protein delivery**. Direct introduction of the genetic material into the cells of the body (mice genetically fluorescently labeled) is also possible. Certain therapeutic cells can be expanded (grown to large numbers) outside of the body for transplantation, to treat a disease (such as cancer), or to form new tissues biocompatible polymers may be important in cell transplantation by serving as a scaffold for attachment of the genetically modified cells. For example, hydrogel materials have been used to encapsulate engineered cells as a means of isolating them from direct contact with cells of the immune system.

Competitive Landscape:
A top-level view of the competitive landscape and some of the companies involved in various areas has been developed as follows:

Sustained Release Technology

Nanobiotechnology involves the design of materials with nano-scale dimensions, which provides them with unique abilities to interact with biological systems. In the heart of nanotechnology are geometry and minimization.

A number of companies are working in this arena with technologies varying from ultrasonic de-aggregation, to heat vaporization of the drug.

Meeting New Challenges:
Meeting the challenges that are presented by emerging drug technologies and the requirement for improved stakeholder benefits, including the impact of the aging population, will require some combination of drugs, delivery devices and mechanisms currently under development, as well as the identification and integration of new yet to be defined technologies. Complicating the need for self-administered, targeted, sustained release with increased bioavailability is the need to improve patient compliance. To achieve improved compliance will require further simplification of the user experience. The next step can easily be envisioned as involving further integration of devices and drugs to provide means to deliver multiple therapies in a simple, pain free, unobtrusive, and targeted sustained release device. The proper combination of technology portfolios, intellectual property, market and stakeholder understanding required achieve this next step is the challenge on the horizon. Making sense of this complex interaction of competing companies, intellectual property, core competencies, stakeholder needs, and technology trends, in a manner that will meet the corporate goals requires a structured methodology, such

as the Innovation Genesis framework, to drive corporate planning and decision-making. Based on the corporate strategies, portfolio investments can be managed to meet the appropriate mix of high and low risk activities for the company. By establishing a deep understanding of convergent trend (and the conditions and drivers underlying the trends), and by maintaining knowledge of emerging technologies outside the core competencies of the firm, IP and technology portfolio strategies can be optimized. Visibility of long-range evolution scenarios enables actionable short and mid-range activities and decisions that are aligned with the long term goals. In short, this structured methodology enables Strategic Innovation. The approach enables informed technology investments that deliver meaningful business consequences, and the development of new ideas that fundamentally change the basis of competition within the drug delivery industry.

Emphasizing Bioengineering Aspects to Drug Delivery: Achieving Precision

The expanding arena of emerging drugs combined with increased sensitivity to clinical outcomes and healthcare costs are driving the need for alternative drug delivery methods and devices. More and more, the development of drugs and the development of delivery systems are being integrated to optimize the efficacy and cost effectiveness of the therapy. As new methods are developed and patented, the intellectual property landscape becomes congested. Some companies may find themselves locked out and in a position of having few attractive paths to market. Possibly worse, is the potential for a pharmaceutical company to find itself in the unenviable position discovering a potential blockbuster drug that requires a delivery method that is thoroughly protected by a competitor's patents? For the companies in the drug development and delivery business, **strategic innovation** is required to ensure complex and competing technology investments are aligned with emerging market needs and support corporate goals. Gaining an understanding of long-term trend convergence in drug development and delivery methods, the intellectual property landscape, and emerging technologies will position a company to make the best possible investment decisions. Through the rigorous application of a strategic innovation analysis framework, actionable investment criteria may be identified that both support the short-term financial responsibilities to shareholders while securing the future potential for the firm to participate in major growth opportunities.

So far, in advanced drug delivery arena there are two innovative drug control delivery systems:

1. *Novel drug delivery* and
2. *Microchip control* drug delivery

About 15 years ago, MIT professors Robert Langer and Michael Cima came up to interesting idea to develop a **programmable, wirelessly controlled microchip that would deliver drugs after implantation in a patient's body**. Finally, the MIT researchers and scientists from MicroCHIPS, Inc., reported that they have successfully used such a chip to administer daily doses of an *osteoporosis drug* normally given by injection.

According to Dr. Langer, the results, published in the Feb. 16 online edition of Science Translational Medicine, represent the first successful test of such a device and could help usher in a new era of telemedicine—delivering health care over a distance. **That means literally have a pharmacy on a chip.** According to these scientists, you can do **remote control delivery**, you can do pulsatile drug delivery, and you can deliver multiple drugs. In the new study, funded and overseen by MicroCHIPS, scientists **used the programmable implants** to deliver an osteoporosis drug called teriparatide to seven women aged 65 to 70. The study found that the device delivered dosages comparable to injections, and there were no adverse side effects. These programmable chips could dramatically change treatment not only for osteoporosis, but also for many other diseases, including cancer and multiple sclerosis. The scientists say that patients with chronic diseases, regular pain-management needs or other conditions that require frequent or daily injections could benefit from this technology. This also avoids the compliance issue completely, and points to a future where you have fully automated drug regimens, approach much more comfortable for the patient.

In 1999, the MIT team published its initial findings in *Nature*, and MicroCHIPS was founded and licensed the microchip technology from MIT. The company refined the chips, including adding a hermetic seal and a release system that works reliably in living tissue. *Teriparatide* is a **polypeptide** and therefore much less chemically stable than small-molecule drugs, so sealing it hermetically to preserve it was an important achievement. The human clinical trial began in Denmark in January 2011. Chips were implanted during a 30-min procedure at a doctor's office using local anesthetic, and remained in the patients for four months. The implants proved safe, and patients reported they often forgot they even had the implant.

Another approach is the work on nanoparticles with a reservoir under remote control in order to control the drug delivery into targeted tissues or organs. Maybe two the most interesting examples are coming from MIT (Langer's group) and the group of Dr. Sakhrat Khizroev from FIU [1, 2].

Langer: A reservoir that could be remotely triggered to release a drug would enable the patient or physician to achieve on-demand, reproducible, repeated, and tunable dosing [1]. Devices that release a drug in response to a remote trigger would enable on-demand control of the timing and dose of drug released. They would allow the patient or physician to adjust therapy precisely to a target effect, thus improving treatment and reducing toxicity. Langer's group has developed **implantable reservoirs** that release a drug when irradiated with **near-infrared laser light**. The release rate was correlated to laser intensity, with negligible leakage between doses. Devices containing **aspart**, a fast-acting analog of insulin, were implanted in diabetic rats and were able to achieve glycemic control upon irradiation. Such devices can be loaded with a wide range of drugs to treat a variety of clinical indications [1].

Khizroev: Dr. Khizroev's group has fabricated **magnetic nanoparticles** that have targeted ovarian carcinoma cells, controlled by **a low energy remote magnetic field, from outside**. They confirmed through kinetic studies that that the drug penetrated through the tumor cell membrane and eradicated the majority of tumor cells within a 24-h period without affecting surrounding healthy cells [2]. **It would be very interesting to target Cancer stem cells only in order to see how the concept of targeted Cancer Stem Cell Therapy works.** The future work will show how relevant these two interesting approaches are for therapeutic purposes.

References

1. Timko, B.P., Arruebo, M., Shankarappa, S.A., McAlvin, J.B., Okonkwo, O.S., Mizrahi, B., Stefanescu, C.F., Gomez, L., Zhu, J., Zhu, A., Santamaria, J., Langer, R., Kohane, D.S.: Near infrared-actuated devices for remotely controlled drug delivery. Proc. Natl. Acad. Sci. U. S. A. **111**(4), 1349–1354 (2014)
2. Guduru, R., Liang, P., Runowicz, C., Nair, M., Atluri, V., Khizroev, S.: Magneto-electric nanoparticles (MENs) to enable field-controlled high-specificity drug delivery to eradicate ovarian cancer cells. (2013) Nature.com: Scientific reports online: pp. 1–13
3. Cirillo, G., Lemma, F., Puoci, F., Parisi, O.I., Curcio, M., Spizzirri, U.G., Picci, N.: Imprinted hydrophilic nanospheres as drug delivery systems for 5-fluorouracil sustained release. J. Drug Target. **17**(1), 72–77 (2009)
4. Cirillo, G., Hampel, S., Spizzirri, G.U., Paris, O., Picci, N., Lemma, F.: Carbon Nanotubes Hybrid Hydrogels in Drug Delivery: A Perspective Review. Hindawi Publishing Corporation BioMed Research International Volume (2014), Article ID 825017, 17 pages
5. Daum, N., Tscheka, C., Neumeyer, A., Schneider, M.: Novel approaches for drug delivery systems in nanomedicine: effects of particle design and shape. Wiley Interdisciplinary Reviews **4**(1), 52–65 (2012)
6. Al-Sarraf, M., Martz, K., Herskovic, A., Leichman, L., Brindle, J.S., Vaitkevicius, V.K., Cooper, J., Byhardt, R., Davis, L., Emami, B.: Progress report of combined chemoradiotherapy versus radiotherapy alone in patients with esophageal cancer: an intergroup study. J. Clin. Oncol. **15**(1), 277–284 (1997)
7. Brannon-Peppas, L., Blanchette, J.O.: Nanoparticle and targeted systems for cancer therapy. Adv. Drug Deliver. Rev. **56**(11), 1649–1659 (2004)
8. Citron, M.L., Berry, D.A., Cirrincione, C., Hudis, C., Winer, E.P., Gradishar, W.J., Davidson, N.E., Martino, S., Livingston, R., Ingle, J.N., Perez, E.A., Carpenter, J., Hurd, D., Holland, J.F., Smith, B.L., Sartor, C.I., Leung, E.H., Abrams, J., Schilsky, R., Muss, H.B., Norton, L.: Randomized trial of dose-dense versus conventionally scheduled and sequential versus concurrent combination chemotherapy as postoperative adjuvant treatment of node-positive primary breast cancer: first report of Intergroup Trial C9741/Cancer and Leukemia Group B Trial 9741. J. Clin. Oncol. **21**(8), 1431–1439 (2003)
9. Collins, A.T., Berry, P.A., Hyde, C., Stower, M.J., Maitland, N.J.: Prospective identification of tumorigenic prostate cancer stem cells. Cancer Res. **65**(23), 10946–10951 (2005)
10. Davis, M.E.: Nanoparticle therapeutics: an emerging treatment modality for cancer. Nat. Rev. Drug Discov. **7**(9), 771–782 (2008)
11. Delaney, G., Jacob, S., Featherstone, C., Barton, M.: The role of radiotherapy in cancer treatment. Cancer **104**(6), 1129–1137 (2005)
12. Farokhzad, O.C., et al.: Nanoparticle-aptamer bioconjugates a new approach for targeting prostate cancer cells. Cancer Res. **64**(21), 7668–7672 (2004)

13. Fisher, B., Wolmark, N., Rockette, H., Redmond, C., Deutsch, M., Wickerham, D.L., Fisher, E.R., Caplan, R., Jones, J., Lerner, H., et al.: Postoperative adjuvant chemotherapy or radiation therapy for rectal cancer: results from NSABP protocol R-011. J. Nat. Cancer I. **80**(1), 21–29 (1988)
14. Gagliardi, G., Lax, I., Ottolenghi, A., Rutqvist, L.E.: Long-term cardiac mortality after radiotherapy of breast cancer—application of the relative seriality model. Brit. J. Radiol. **69**(825), 839–846 (1996)
15. Gaitanis, A., Staal, S.: Liposomal doxorubicin and nab-paclitaxel: nanoparticle cancer chemotherapy in current clinical use. Cancer Nano. **624**, 385–392 (2010)
16. Gil, J., Stembalska, A., Pesz, K.A., Sasiadek, M.M.: Cancer stem cells: the theory and perspectives in cancer therapy. J. Appl. Genet. **49**(2), 193–199 (2008)
17. Gradishar, W.J., Tjulandin, S., Davidson, N., Shaw, H., Desai, N., Bhar, P., Hawkins, M., O'Shaughnessy, J.: Phase III trial of nanoparticle albumin-bound paclitaxel compared with polyethylated castor oil-based paclitaxel in women with breast cancer. J. Clin. Oncol. **23**(31), 7794–7803 (2005)
18. Hillner, B.E., Smith, T.J.: Efficacy and cost effectiveness of adjuvant chemotherapy in women with node-negative breast cancer: a decision-analysis model. New Engl. J. Med. **324**(3), 160–168 (1991)
19. Stupp, R., Wong, E.T., Kanner, A.A., Steinberg, D., Engelhard, H., Heidecke, V., Kirson, E.D., Taillibert, S., Liebermann, F., Dbalý, V., Ram, Z., Villano, J.L., Rainov, N., Weinberg, U., Schiff, D., Kunschner, L., Raizer, J., Honnorat, J., Sloan, A., Malkin, M., Landolfi, J.C., Payer, F., Mehdorn, M., Weil, R.J., Pannullo, S.C., Westphal, M., Smrcka, M., Chin, L., Kostron, H., Hofer, S., Bruce, J., Cosgrove, R., Paleologous, N., Palti, Y., Gutin, P.H.: NovoTTF-100A versus physician's choice chemotherapy in recurrent glioblastoma: a randomised phase III trial of a novel treatment modality. Eur. J. Cancer **48**, 2192–2202 (2012)

Chapter 11
Engineering Balances

In questions of science, the authority of a thousand is not worth the humble reasoning of a single individual.

Galileo Galilei (1564–1642)

Analysis of any engineering is based upon essential concepts: the concept of a system under study and system balances. These concepts, after being introduced, can be illustrated in the context of lung, gastrointestinal, cardiovascular and renal physiology. On the other hand, the human body is an elegant machine that requires input for sustained operation. However, the workings of the human bodies—which convert food, air and water into ENERGY and BODY MASS is extremely complex. Biomedical engineers can contribute to understanding relationships between intake of nutrients, air, drugs, toxins, and other molecules and human health by applying physical, chemical and mathematical knowledge where appropriate. Beside introducing measurable values and terms, it also helps with the problem solving attitude in many functional obstructions of systemic significance.

The human body can be envisioned as an elegant machine that requires **input** for **sustained operation.** However, the workings of the human bodies—which convert food, air and water into energy and body mass is extremely complex. Biomedical engineers can contribute to understanding relationships between intake of nutrients, air, drugs, toxins, and other molecules and human health.

M. Pavlovic, *Bioengineering*, DOI 10.1007/978-3-319-10798-1_11

Breaking through: Engineering Systems: Karl Ludwig von Bertalanffy (1901–1972) and James Grier Miller (1916–2002) among the first who developed the System Theory and gave the concepts of open/closed and living systems (Engineering Systems)

Engineering Systems

Analysis of any engineering is based upon essential concepts: the concept of a system under study and system balances. These concepts, after being introduced, can be illustrated in the context of lung, gastrointestinal, cardiovascular and renal physiology. Conceptual illustration of global bioengineering system is given in Figs. 11.1 and 11.2.

This approach is inevitable, since one of central goals of engineering is the analysis and interpretation of the mechanisms, often from complex systems (brain, circulation) in which all of the internal workings of the system are not known. The human machine requires food and water for continued operation, but the relationship between food intake and human health is complex and poorly understood.

Engineering analysis invariably begins by **defining the system under study**. The system might be a supporting beam in the bridge, the human body or individual cell in the body, even molecule of the cell in the body. This analysis is accomplished by identifying system boundaries.

System boundaries are important part of description of the system, e.g. they are **physical sites of intersection** between the system under the study and the rest of the world. There are often multiple choices for them, leading to alternate description of the system under study.

When we decide to study or communicate information about a system, we first have to explicitly define the boundaries to our description and what flows in and out (Table 11.1).

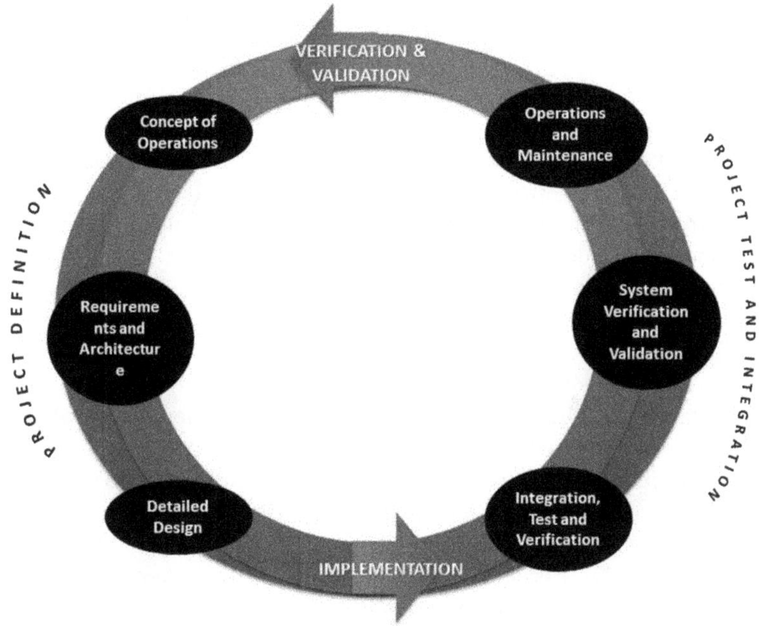

Fig. 11.1 Conceptual illustration of global bioengineering system

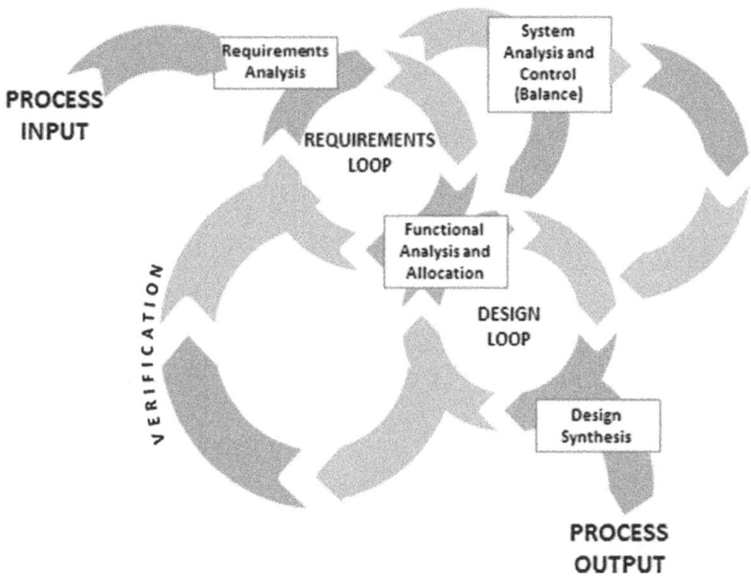

Fig. 11.2 System boundaries

Table 11.1 Open and closed systems (differences)

Open systems	Closed systems
• Can perform work	• Work unachievable
• Not in an equilibrium state	• No change between system and its environment

Open Systems

The concept of equilibrium and steady state conditions need to be clarified before we go further into how open systems operate. In a closed system, equilibrium is achieved when opposing variables in the system are in balance (Miller, 1978). In addition, the equilibrium can be static or dynamic (Fig. 11.3).

The former is commonly found in closed systems while the latter is a property of an open system. Since living systems are open systems, with a recurrent alteration of fluxes of matter, energy, and information, their equilibrium is dynamic. Miller (1978) termed the dynamic equilibrium a 'flux equilibria' or 'steady state' [1]. The term **dynamic equilibrium has, however, also been utilized interchangeably in both closed and open systems** [2, 3]. We argue that both closed and open systems can exhibit equilibrium; however, in the latter case, the equilibrium is 'quasi' rather than being a true one as in closed systems. In some instances, a **steady state was characterized** as a dynamic equilibrium that exists in open systems. According to Kramer and De Smith (1977), a steady state refers to an **open system maintaining an unchanging state even when input and output are still in operation** [4]. This makes the system **appear static to the observer despite the fact that the flow of resources through the system is dynamic and continuous**. A popular example of this is the maintenance of the human body temperature at 37 °C. In this case, the amount of heat generated by the body's metabolism is kept equal to the heat lost to the environment. As a result, a constant body temperature can be maintained.

The most important quality of an open system is that it can perform work, which is unachievable in a closed system in an equilibrium state because a closed system in equilibrium does not need energy for the preservation of its state, nor can energy be obtained from it. In order for it to perform work, it is necessary that **an**

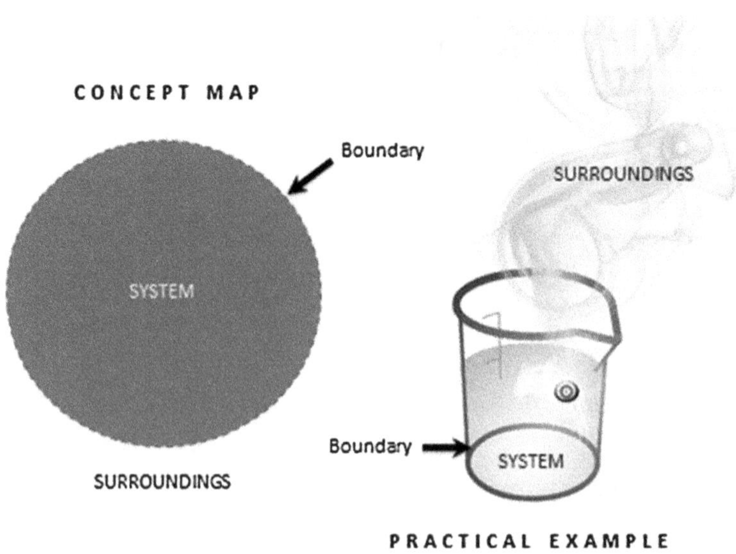

Fig. 11.3 Concept example

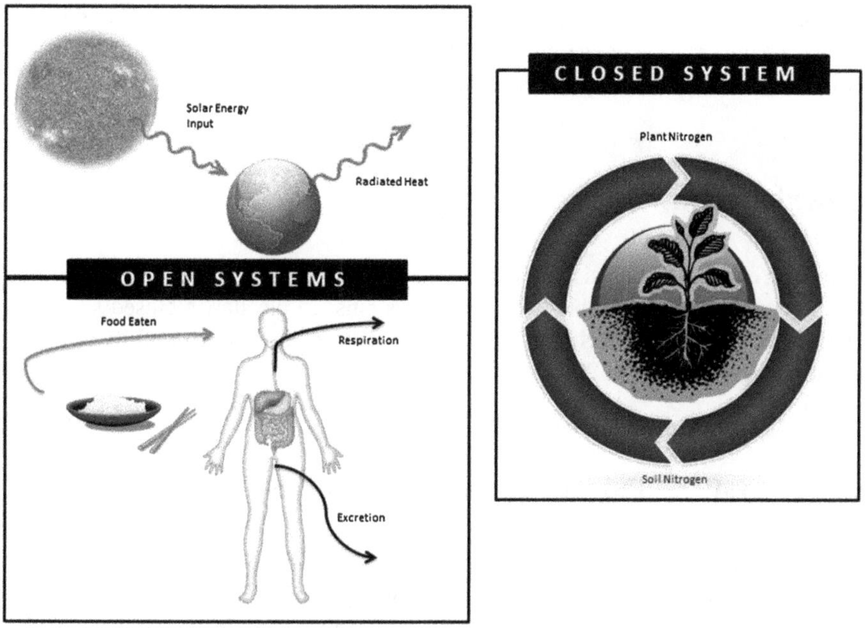

Fig 11.4 Examples of open and closed systems

open system is not in an equilibrium state. Nevertheless, the system has a tendency to attain such a state. As a result, the equilibrium found in an organism (or any open system) is not a true equilibrium, incapable of performing work. Rather, it is a **dynamic pseudo-equilibrium (or quasi-equilibrium) kept constant at a certain distance from the true equilibrium**. In order to achieve this, the continuous importation of energy from the environment is required [2, 3] (Fig. 11.4).

The homology between an open system and human or work organizations can be drawn from the chain of logic mentioned in the previous paragraph. A fictitious organization, which is largely closed to the external environment, will eventually lose its alignment with the environment because only limited or no resources (i.e. materials, energy, and information) from the environment are allowed to cross the boundary into the organization. This leads to a misalignment between organizational strategy-structure and the environment, which results in substandard performance as the acquisition and usage of resources become inconsistent with the demand from the environment. The organization that persistently performs poorly will deteriorate over time and, we argue, is on the way to equilibrium according to the second law. On the other hand, a viable organization needs a continuous inflow of new members for new ideas, skills and innovations, raw materials and energy to produce new products and/or services, and new information for reasonable planning, strategy formulation and coordination. Only the importation of these resources from the environment can keep it away from equilibrium and can allow it to perform its activities in a viable manner. It should be noted at this point that the meaning of equilibrium as it is used here, is 'entropic equilibrium' in which equilibrium is maintained at the expense of structure (Grey, 1974; Van Gigch, 1978) [5]. In other words, the system's structure and organization will deteriorate over time, according to the second law, if there is no importation of energy and materials from the environment and processing of information. Another type of equilibrium will be introduced in the next section.

Closed Systems

Before an extensive analysis of theories of open systems is conducted, it is useful to briefly consider some attributes of closed systems. In physics, a closed system is one **where there is no exchange of matter between the system and its environment** (Cengel and Boles, 2002) [6].

Closed Systems and Organizational Theories

In nonrelativistic classical mechanics, a closed system is a physical system which doesn't exchange any matter with its surroundings, and isn't subject to any force whose source is external to the system [3]. A closed system in classical mechanics would be considered an isolated system in thermodynamics. However, Kramer and De Smith (1977) define a closed system as a system that has no interaction at all with its environment [4]. But they explain further that a system can be deliberately considered as a closed one by researchers if the relations that exist between the system and its environment are disregarded for the sake of simplicity in their analysis. For example, a production or assembly line, which is built on the theory of scientific management and operations research, can be treated as a closed system if it is insulated from fluctuations in demand and supply (environmental contingencies) through the stockpiling of raw materials and finished-goods to keep it in a relatively static environment.

Even though it is impossible to treat a work organization as a completely closed system, in the past several organizational theories have assumed this view (Robbins, 1990; Scott, 1998) [7]. Between 1900 and 1930, the most dominant theories, which were based on closed-rational system models, were Taylor's scientific management approach, Weber's model of bureaucracy, and Fayol's administrative theory. From the 1930s through the 1950s, the most influential theories were based on a new perspective of closed-natural system models, such as Barnard's theory of cooperative systems and Mayo's human relations model. It is reasonable to say that the ideas of scientific management and bureaucracy are rooted in engineering where the system designer believes that, through proper design and without referring to external factors, a purposive system will perform in an efficient and effective manner. This belief has become the foundation of the machine metaphor or the mechanistic organization (Morgan, 1997) [8].

A prevalent example of a management system built on a closed system model is a machine bureaucracy, which is still, to various degrees the prevailing paradigm in most organizations (Brown, 1992; Beetham, 1996; Du Gay, 2000) [9, 10].

The main objective of a bureaucracy is to promote efficiency and control in systems through the following: a fixed division of labor; a hierarchy of offices; a set of general rules that govern performance; a separation of personal from official property and rights; selection of personnel on the basis of technical qualifications; and employment viewed as a career by participants (Scott, 1998). From an engineering viewpoint, it is a superbly designed system based on technical rationality, aimed at maximizing operational efficiency and control. However, the emphasis on internal operational efficiency without referring to external factors can result in system-environment misalignment. In addition, the sole concentration on control without flexibility may well cause poor adaptation, which leads to unsatisfactory performance in the long run.

Closed Systems and Change

One possible explanation for the existence of organizations that continuously remain in a steady state condition is that they reside in a relatively static environment (e.g. some not-for-profit organizations). When the environment is relatively static, stable, and predictable, interactions and relationships between the organization and its environment are trivial and, thus, can be ignored or otherwise managed [11]. The closed system model was universally adopted in management theory development during the early 20th century. **However, the environment has changed dramatically over the past century** and the direction of change is toward an increase in both complexity and dynamism [11, 12]. A model that was valid in the past might not be effective in describing, explaining, and predicting organizational phenomena in a changing context. For example, the Just-in-Time (JIT) inventory system increases the alignment between the production system and its environment, giving a substantial increase in operational efficiency and a reduction in inventory cost [13, 14].

If the human or work organization is assumed to be a closed system, the direction of change should go toward an equilibrium state in which entropy will maximized, according to the second law of thermodynamics. In this case, the organization as a system should deteriorate rather than prosper over time. The increase in entropy suggests that the organization and order of the system will be degraded and the system will run down.

Homeostasis

The living system always tends to remain in a steady state with a help of self-regulatory mechanisms. Such a phenomenon, which **involves maintenance of constant internal environment**, is known as homeostasis. It is the ability or the tendency **of a living organism (or an opened system in general) to maintain a stable internal environment by adjusting its physiological processes**.

Almost every organ in the body contributes to homeostasis at the level of whole body. Some examples of homeostasis from everyday life are:

- When we enter a darkened room from bright light, we are unable to see anything for a few moments. Within a few seconds we are able to adjust our sight to the dim light and move about easily.
- After vigorous exercise, the body temperature rises and there is profuse sweating. Evaporation of the sweat results in cooling which brings down the body temperature to normal.
- In winter the skin contracts to conserve body heat.
- The homeostatic mechanism works through a negative feedback mechanism where the rate of formation or utilization of any product is regulated by the amount of that product available at a given period of time.
- The homeostatic mechanism operates at all levels of hierarchy in the living system, namely cells, tissues, organs, organ systems and organisms. It occurs in almost all forms of life and contributes to a balance in nature.

A "closed system" is one in which the boundary includes all stocks and flows and for which there are no source/sink components. Often the decision of whether or not a system is open or closed requires **a judgment based on the significance of some of the smaller losses or gains and a decision on the time scale of your study**. For example, you might model a forest as a closed system for nutrients ignoring initially small amounts of nitrogen that comes in from rain or lost through streams. The **time scale question is apparent** if, for example, you are studying the gain and loss of species in a city park but are ignoring evolution. The description and diagramming of a systems model should attempt to make these boundaries very clear (Fig. 11.5).

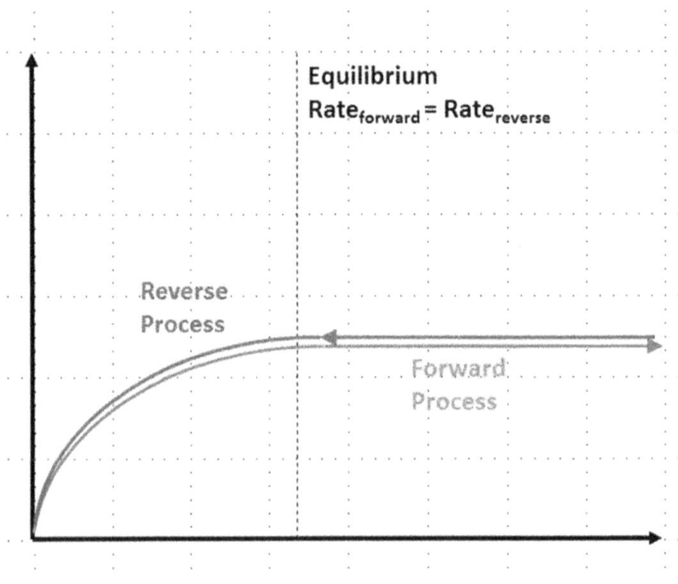

Fig. 11.5 Essentials for forward process reverse process and dynamic equilibrium

Mass Balances

Methods for accounting for **material entering and leaving** a system based on the conservation of mass principle (matter cannot be created or destroyed).

Steady State

An unvarying condition (all state variables are constant) despite ongoing condition to change the state. Systems at steady state do not change appreciably at time. The inflow to and outflow from a stock can create a situation where steady state is possible. If the input and output are equal then the value of the stock will not change with time. Slight increases in the input create or a slight decrease in the output rate can lead to an increasing stock. Other examples of steady state conditions are the CO_2 concentration in the atmosphere (currently not in steady state), use and replenishment of natural capital, or the human population at zero population growth. The conditions that lead to steady state are important to understand because the steady state may be the consequence of a very slow input and very slow output, in which case not much will ever happen very quickly. Conversely, the steady state could be a very tenuous balance between rapid input and output [15–21].

With rapid fluxes, slight disturbance in one rate could have dramatic consequences. A good example of this delicate balance is a pond in which a large amount of algae growth is growing and contributing oxygen to the water, but then with a slight change in temperature the large amount of algae turns from a net oxygen producer to a net oxygen consumer. These ponds crash into a scummy mass very quickly and then start to stink. The simpler the system and the controls on the system, the more likely that these rapid fluxes can flip flop [15–21].

Equilibrium and Dynamic Equilibrium

From thermodynamic point of view, equilibrium is a state in which opposing or influencing forces are **perfectly balanced. Dynamic equilibrium** however, mostly relates to chemical systems and defines the state in which the rates (but not necessarily concentrations) of the forward and reverse reactions are equal. The system is dynamic because individual molecules react continuously. It is at equilibrium because no net change occurs and the speed of reactants and products are the same at that point although the concentrations might not be [1].

For the purposes of this book, we are going to use a specific definition of **how to think of a problem as a system**. This isn't the only definition of a "system" and might be different than the one (or ones) that you are familiar with. The purpose of our **limited definition is to understand that we need to provide a short list**

of characteristics of a system and then try to describe many different structures and behaviors using just this short list. This intellectual process is both a good way to start looking at problems and will help us see **similarities** between different systems.

References

1. Miller, J.G.: Living Systems. McGraw-Hill, New York (1978). ISBN 0-87081-363-3
2. Von Bertalanffy, L.K.: An outline of general systems theory. Brit. J. Philosophical Sci. **1**, 134–165 (1950)
3. Von Bertalanffy, L.K.: Perspectives on general systems theory. In: Taschdjian, E. (ed.) Scientific-Philosophical Studies. George Braziller, New York (1976). ISBN 0-8076-0797-5
4. Kramer, J.T.A.N., de Smith, N.: Systems Thinking: Concepts and Notions. Springer, New York (1977). ISBN 9020705873
5. Van Gigch, J.: Applied General Systems Theory. Addison Wesley, Reading (1978). ISBN 9780060467760
6. Cengel, M.A., Boles, Y.A.: Thermodynamics: An Engineering Approach. McGraw-Hill, New York (2002)
7. Scott, J.: SEEING LIKE A STATE. How Certain Schemes to Improve the Human Condition Have Failed. Yale, New Haven (1998)
8. Morgan, D.L.: Focus Groups as Qualitative Research. Sage, London (1997)
9. Beetham, D.: Bureaucracy, p. 120. University of Minnesota, Minneapolis (1996)
10. du Gay, P.: In Praise of Bureaucracy: Weber-Organization—Ethics. Sage, London (2000)
11. Robbins, S.P.: Organization Theory: Structure, Design, and Applications. Prentice Hall, Englewood Cliffs (1990)
12. Clarck, J., Neumann, J.: The Managerial State. Sage, London (1997)
13. Gaither, N., Frazier, G.: Production and Operations Management. Litten Editing and Production, Stow (1999)
14. Wagner-Tsukamoto, S.: Human Nature and Organization Theory. Edward Elgar, Cheltanham (2003)
15. Hatch, M.J.: Organization Theory: Modern, Symbolic, and Postmodern Perspectives. Oxford University, Oxford (1997)
16. Nickelson, J.A., Todd, R.Z.: Being efficiently fickle: a dynamic theory of organizational choice. Organizational Science **13**, 547–566 (2002)
17. Pfeffer, J.: New Directions for Organization Theory: Problems and Prospects. Oxford University Press, Oxford (1997)
18. Putnam, L.L., Jablin, F.M.: New Handbook of Organizational Communications: Advances in Theory, Research, and Methods. Sage, Newbury Park (2004)
19. Brown, L., Holme, T.: Chemistry for engineering students. Mary Finch, Books/Cole, Cengage learning (2011) ISBN: 13:078-1-4390-4791-0
20. Gray, R.M., Wyner, A.D.: Source coding over simple networks. Bell Syst. Tech. J. **53**(9), 1681–1721 (1974)
21. Greene, J.C., Caracelli, V.J.: Advances in Mixed-Method Evaluation: The Challenges and Benefits of Integrating Diverse Paradigms. Jossey-Bass, San Francisco (1997). ISBN 0787998222, 9780787998226

Chapter 12
Respiration and Digestion: Bioengineering Basics

Science is a way of thinking much more than it is a body of knowledge.

Carl Sagan (1934–1996)

Respiration is exchange of oxygen through the lungs and circulatory system. Oxygen diffuses into capillaries, which surround the alveolus. The circulatory system carries the oxygen to all of the cells within the body. Mitochondria inside of cells use this oxygen to produce cellular energy. The concentration of oxygen in the external air is measured in units of partial pressure—it is usually about 150 mmHg.

The digestion is a complex process of breaking food into smaller molecules elaborated by digestive system and its enzymes. The human diet is complex: so is the metabolism and metabolic control taking place in digestive system and being controlled from local and higher CNS and autonomic nervous system levels. The digestive system is an open system which intakes the food and water, digests it, and eliminates what is not necessary in the external environment. Most of the nutrients are absorbed into the circulation through the wall of digestive tract and its lymphatic and blood vessels. In the cell, these nutrients will be metabolized and stored as energy deposits or used as building material. The digestive system has a number of actions that are familiar to engineers; storage and controlled emptying, mixing, secretion, digestion, and adsorption. In its simplest version it is an 8 m of tubing connecting the mouth to the anus.

A. Respiratory tract

- **Respiratory physiology**

In many aspects, respiration is the basis of human life, since it is one of essential vital functions. Remember mitochondria and chloroplasts and the sense of the life on the Earth? Respiration is the process which enriches the atmosphere with CO_2 necessary for life of the plants [1]. The concept of respiration involves two related concepts: internal and external respiration [1].

M. Pavlovic, *Bioengineering*, DOI 10.1007/978-3-319-10798-1_12

- **External and internal respiration**

External respiration is known as *ventilation* and it plays a role in gas exchange with the surrounding/external atmosphere [1, 2]. The process of internal respiration takes place in the cells within the body when the cells consume oxygen producing carbon-dioxide as a by-product. The lung, in the process cold external respiration, accomplishes this intake and expulsion of oxygen and carbon-dioxide.

- **Ventilation (inspiration and expiration)**

External and internal ventilation are physically linked by circulatory system, which is different from respiratory (open), e.g. closed system. Gaseous exchange in alveolus of man is shown on the Fig. 12.1. Oxygen diffuses into capillaries, which surround the alveolus and haemoglobin from blood, binds the oxygen. The circulatory system carries the oxygen to all of the cells within the body, where it will be released at the capillary bed into intercellular space and then reach the cells and enter mitochondria on the basis of concentration gradient. Mitochondria inside of cells use now this oxygen to produce cellular energy in Kreb's cycle and Respiratory chain reaction-cellular, aerobic, respiration [2]. The concentration of oxygen in the external air is measured in units of partial pressure-it is usually **about 150 mmHg** [1] (Fig. 12.2).

Concentration is lower within the alveoli because of the action of capillary blood flow, which removes oxygen [1, 2]. Oxygen concentration is lower in cells because of the chemical reactions that consume oxygen in generation of energy [1].

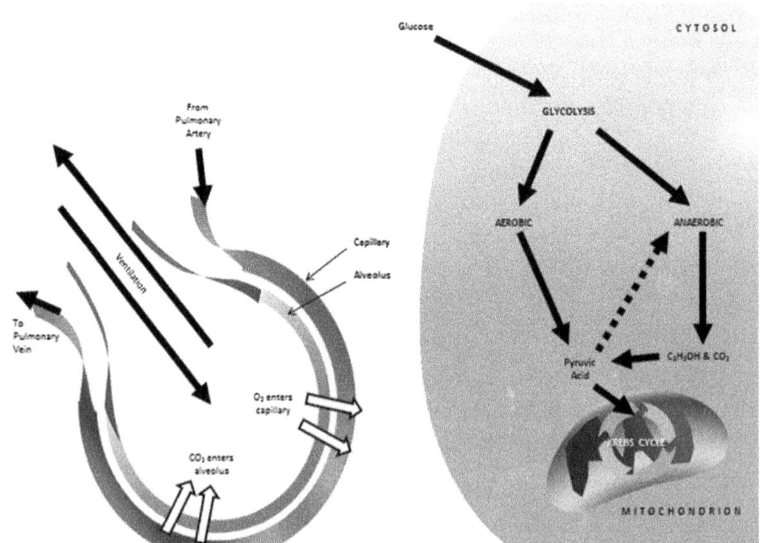

Fig. 12.1 Visualization of ventilation and cellular respiration. Flow of oxygen and carbon dioxide through alveolus and alveolar cells

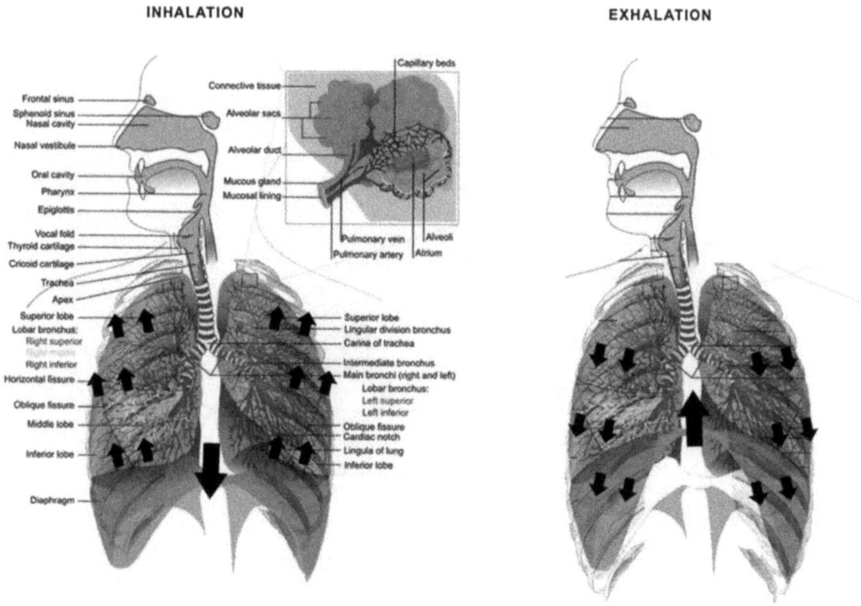

Fig. 12.2 Ventilation: inhalation and exhalation

Human lung is an elegant open engineering system. It accomplishes external respiration, or ventilation. The exchange of the air between alveoli and atmosphere is called **ventilation**. It occurs as a diffusion of gases through about 300 million small sack-like units that make up the lung and through which the exchange of oxygen and carbon dioxide occurs by diffusion across the complex membrane that separates following blood from gases (Fig. 12.3).

Alveoli are connected to the last branches of the bronchial network and elaborate tree-like structure of ever finer branches that grows from each end of the **left and right main bronchus**, branching from trachea the tube after epiglottis. Diameter of alveolus **is 300 μm** [1]. They contain **3,000 mL of** gas, while the conducting air volume **is about 150 mL** [1] (Fig. 12.4).

- **Lung volumes (tidal, functional residual capacity) can be measured by the device known as SPIROMETER.**

VOLUMES

Tidal Volume (TV). The amount of gas inspired or expired with each breath.

Inspiratory Reserve Volume (IRV). Maximum amount of additional air that can be inspired from the end of a normal inspiration.

Expiratory Reserve Volume (ERV). The maximum volume of additional air that can be expired from the end of a normal expiration.

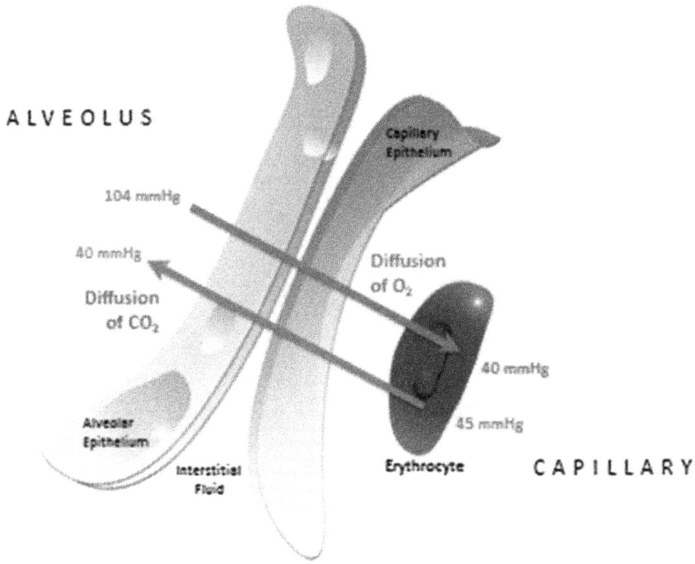

Fig. 12.3 Exchange of gases in the lungs. Diffusion of gasses

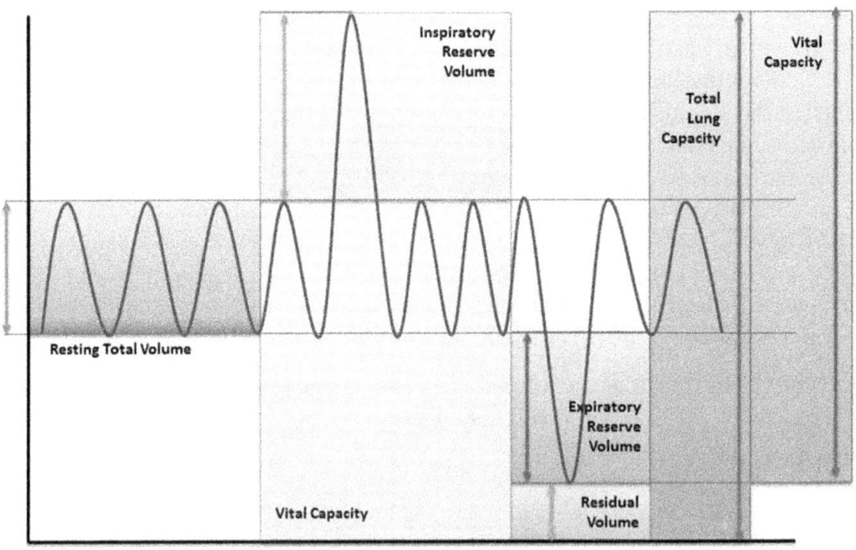

Fig. 12.4 Lung volume subdivisions: *There are four lung volume subdivisions which*: (**a**) do not overlap, (**b**) can not be further divided, (**c**) when added together equal total lung capacity (TLC)

Residual Volume (RV). The volume of air remaining in the lung after a maximal expiration. This is the only lung volume which cannot be measured with a spirometer. It is about 2 L in most subjects. Lung capacities are subdivisions of total volume that include two or more of the 4 basic lung volumes [1, 3].

CAPACITIES

Total Lung Capacity (TLC). The volume of air contained in the lungs at the end of a maximal inspiration. Called a capacity because it is the sum of the 4 basic lung volumes [1, 3].

$$\mathbf{TLC = RV + IRV + TV + ERV}$$

Vital Capacity (VC). The maximum volume of air that can be forcefully expelled from the lungs following a maximal inspiration. Called a capacity because it is the sum of inspiratory reserve volume, tidal volume, and expiratory reserve volume.

$$\mathbf{VC = IRV + TV + ERV = TLC - RV}$$

Functional Residual Capacity (FRC). The volume of air remaining in the lung at the end of a normal expiration. Called a capacity because it equals residual volume plus expiratory reserve volume. FRC = RV + ERV

Inspiratory Capacity (IC). Maximum volume of air that can be inspired from end expiratory position. Called a capacity because it is the sum of tidal volume and inspiratory reserve volume. This capacity is of less clinical significance than the other three.

$$\mathbf{IC = TV + IRV}$$

- **Resistance to flow**
- **Ventilation rates** (15 breaths/min). Faster: tachi, slower: bradi
- **Anatomical dead space**
- **Partial oxygen pressure**

Air contains **only 0.03 % CO_2**; therefore, under normobaric conditions, air inspired into the lungs is almost devoid of CO_2. This creates a large difference in the partial pressure of CO_2 (PCO_2) between blood and inspired air, promoting CO_2 to diffuse rapidly from blood into the gas phase of the lungs. At rest, ventilation is controlled by the PCO_2 in the ventilatory control center of the brain. The nervous system adjusts ventilation to maintain arterial blood PCO_2 ($PaCO_2$) constant, which at rest ranges from 35–45 mmHg (average 40 mmHg) [1]. Venous blood entering the lungs has a CO_2 partial pressure ($PvCO_2$) approximately 5 mmHg higher than arterial blood, or 45 mmHg. Because CO_2 is very soluble in blood, a large volume of CO_2 exists in a dissolved state in blood. This means that to lower blood PCO_2 any given amount, a large amount of CO_2 must be removed. As CO_2 diffuses into the gas space (alveoli) of the lungs, an equilibrium is established when the alveolar gas phase partial pressure of CO_2 ($PaCO_2$) and blood PCO_2 reach 40 mmHg [2]. The volume of gas breathed per minute (minute ventilation) controls removal of CO_2 from the blood perfusing the lungs. When CO_2 production increases during exercise at 1 ATA, minute ventilation also increases to maintain $PaCO_2$ constant. With severe exercise at 1 ATA, $PaCO_2$ may decrease slightly. During exercise, if minute ventilation does not increase to match the increase in CO_2 production, then arterial PCO_2 will increase [1].

- **Oxygen carriage in the blood**

The theoretical maximum oxygen carrying capacity is 1.39 ml O_2/g Hb, but direct measurement gives a capacity of 1.34 ml O_2/g Hb. 1.34 is also known as **Hüfner's constant** [1]. The oxygen content of blood is the volume of oxygen carried in each 100 mL blood. It is calculated by: (O_2 carried by Hb) + (O_2 in solution) = $(1.34 \times Hb \times SpO_2 \times 0.01) + (0.023 \times PaO_2)$.

The sigmoid shape of the oxygen dissociation curve is a result of the cooperative binding of oxygen to the four polypeptide chains. Cooperative binding is the characteristic of a haemoglobin to have a greater ability to bind oxygen after a subunit has bound oxygen. Thus, haemoglobin is most attracted to oxygen when three of the four polypeptide chains are bound to oxygen. For a normal adult male the oxygen content of arterial blood can be calculated:

Given arterial oxygen saturation (SpO_2) = 100 %, Hb = 15 g/100 mL and arterial partial pressure of oxygen (PaO_2) = 13.3 kPa, then the oxygen content of arterial blood (CaO_2) is: $CaO_2 = 20.1 + 0.3 = 20.4$ mL/100 m. The figures shown above illustrates that if the level of haemoglobin is halved, the oxygen content of arterial blood will be halved. Carbon monoxide (CO) interferes with the O_2 transport function of blood by combining with Hb to form carboxyhaemoglobin (COHb). CO has about 240 times the affinity of O_2 for Hb. For this reason, small amounts of CO can tie up a large proportion of the Hb in the blood, thus making it unavailable for O_2 carriage. If this happens, the Hb concentration and PO_2 of blood may be normal, but its O_2 concentration is grossly reduced (Fig. 12.5).

The presence of COHb also shifts the O_2 dissociation curve to the left, thus interfering with the unloading of O_2. This is an additional feature of the toxicity of CO (Fig. 12.6).

- **The role of carbonic anhydrase in red blood cells, carbon-dioxide carriage and acid–base balance in the blood**

Similarly the oxygen content of mixed venous blood can be calculated. Given normal values of mixed venous oxygen saturation (SvO_2) = 75 % and venous partial pressure of oxygen (PvO_2) = 6 kPa, so:

$$\mathbf{CvO_2 = 15.2 + 0.1 = 15.2\,mL / 100\,mL}$$

- **Diffusion (Respiratory membrane)**

Red blood cell membrane, blood capillary wall, basal membrane, basal epitel of alveoli, surfactant, are all the components of the respiratory membrane (Fig. 12.7).

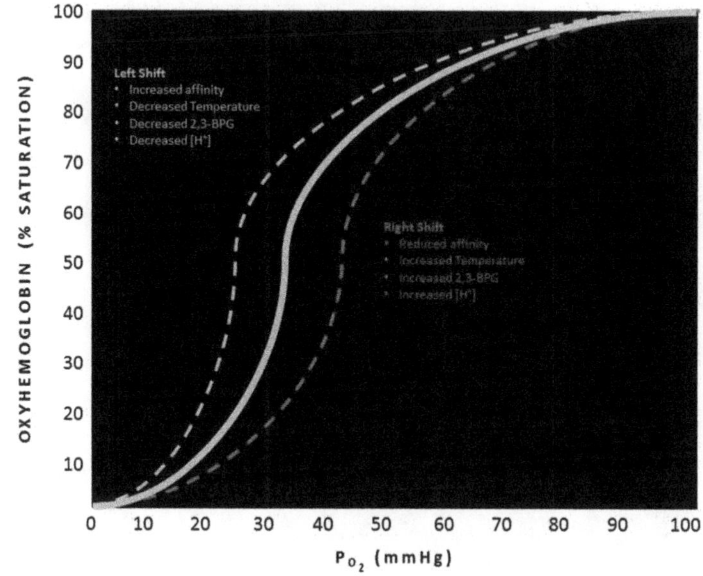

Fig. 12.5 Shift in oxygen binding curve due to partial pressure of oxygen vs haemoglobin. Where: ***SO2 = percentage saturation of Hb with oxygen. Hb = haemoglobin concentration in grams pre 100 mL blood. PO2 = partial pressure of oxygen (0.0225 = mL of O2 dissolved per 100 mL plasma per kPa, or 0.003 mL per mmHg)***

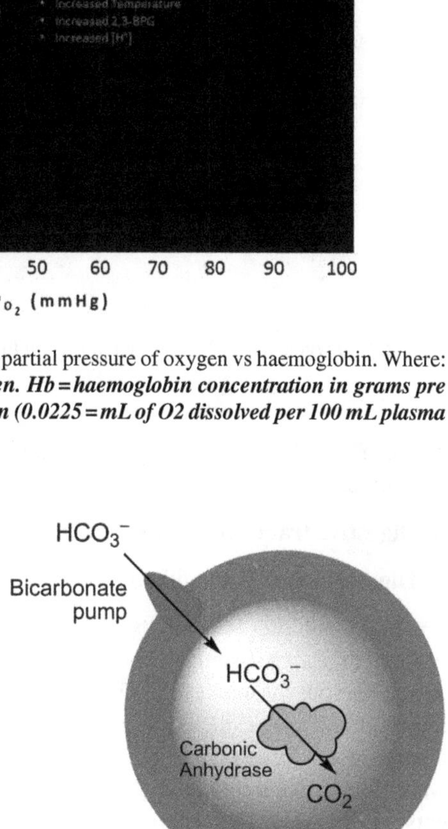

Fig. 12.6 The role of carboxyl-anhydrase in red blood cells

- **Henry's Law**

  The special expression of the O_2 partial pressure: $\mathbf{CO_2 = HPO_2}$

- **Fick's law of diffusion (determines diffusing capacity in humans)**

$$D_{L_{gas}} = \frac{V_{gas}}{P_{A_{gas}} - P_{C_{gas}}}$$

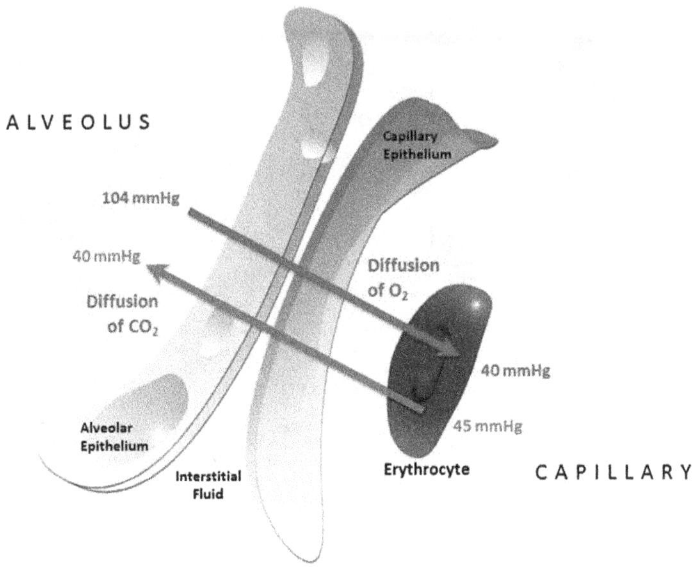

Fig. 12.7 Oxygen and carbon dioxide diffusion through (in/out) capillaries and red blood cells

where D = diffusing capacity; L = lung; A = alveolar; c = mean capillary value; P = pressure

B. Digestive tract, gastrointestinal tract (GIT)

- **Digestion and metabolism**

The human diet is complex: so is the metabolism and metabolic control taking place in digestive system and being controlled from local and higher CNS and autonomic nervous system levels. The digestive system is an open system which intakes the food and water, digest it, and eliminate what is not necessary in the external environment. Most of the nutrients are absorbed into the circulation through the wall of digestive tract and its lymphatics and blood vessels. In the cell, these nutrients will be used, and with oxygen brought up in there by the action of lungs and circulation, will be metabolized and stored as energy deposits or used as building material. The digestive system has a number of actions that are familiar to engineers; storage and controlled emptying, mixing, secretion, digestion, and adsorption. In its simplest version the GIT **is an 8 m of tubing connecting the mouth to the annus.**

- **Organs of the digestive system and their function**

The organs of GIT are given in Fig. 12.8:

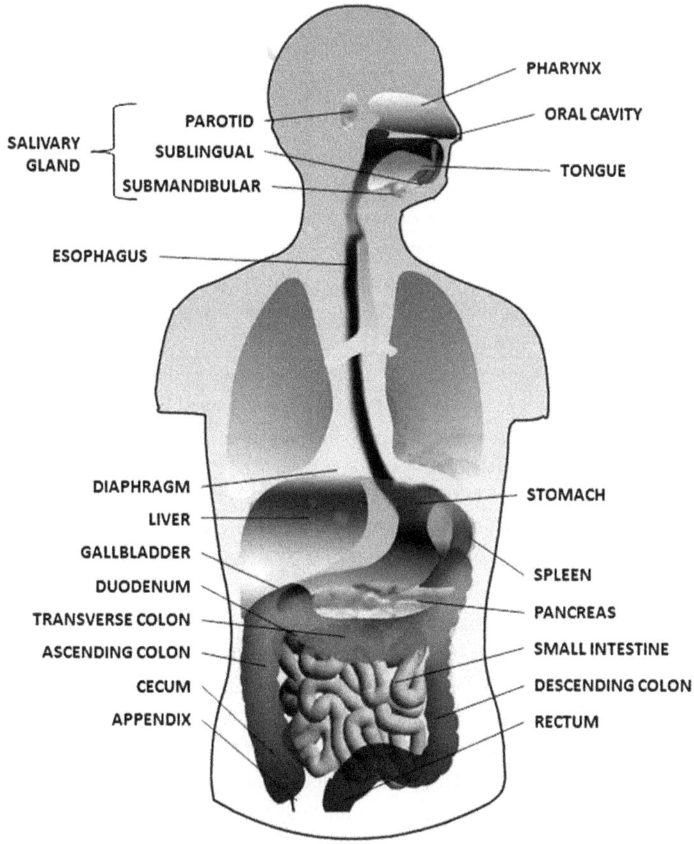

Fig. 12.8 Digestive physiology: Digestion and metabolism

Emphasizing Bioengineering Aspects to Respiratory and Digestive Tract

Modeling the Digestive Tract

The chemical reactions of digestion can be understood in terms of mathematical models of chemical reactors (Table 12.1).

It can be modeled as:

- Chemical reactor
- Batch reactor (only for closed systems explanation)
- Ideal reactor models: plug flow and stirred tank reactors

Table 12.1 Digestive system

Digestive system	
Teeth	Physically breaks down food by chewing
Tongue	Assists in chewing and helps in swallowing
Salivary glands	Secretes salivary amylase to break down complex carbohydrates
Stomach	Mixes salivary amylase, food and gastric juices
	Secretes hydrochloric acid which begins protein digestion
	Begins digestion of fats
Small intestines	Averages 3 m in length
	Proteins and fats continue to be broken down
	Carbohydrates are broken down
	Approximately 90 % of absorption takes place
Pancreas	Secretes digestive enzymes into the small intestine
Gallbladder	Secretes bile into the small intestine, which emulsifies fats (makes the globules smaller)
Liver	Makes bile and sends to the gallbladder
	Maintains normal blood glucose levels
	Metabolizes protein for use elsewhere in the body
	Metabolizes fats
	Stores fat soluble vitamins A, D, E and K
	Stores vitamin B12 and the minerals iron and copper
Large intestines	Approximately 1.5 m in length and 6.5 cm in width
	Absorbs water
	Produces bacterial, vitamin K and some 8 vitamins
	Produces feces *via* chemical breakdown and water reabsorption
	Evacuates feces from the body though the rectum & anus

References

1. Hall, G.E., Guyton, A.C.: Textbook of Medical Physiology. Saunders Elsevier, Philadelphia (2011). ISBN: 978-1-4160-4574-8
2. Saltzman, M.W.: Biomedical Engineering. Cambridge University, New York (2009). ISBN: 978-0-521-840099-6 (hardback)
3. Andjus, R.K.: General Physiology and Biophysics. Modules 1–7. Center for Multidisciplinary Studies. University of Belgrade, Belgrade, Serbia (2002). ISBN: 86-80109-11-8

Chapter 13
Circulation and Lungs

Measure what can be measured, and make measurable what cannot be measured.

Galileo Galilei (1564–1642)

The principal function of the blood flow in the cardiovascular system is to provide oxygen (O_2) and nutrients to the tissues of the body and to remove carbon dioxide (CO_2) and waste products. The flow of blood through the cardiovascular system follows physical law's known from fluid mechanics. Oxygen diffuses into capillaries, which surround the alveolus. The circulatory system carries the oxygen to all of the cells within the entire body. Mitochondria inside of bodily cells use this oxygen to produce cellular energy, adenosine-triphosphate (ATP), endogenous water and CO_2, which is returned by venous blood into the lungs and exhalated into surroundings, during ventilation.

Circulation (Circulating Fluid, Blood Vessels, and Pump)

The circulatory system is a closed system in which the heart is the pump, and blood vessels (arteries, veins and capillary bed) are the tubes of different diameter connected to each other and enabling the blood to flow through [1]. Blood itself, has its own physic-chemical properties. It is composed of water, salts, proteins, other metabolites, and cells. Due to that, it is *viscouse liquid* [1].

Viscosity of Blood

Viscosity is the *inner friction in the fluid,* which is due to the interaction between molecules and particles/cells in the blood passing a cylindrical vessel [2]. Telescope cylinders (laminae) of blood sliding against each other can illustrate this inner

M. Pavlovic, *Bioengineering*, DOI 10.1007/978-3-319-10798-1_13

friction. The outermost blood cylinder rests against the vessel wall (velocity is zero), and the central cylinder moves (laminar flow) with the greatest velocity (v). The velocity profile is parabolic [2]. The *velocity gradient,* with the distance x from the center of the blood vessel towards the outermost blood cylinder, is called the *shear rate* (dv/dx). The tangential force (F) between these blood cylinders depends upon the area (A) sliding against each other, and the relation to viscosity (h) is given by the equation: $\mathbf{F/A = h \times dv/dx}$. The viscosity (h) *one Pascal sec* (1 Pa s) is the tangential force, working on 1 m^2 of surface area, when dv/dx is (s-1) [2].

Essential breakthrough: William Harvey (1578–1657) developed a coherent theory of blood circulation and provided experimental-if partial-proof, contradictied to Galen (Roman physician c.129-216 AD) who propounded that the blood was produced in the liver and used up as it reached the tissues.

Harvey was also the first to suggest that humans and other mammals reproduced via the fertilization of an egg by sperm. It took a further two centuries before a mammalian egg was finally observed, but nonetheless Harvey's theory won credibility during his lifetime.

The principal function of the blood flow in the cardiovascular system is to provide oxygen (O_2) and nutrients to the tissues of the body and to remove carbon dioxide (CO_2) and waste products [1]. The flow of blood through the cardiovascular system follows physical law's known from fluid mechanics (see principles).

Strictly speaking, *Poiseuille's law* has validity in a circulatory system, only when the fluid flow is laminar and non-pulsating in horizontally situated cylindrical vessels of constant dimensions. The resistance for laminar flow of a Newtonian fluid is only dependent on the dimensions of the vessel and the viscosity of the fluid. Resistance varies inversely as the fourth power of the radius of the vessel [1, 2].

For *resistances in parallel,* the total resistance is less than that of any individual resistance. Although the total cross sectional area of all arterioles is much larger than that of all arteries, their resistance to blood flow is much greater than that of the arteries. The number of daughter vessels is not high enough to balance the decrease in vessel diameter. The resistance is highest in the capillaries and it diminishes as the vessels increase in radius. For *resistances in series*, the total resistance equals the sum of the individual resistances [1, 2].

In contrast to Poiseuille's conditions, the blood flow in the human circulation is pulsating and sometimes turbulent, and its blood vessels are not horizontally located, cylindrical or inflexible. Neither is the blood viscosity constant nor independent of vessel diameter and flow. Viscosity is inner friction of the blood due to molecular and cellular movement in circulation [1–3]. The viscosity of non-Newtonian fluids decreases with increasing shear rate according to the equation above. Blood is namely not homogenous with a viscosity that is independent of shear rate. On the

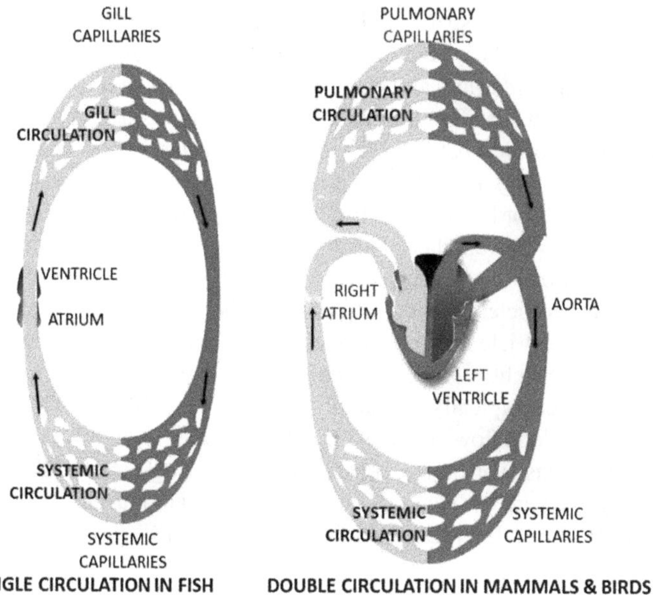

Fig. 13.1 Comparative circulatory systems in fish and mammals or birds

contrary, at low shear rates (low blood flow); the viscosity of blood can be tenfold higher than normal. The typical normal viscosity of body warm blood is 5 centi-Poise equal to 5 milli-Pascal seconds or 5 (mPa*s). Blood viscosity depends upon the concentration of red cells (the hematocrit) [1, 2] (Fig. 13.1).

At rest the mean red cell velocity in the capillaries is observed to be approximately 1 mm in one second (s); this provides ample time for gas exchange. Since the circulating blood moves continuously, the cardiac output must pass a cross section of all open capillaries. At rest a cardiac output of 5,000 mL per min is a reasonable estimate; when changed into volume rate per s, the cardiac output is equal to 10^{-4} m^3 s^{-1} [1]. Hence, it is possible to calculate the large cross sectional area of all open capillaries in a resting person. The total blood volume is approximately 5 L in a healthy adult. The *right atrium* receives venous blood from the caval veins, and the left atrium receives oxygenated blood from the pulmonary veins. The two atria function as thin walled reservoirs and conduit organs for the blood. On average, atrial systole contributes only about 15 % of the total ventricular filling, but in cardiac insufficiency the atrial contribution may increase significantly [1]. The left and right ventricles provide most of the energy needed to transport the blood through the circulation. The left ventricle accelerates the blood into the systemic or peripheral high-pressure system, and its walls are thick in contrast to the thin, weak right ventricle, which pump blood into the low-pressure pulmonary system [1].

The *left ventricle* consists of cardiac muscle fibers originating from the fibrous rings at the base of the heart and the fibers are twitching towards the apex [3]. The orifice between the left atrium and the left ventricle carries two valve cusps, and this valve is called the *bicuspid* or mitral valve b [1]. Three cusps form the *tricuspid* valve closing the orifice between the right atrium and ventricle during systole. Strong filaments (*chordae tendineae*) arise from the *papillary muscles* of the ventricles [1]. These chordae are attached to the free edges of the atrio-ventricular valves and normally prevent the valves from bulging into the atria during ventricular systole. The two *atrio-ventricular valve* systems prevent the leakage of blood backward from the ventricles into the atria. Two other valve systems are interposed between the left ventricle and the aorta (the aortic valves) and between the pulmonary artery and the right ventricle (the pulmonary valves).

At rest the athlete typically has an oxygen uptake of 250 mL STPD per min [3]. The total muscle blood flow at rest is $(35{,}000/100) \times 3 = 1{,}050$ mL of blood per min [1–3]. The total muscular oxygen uptake at rest is $(1{,}050 \times 50/1{,}000) = 53$ mL per min. During maximal dynamic activity the total muscle blood flow is: $(35{,}000/100) \times 75 = 26{,}250$ mL/min or 26.25 L per min [3].

Circulating Fluid

Distribution of blood and its flow. The total blood volume (5 L) is distributed with 60–75 % in veins and venules, 20 % in arteries and arterioles, and only 5 % in capillaries at rest [1]. Of the total blood volume only 12 % are found in the pulmonary low-pressure system [1]. The distribution of the cardiac output to the top athlete can show a 6-fold increase in cardiac output from 5 to 30 L of blood each min, when going from rest to maximal dynamic exercise. The heart rate increases from 60 to 180–200 beats per min. The muscle blood flow can rise from 3 to 75 mL per min per 100 g of muscle tissue (FU) or factor 25 in a total muscle mass of 35 kg. The muscular arterio-venous-O_2 content difference can rise from the resting level (200–150) = 50 mL STPD per l of blood to (200–40) = 160 mL STPD per l (Fig. 13.2).

The total muscular oxygen uptake is increased to $(160 \times 26.25$ L/min$) = 4{,}200$ mL STPD per min. Accordingly, the total muscular oxygen uptake rises by a factor of (4,200/53) almost 80 from rest to exercise. At the start of exercise, signals from the brain and from the working muscles bombard the cardiopulmonary control centers in the brainstem. Both cardiac output and ventilation increase, the β-adrenergic tone of the muscular arterioles falls abruptly, whereas the vascular resistance increases in inactive tissues. The systolic blood pressure increases, whereas the mean arterial pressure (**MAP**) only rises minimally during dynamic exercise. The total peripheral vascular resistance (**TPVR**) falls during exercise towards 30 % of the level at rest, because of the massive vasodilatation in the muscular arterioles of almost 35 kg muscle mass. This is why the major portion of the cardiac output passes through the skeletal muscles and why the diastolic pressure often decreases during exercise. At moderate exercise the skin blood flow and heat dissipation is increased. The coronary blood flow increases from rest to **exertion**.

Fig. 13.2 Blood fractions after centrifugation

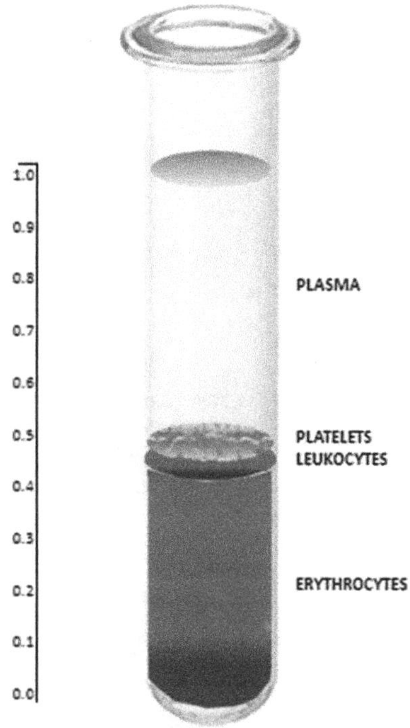

Table 13.1 Distribution of flow in % of the cardiac output

Organ system	Distribution	A-v difference	O_2 uptake	Bloodflow
	Flow%	ml STPD* l^{-1}	ml STPD*min^{-1}	ml*min^{-1}
Splanchnic	27 (2)	40 (80)	60 (40)	1,500 (500)
Kidneys (300 g)	22 (2)	12–14 (28)	16 (17)	1,200 (600)
CNS	14 (1)	60 (120)	45 (36)	750 (300)
Myocardium (250 g)	4.5 (6.7)	140 (190)	35 (380)	250 (2,000)
Muscle (35 kg)	19 (88)	50 (160)	53 (4,200)	1,050 (26,250)
Other organs	14 (1–2)	50 (100)	38 (35)	750 (350)
Total body	100 (100)	50 (150)	250 (4,500)	5,500 (30,000)

***F/A = h × dv/dx**. The viscosity (h) one Pascal sec (1 Pa s) is the tangential force, working on 1 m^2 of surface area, when dv/dx is 1 (s^{-1})

Distribution of flow in % of the cardiac output, arterio-venous oxygen content difference, oxygen uptake and absolute blood flow at rest. The same variables are given for maximal exercise (in brackets)

This simplified description is valid for water, gas, and other homogenous fluids that are Newtonian fluids. *Newtonian fluids* are defined as *fluids with a viscosity that is independent of the shear rate.* Newtonian fluids move streamline or with so-called ideal laminar flow (Table 13.1).

The viscosity of non-Newtonian fluids decreases with increasing shear rate, according to the equation above. Blood is namely not homogenous with a viscosity

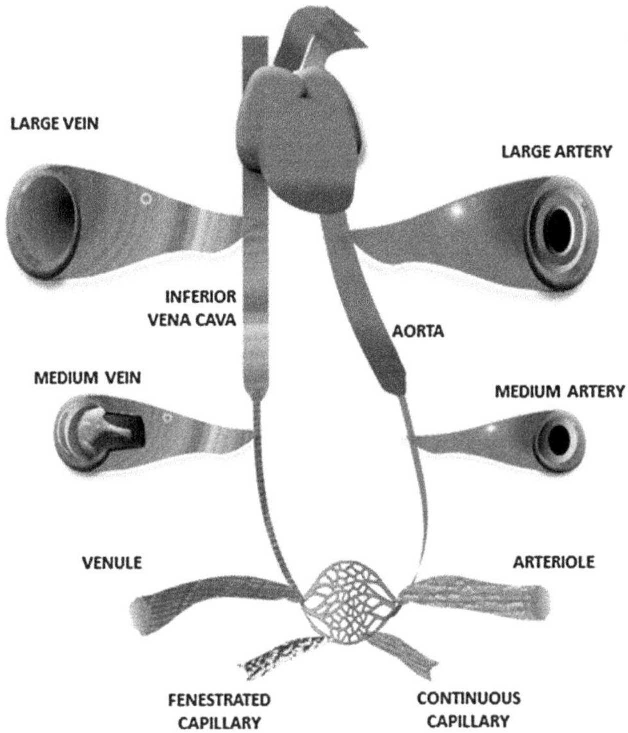

Fig. 13.3 Blood flow through a cylindrical vessel

that is independent of shear rate. On the contrary, at low shear rates (low blood flow); the viscosity of blood can be tenfold higher than normal. The typical normal viscosity of body warm blood is 5 centi-Poise equal to 5 milli-Pascal seconds or 5 (mPa*s).

Blood viscosity depends upon the concentration of red cells (the hematocrit) (Figs. 13.3 and 13.4).

- **Arteries and veins**

Arteries and veins are main blood vessels branching to smaller arterioli and venuli and ending with the capillary bed at the bottom and top of the body closing the general circulation. The architecture of arterial wall is different than in veins due to higher blood pressure that they have to deal with. Figure 13.5 shows three layers of arterial wal (endothelial, smooth muscle and serosal) where muscle layer is thick compared to the one in viens. However, Fig. 13.6 shows that this intermediate layer is thinner in viens and that viens have valves in order to keep the blood against the gravitation force and prevent it from regurgitation into lower parts of vascular bed.

- **Capillary function**

The role of capillaries is to facilitate the exchange of the gasses (O_2 and CO_2). The capillaries are built up of one layer of fenestrated endothelial cells and their

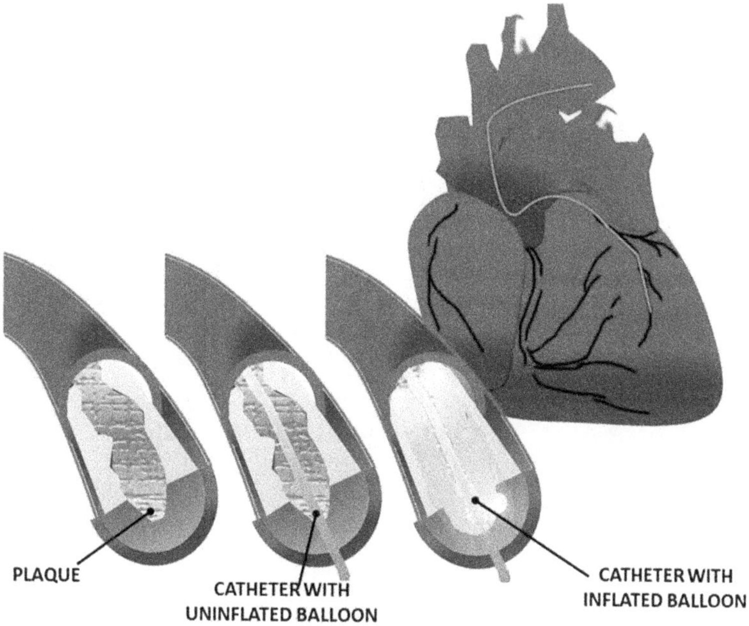

Fig. 13.4 Angioplasty of atherosclerotic plaque in cardiac artery

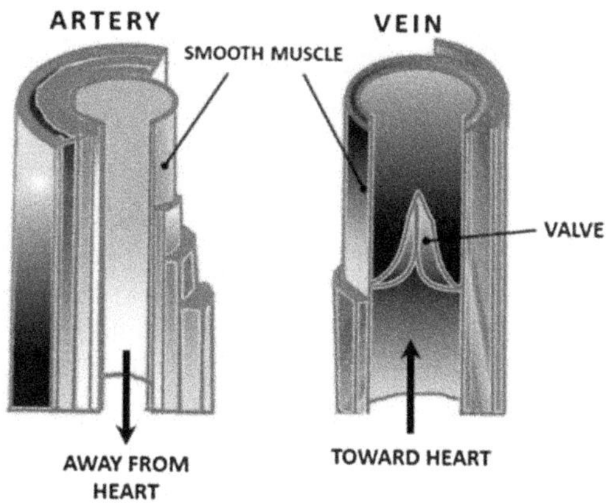

Fig. 13.5 Gross comparison of artery and vein physical features

oxygen pressure in arterial part is higher than in tissues. Thus, oxygen diffuses into the extracellular tissue and then into the cells and mitochondria where it will enter metabolic cycles such as Krebs cycle and respiratory chain and CO_2 produced within these cycles will be expelled from the cell on the basis of higher partial pressure, consecutively bound to hemoglobin at venous part of capillaries, and brought up *via* systemic venous circulation into the lungs (alveoli). In that way two gases will exchange the environment and binding to hemoglobin will help CO_2 to be expelled with exhalation during lung ventilation from the alveoli of the lungs.

Fig. 13.6 Capillary bed

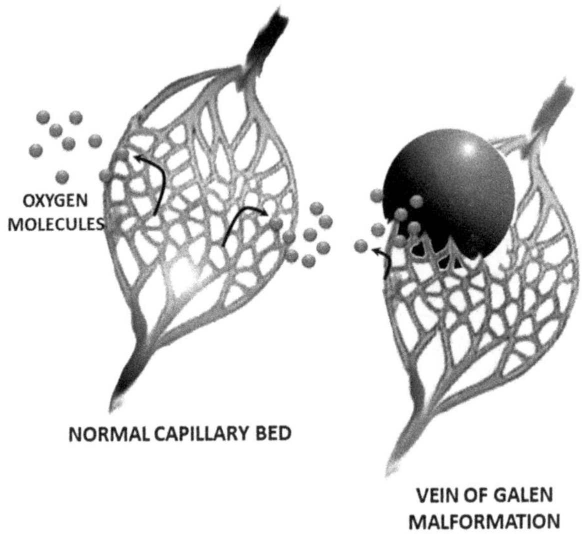

Heart Muscle Cells

Cardiac muscles are self-contracting; astronomically regulated and must continue to contract in rhythmic fashion for the whole life of the organism. Hence, they have special features that give them characteristics of excitable tissues [1].

- **The heart (pump)**

The heart is composed of working musculature (myocardium/muscle cells/tissue) and conductive system which creates, and convey electrical signal for proper work of working musculature. Heart itself has two atriums and two ventricles and is "divided" into right and left heart. The right one is pumping the blood into pulmonary blood vessels while the left one into systemic circulation. The heart muscle has morphological features similar to striated skeletal muscle but functional are similar to smooth muscle. The cells of working part of heart themselves, are known as cardiomyocytes (Fig. 13.7).

The cells are Y shaped and are shorter and wider than skeletal muscle cells [1, 2]. They are pre-dominantly, mono-nucleated. The arrangement of *actin* and *myosin* is similar to skeletal striated muscle. Some of the cardiac muscle cells are autorhythmic, i.e. they contract even in the absence of neuronal innervation (known as **pacemaker cells**). Intercalated disks are located between cardiac muscles cells. These contain gap junctions which provide **communicating channels between cells.** The intercalated disks allow waves of depolarization to sweep across the cells thus synchronizing muscle contraction. Depolarization of cardiac muscle cells differs from that of other muscle cells [1]. Repolarization takes much longer to occur and thus cells cannot be stimulated at high frequency [1]. The advantage is that cardiac muscle is prevented from going into tetanus (Fig. 13.8).

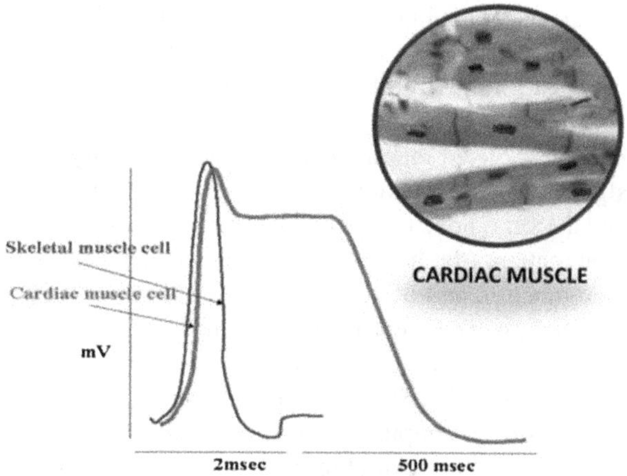

Fig. 13.7 Cardiac muscle contractility versus skeletal muscle contraction

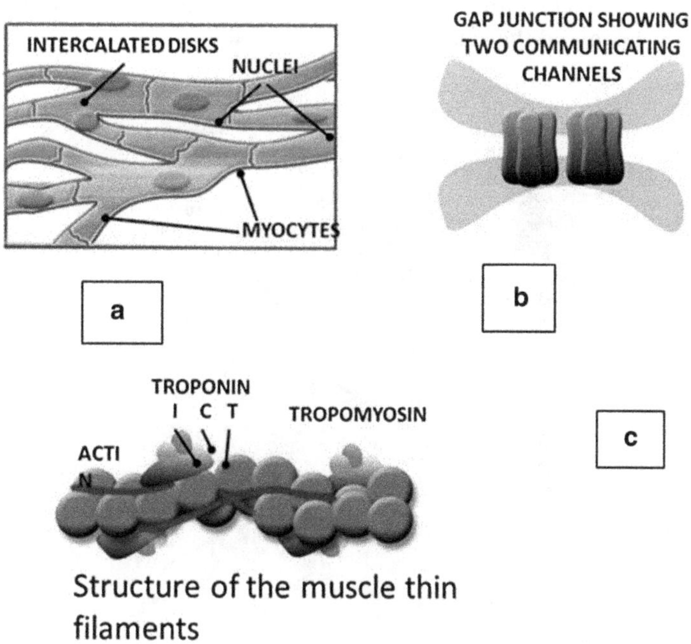

Fig. 13.8 (**a**, **b**, **c**) Microscopic view of physiological relevant structures in various muscle tissues

The normal heart is characterized by an electrical insulation between the atria and the ventricles mainly due to the fibrous ring (*annulus fibrosus*) [1]. As indicated, the heart possesses a *specialized electrical system,* the cardiac *conduction system* that leads the electrical signal from the atria to the ventricles. The conducting system consists of *modified myocardial cells*. An optimal timing of atrial and ventricular pumping allows the emptying of the atria to be completed before the ventricular contraction. This allows the heart to pump the required cardiac output. The heart normally has a self-firing unit, located in the right atrium, called the *sinoatrial node* or sinus node [1]. The *sinus node* contains round cells (*pacemaker cells*), elongated intermediary cells and ordinary atrial cells [1–3]. The electrical signal that automatically originates from the sinus node has the highest frequency, and the sinus node is thus the *natural pacemaker* of the heart. Even a cardiac transplant patient (the heart is totally de-nervated) adapts to the altered needs for cardiac function and of course initiates new heart beats as long as the transplant is functioning. The electric signal from the sinus node activates the atrial walls to contraction, and then reaches the main conduction system at the level of the *atrio-ventricular node* (AV node). The AV node consists of the same cell types as the sinus node. The impulse is delayed in the AV node, and this delay allows the atrial systole to squeeze extra blood into the ventricles just before the ventricular systole occurs (Fig. 13.9).

The cardiac conduction system (left) is the only electrical connection between the atria and the ventricles of the normal heart.

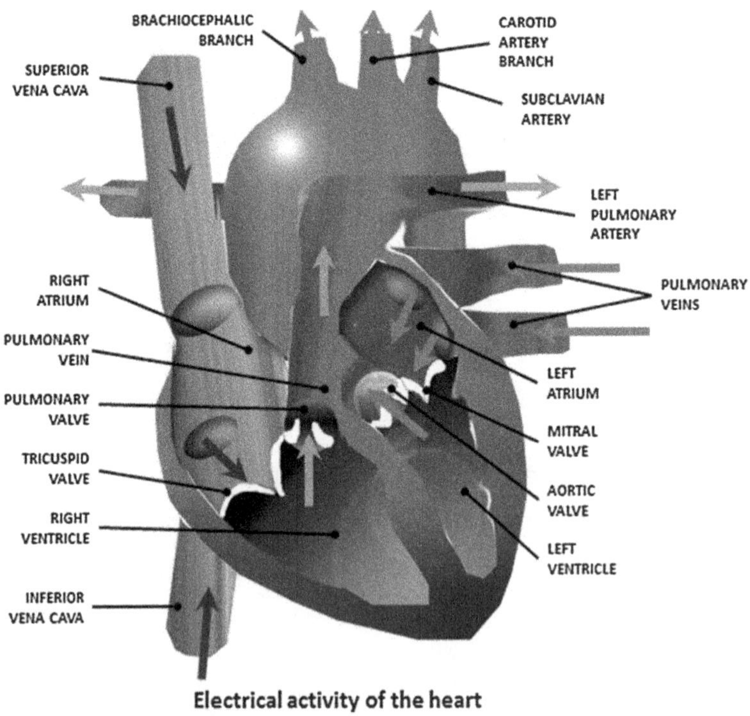

Fig. 13.9 Structure of heart and major blood vessels

ECG

From the *bundle of His*, the signal is transmitted down a rapid conduction pathway, composed of the right and left *bundle branches,* to stimulate the right and the left ventricle and cause them to contract. The right bundle branch proceeds down the right side of the ventricular septum, and the large left bundle branch perforates the septum and divides into an anterior and a posterior division. These bundle branches divide into a network of conducting *Purkinje fibers* just below the endocardial surface. Purkinje fibres are large diameter cells without T-tubules, and with a long refractive period, so they can block premature depolarization waves from the atria. The propagation wave spreads in the septum from both branches with the thick left bundle branch being dominant. The spread along the Purkinje fibres is rapid, whereas the spread from the endocardium to the epicardium is slow (Fig. 13.10).

Ectopic foci become pacemakers, when the normal dominant pacemakers fail by blockade or depression: In the AV node, the atria, and the Purkinje fibres or in ischemic ventricular fibres.

Vascular Compliance and Stiffness

Distensibility or compliance is the increase of volume per unit of transmural pressure increase (DV/DP_t). The specific compliance is the relative increase in volume per unit of pressure increase. The elastance or stiffness is the reciprocal value of the compliance. The compliance of the venous system can be 30 times as large as that of the arterial system. The venous system can be expanded to contain more than 75 % of the total blood volume. The veins function as capacitance vessels, and

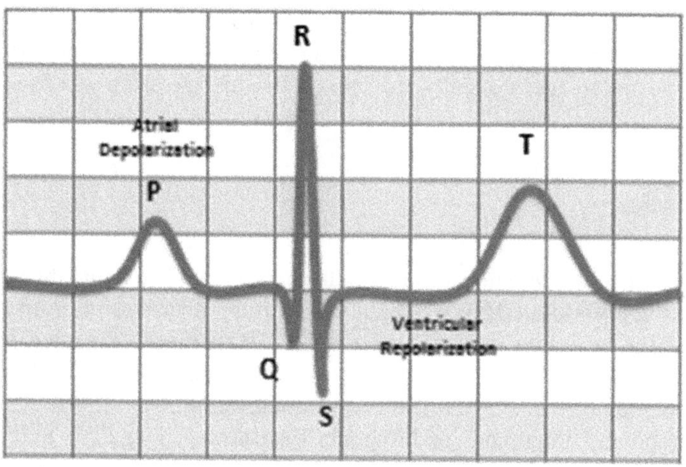

Fig. 13.10 ECG: Electrical activity of the heart

become much distended when blood is given in transfusions, in heart insufficiency, or during a heart attack. Severe exercise and loss of blood cause an increase in venous tone, which for a period actually can increase the circulating blood volume. During hard work the muscular venous pump provides up to 1/3 of the energy required for blood circulation (the peripheral venous heart). The venous system also plays an important role by its graded venous return to the heart.

Blood Pressure

MAP, is usually being defined as the diastolic pressure plus 1/3 of the pulse pressure. MAP is about 12 kPa (=90 mmHg) in the arteries. Notice the *fall* in MAP from the abdominal aorta to the femoral artery, whereas the systolic pressure *increases*. The arterial mean pressure falls to a mean value around 2.4 kPa (18 mmHg) in the capillaries. The *arterial pulse pressure* is the difference between the systolic and the diastolic arterial pressure. At a heart rate of 75 beats/min at rest, the cardiac cycle length is 0.8 s with 0.3 s systole and 0.5 s diastole. A stroke volume of 70 mL is deposited in the aorta and the larger elastic arteries during systole. During the systolic period 26 ml of blood ($70 \times 3/8$) is streaming through the resistance vessels, leaving the arterial system, so the systolic *volume expansion* is 44 mL of blood. A young healthy subject has an arterial distensibility or compliance of 1 mL of blood per mmHg, which creates a pressure rise during systole (pulse amplitude) of (70–26) = 44 mmHg (Fig. 8.8). With a diastolic pressure of 70 mmHg, this implies a systolic pressure of 114 mmHg, conventionally written 114/70 mmHg or 15.2/9.3 kPa.

Aging and arteriosclerosis increase the stiffness (reduce the distensibility) of the elastic arteries, causing the arterial compliance to fall from 1 (one) to 0.5 mL of blood per mmHg. In this case, a systolic volume expansion of 44 mL of blood increases the pulse pressure amplitude to 88 mmHg (44/0.5 = 88), and the blood pressure to perhaps 180/92 mmHg. This is a likely process in an otherwise healthy person of advanced age. Typically, the average diastolic pressure will rise with age.

Wall Tension

For a *thin-walled organ* with two main radii, Laplace predicted that the transmural pressure at equilibrium (DP_t), was identical with the fiber tension in the wall (T) divided by the two main radii: $DP = T / (r_1 + r_2)$. This model has often been used (with modifications for wall thickness, w) for the relaxed ventricle (Fig. 13.11). In the thin filaments of skeletal and cardiac muscles, proteins called actin, tropomyosin, troponin I, troponin C, and troponin T, are arranged regularly to form a filamentous protein complex. The regulatory mechanism of muscle contraction lies within this complex.

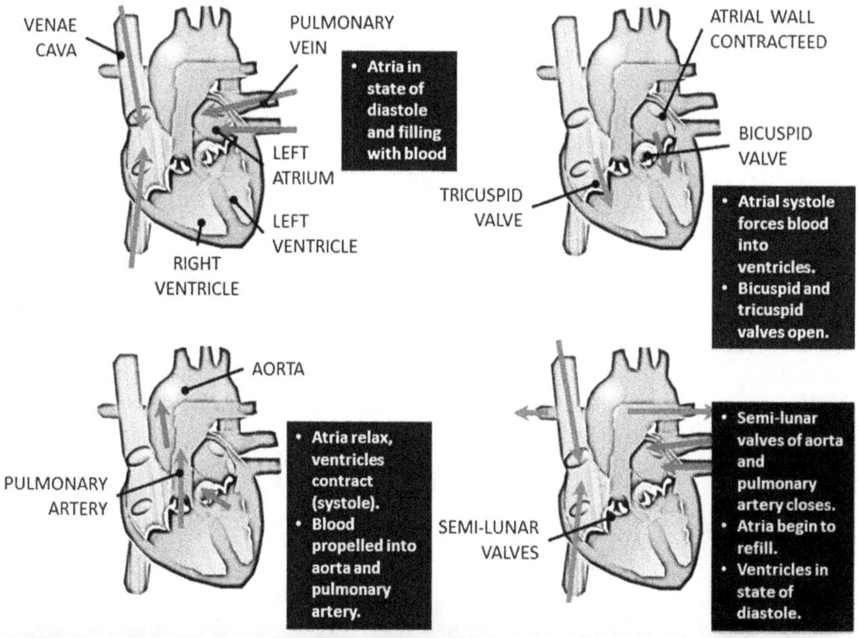

	Atrial Systole	Early Ventricular Systole ⟷ Late Ventricular Systole	Early Ventricular Systole ⟷ Late Ventricular Systole
Atria	CONTRACT	RELAX	RELAX
Ventricles	RELAX	CONTRACT	RELAX
AV Valves	OPEN	CLOSED	OPEN
Semilunar Valves	CLOSED	OPEN	CLOSED

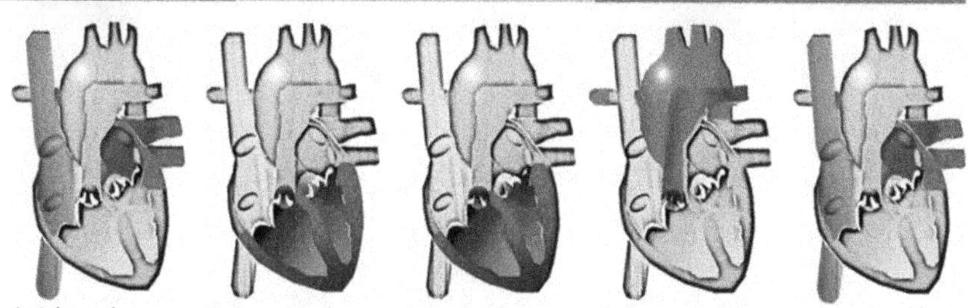

Fig. 13.11 Cardiac cycle: visual presentation

Emphasizing Bioengineering Aspects to Circulation and Lungs:
Bioengineering work on heart/circulation and lungs is equally important given the fact that both systems are of vital significance for the fundamental functions in the body. On average, 18 people die each day awaiting an organ donation—one person is added to the waiting list for organ transplants every ten minutes [4, 5]. Although nearly 80 people receive an organ transplant per day according to *OrganDonor.gov*, the gap between donations made and those awaiting organs is devastating for those on the waiting list. There is good news on the horizon for the 100,000-plus people awaiting a transplant, however. Researchers in Texas and across the globe are working on a way to create organs from the patient's own body. While 3-D printed versions of large organs such as kidneys are not yet successful, it is good for growth of the bladder, for example [5]. The techniques have been developed in order to help the higher evolution of the organ development in the case they are needed. Several are already working at least in experimental conditions.

Building New Organs

Custom-made hearts, lungs, kidneys, and other organs could revolutionize organ transplantation. Scientists are learning how to grow custom-made body parts so they can be ready when you—and your vital organs—start falling apart [4–7]. At the University of Minnesota, Doris Taylor and her colleagues strip organs of their cells, reseed the organ "skeletons" with living cells, and watch as the organs start working right in front of their eyes [4]. Doris Taylor and her team are building new organs, hoping to reverse disease, maybe even the aging process. It sounds like science fiction, but it is isn't. On the ninth floor of the Texas Heart Institute's Denton Cooley building, Doris Taylor and her team are building human hearts, with help from pigs and stem cells. "We think a pig heart is a perfect scaffold for a human heart, based on its structure and size," says Taylor, a passionate scientist with a PhD in pharmacology. One recent morning, a pig heart hung suspended in a clear homemade tank in the lab built for Taylor and her team.

Growing/Replacement Organs

Repairing cardiac tissue in the aftermath of a heart attack [7–11]. These ideas have belonged to the realm of science fiction for generations, but all are now realistic goals, thanks to groundbreaking research in the field of adult stem cells [12]. How promising is the adult stem cell field? In awarding the Nobel Prize to the UK's John Gurdon and Japans Shinya Yamanaka, the Nobel committee said the duo and this type of research have completely revolutionized science. And science is only the first thing it promises to revolutionize. The best part from our perspective? Texas has a realistic opportunity to become the center of this rapidly emerging industry.

References

1. Hall, G.E., Guyton, A.C.: Textbook of Medical Physiology. Saunders Elsevier, Philadelphia (2011). ISBN 978-1-4160-4574-8
2. Andjus, R.K.: General Physiology and Biophysics. Modules 1–7. Center for Multidisciplinary Studies, University of Belgrade, Belgrade (2002)
3. Saltzman, M.: Biomedical Engineering. Cambridge University, New York (2009)
4. Sjöqvista, S., Jungebluth, P., Lim, M.L., Haag, J.C., Gustafsson, Y., Lemon, G., Baiguera, S., Burguillos, M.A., Del Gaudio, C., Simonsson, O.E., Rodriguez, A.B., Sotnichenko, A., Kublickiene, K., Ulman, H., Kielstein, H., Damberg, P., Bianco, A., Joseph, B., Ribatti, D., Heuchel, R., Ibarra, C., Zhao, Y., Taylor, D.A., Macchiarini, P.: Experimental orthotopic transplantation of a tissue engineered esophagus in rats. Nature Comm 2014. (accepted)
5. Atala, A.: Principles of regenerative medicine (2011). ISBN: 978-0-12-369410-2
6. Robertson, M.J., Dries-Devlin, J.L., Kren, S.M., Burchfield, J.S., Taylor, D.A.: Optimizing recellularization of whole decellularized heart extracellular matrix. PLoS One **9**(2), e90406 (2014). PMID:24587354
7. Pavlovic, M., Balint, B.: The use of stem cells to repair cardiac tissue. Anest. Reanim. Transfuziol. **34**, 129–150 (2006)
8. Pavlovic, M., Balint, B.: Stem Cells and Tissue Engineering. Springer, New York (2013). ISBN: ISBN:978-1-4614-5505-9 (eBook)
9. Vunjak-Novakovic, G., Scadden, D.T.: Biometric platforms for human stem cell research. Cell Stem Cell **8**, 252–261 (2011)
10. Balint, B., Stamatovic, D., Todorovic, M., Jevtic, M., Ostojic, G., Pavlovic, M., Lojpur, Z., Jocic, M.: Stem cells in the arrangement of bone marrow repopulation and regenerative medicine. Vojnosanit Pregl. (VSP) **64**(7), 481–484 (2007)
11. Balint, B., Pavlovic, M., Jevtic, M., Hrvacevic, R., Ignjatovic, L.J., Mijuskovic, Z., Blagojevic, R., Trkuljic, M.: Simple "Closed-cirquit" group-specific immunoadsorption system for ABO–Incompatible kidney transplants. Transfus. Apher. Sci. **36**(3), 225–233 (2007)
12. Harvey, W.: On the motion of the heart and blood in animals. In: Carmichael, A.G., Ratzan, R.M. (eds.) Medicine, a Treasury of Art and Literature, pp. 100–103. Harkaway, New York (1991)

Chapter 14
Waste Disposal from the Body

The opposite of a correct statement is a false statement. But the opposite of a profound truth may well be another profound truth.

Neils Bohr (1885–1962)

Each person ingests a large number of molecules per day with meals and snacks. A similarly large number of molecules enter the body through the respiration. Body processes – such as building proteins, producing energy and replenishing lost nutritional stores – use many of these molecules. However, there is a substantial number of molecules that are neither usable nor needed to organism and therefore, have to be eliminated from the body. Waste products are also generated during metabolic processes in organism and since they might be toxic if accumulate within the body, they should be eliminated, as well. The diversity of molecules that can be ingested is numerous and therefore, the processes of elimination, very complex.

Introduction

Body processes—such as building proteins, producing energy and replenishing lost nutritional stores—use many of ingested or inhaled molecules [1]. However, there is a substantial number of molecules that are neither usable nor needed to organism and therefore, have to be eliminated from the body. Waste products are also generated during metabolic processes in organism and since they might be toxic if accumulate within the body, they should be eliminated, too [1]. The diversity of molecules that can be ingested is numerous and therefore, the processes of elimination, very complex.

M. Pavlovic, *Bioengineering*, DOI 10.1007/978-3-319-10798-1_14

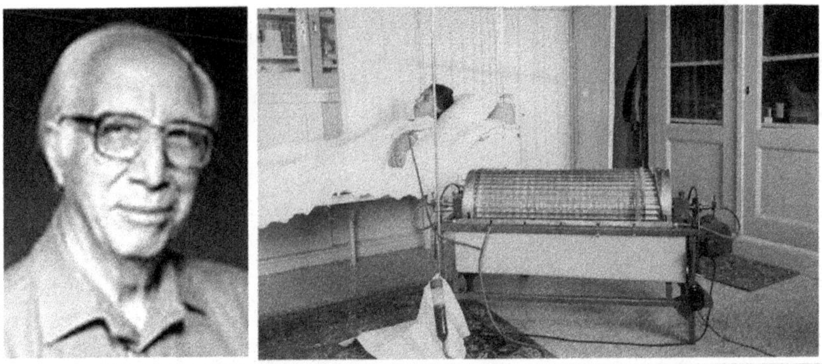

Bioengineering breakthrough: Willem Johan "Pim" Kolff (1911–2009) was a pioneer of hemodialysis as well as in the field of artificial organs. Willem is a member of the Kolff family, an old Dutch patrician family. He made his major discoveries in the field of dialysis for kidney failure during the Second World War.

The Role of Excretory Systems (Kidney and Liver) in Eliminating Wastes and Toxins and Maintaining the Body Balance

Excretion of the molecules by the liver and kidney, and **biotransformation of compounds in the liver**, are responsible for elimination of wastes (such as urea), elimination of toxins (such as drugs), and maintenance of homeostasis. The chemical composition of body fluids is important for the well-being of the cells of the body. The circulatory system is mainly responsible for the physical transport of fluids but not for the composition of those fluids. This function is largely the responsibility of the kidneys. Although they help with various physiological functions, the kidneys' main roles are the r*emoval of wastes* and the *maintenance of the body's water balance*. The function (s) of the kidneys is a very important event to the normal functioning of the entire body, and can be summarized as follows:

1. Control of the body's water balance.
2. Regulation of blood pressure via the renin–angiotensin–aldosterone system.
3. Regulation of blood electrolyte balance - Na^+, Ca^{2+}, K^+ etc.
4. Excretion of metabolic wastes such as urea, creatinine and foreign substances such as drugs and the particular end—metabolic products of chemicals we ingest with our food.
5. Help in the regulation of the body's acid base balance.
6. Regulation of red blood cell production *via* the hormone erythropoietin.
7. Help in the production of vitamin D.

The Concept of Biotransformation

It is the alteration of the substance, such as drug, by chemical reactions, within the body usually from a toxic state to a less toxic state [1, 2]. Biotransformation occurs primarily within the **liver** and excretion of the products is occurring thanks to the work of **kidneys** (see Fig. 14.1).

Cells produce water and carbon dioxide as by-products of metabolic breakdown of sugars, fats, and proteins [1]. Chemical elements such as nitrogen, sulfur, and phosphorous must be stripped, from the large molecules to which they were formerly attached, as part of preparing them for energy conversion [1]. The continuous production of metabolic wastes establishes a steep concentration gradient across the plasma membrane, causing wastes to diffuse out of cells and into the extracellular fluid.

Single-celled organisms have most of their wastes diffuse out into the outside environment. Multicellular organisms, and animals in particular, must have a *specialized organ system to concentrate and remove wastes* from the circulation into the blood capillaries and eventually deposit that material at a collection point for removal entirely from the body (lungs, kidneys, liver).

Urine is produced in the *kidneys from water and wastes extracted from the blood.* The rest of the urinary system is concerned with the storage and ducting of the urine to the outside of the body.

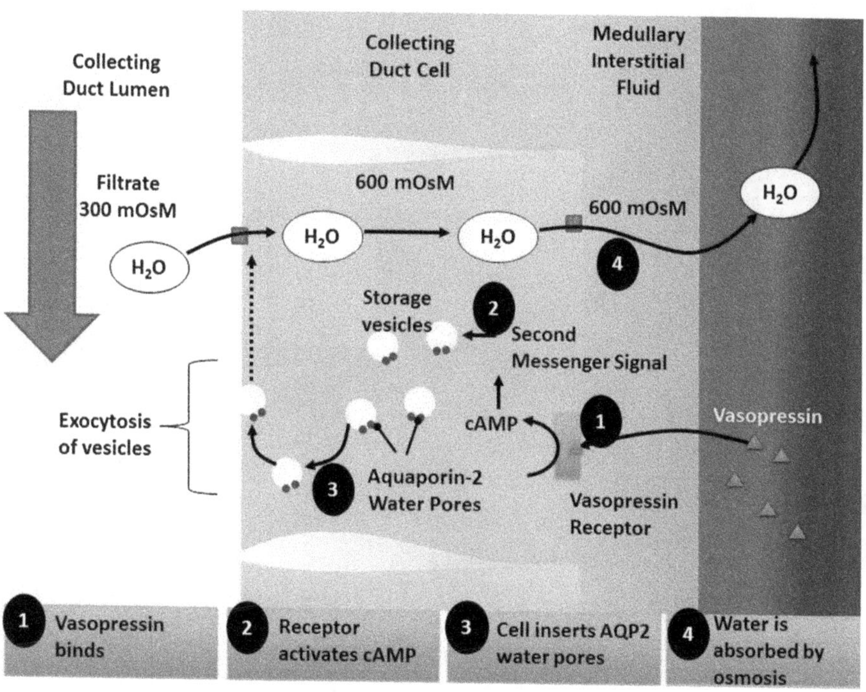

Fig. 14.1 Elimination of molecules by the kidneys: filtration, reabsorption and secretion

Structure of the Renal System

The kidneys are large, bean shaped organs which lie on the dorsal side of the visceral cavity, roughly level with the waistline [1]. Blood is supplied to the kidneys by the renal arteries which branch off the aorta. The kidneys are drained by the renal veins into the inferior vena cava. From the kidneys, urine passes to the urinary bladder via the ureters. Urine is passed to the outside environment *via* the urethra (this is routed differently in males and females).

The kidneys are protected by a tough fibrous coat called the renal capsule. Under the capsule, the arrangement of nephrons (tubular formations) and capillaries (circulatory system) in the kidney produce the appearance of distinct regions when viewed in longitudinal section. The outer cortex region surrounds darker triangular structures called pyramids which collectively form the medulla. The inner part of the kidney, the renal pelvis, collects the urine draining from the nephron tubules and channels it into the ureter (Fig. 14.2).

Microstructure of the kidney. The basic functional unit of the kidney **is the nephron**. There are over one million nephrons in each human kidney and together they are responsible for the complex water regulation and waste elimination functions of the kidneys. The heads of the nephrons are in the cortical region and the tubular component then descends through the medulla and eventually drains into the renal pelvis (Figs. 14.2 and 14.3).

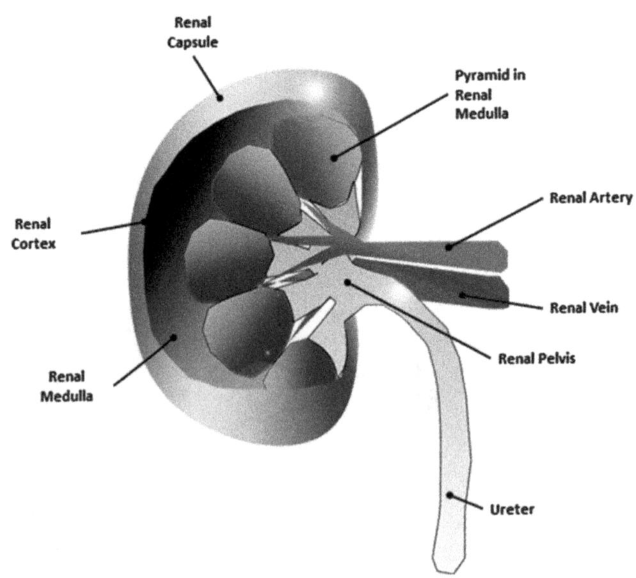

Fig. 14.2 Sectioned view of the kidney

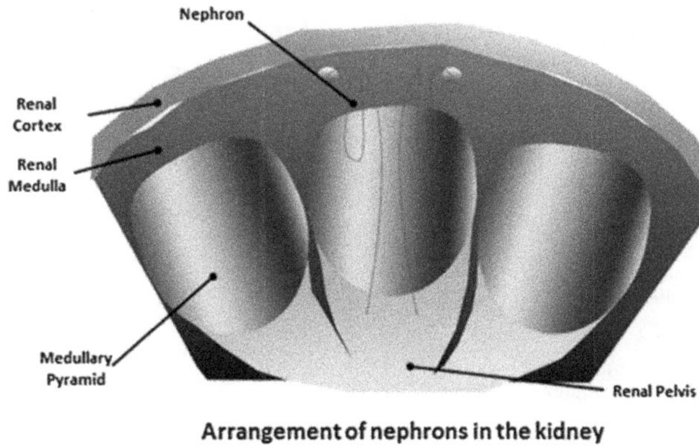

Fig. 14.3 Arrangement of nefrons in the kidney

Regulation of Filtration in Glomerulus

The filtration of the blood in **glomerulus** and reabsorption/secretion in tubular system are controlled and very well regulated processes. The key area of interface between the circulatory system and the tubular part of the kidney is the knot of glomerular capillaries in the **Bowman's capsule**. Those liquid parts of the blood that are able to cross through the filtration membrane of the capillaries, pass into the Bowman's capsule, and then into the tubular section of the nephron (Fig. 14.3). The filtration membrane only allows water to pass through it and small molecules that will dissolve in water such as waste (urea, creatinine, etc.) glucose, amino acids and ions. Large proteins and blood cells are too large to be filtered and remain in the blood.

The filtered fluid or filtrate enters the proximal tubule and then into the loop of Henle which is the part of the nephron which dips in and out of the medulla. **Loop of Henle is** a U-shaped loop between the proximal and distal tubules in the kidney. From the loop of Henle, the filtrate travels through the distal tubule and then into a common collecting duct which passes through the medulla and into the renal pelvis (Figs. 14.2 and 14.3).

Phase I: Filtration in Bowman's Capsule

The glomerulus consists of a cuplike structure, **Bowman's capsul**e, within which lies a cluster of capillaries, and a hairpin-shaped tubule that runs from Bowman's capsule into the medulla of the kidney and to collecting ducts in the medulla (Figs. 14.4, 14.5 and 14.6). The capillaries are extremely thin-walled, and significantly more permeable to plasma than ordinary capillaries. Moreover, the diameter of the arteriole as it leaves Bowman's capsule is less than its entering diameter.

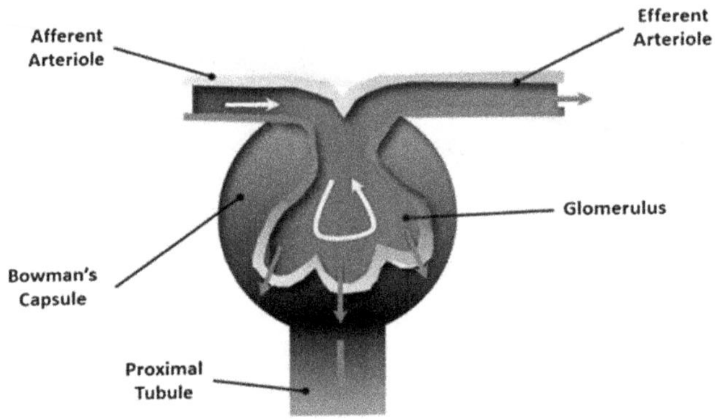

Fig. 14.4 The Bowman's capsule and glomerulus

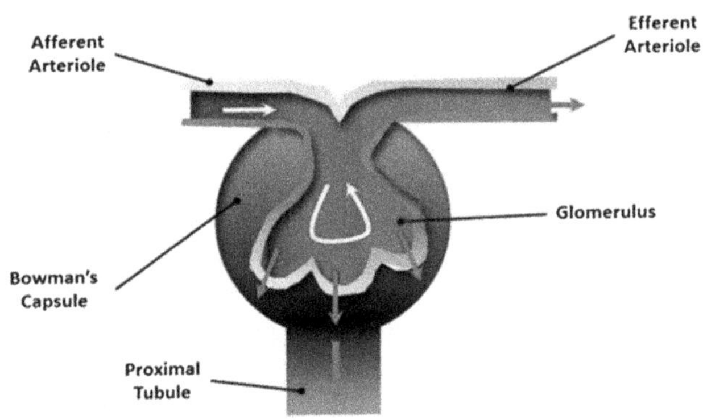

Fig. 14.5 Glomerular filtration

Surrounding the capillaries are cells known as podocytes—they create a network of cytoplasmic extensions which aid in filtration. All of these factors combine to increase blood pressure and force large quantities of plasma out of the capillaries into the Bowman's capsule and down the tube of the nephron. Small solutes (especially salts and urea, but also other water soluble molecules) are also forced out of the bloodstream with the plasma.

Larger proteins and cells remain in the capillaries, creating a very hypertonic solution.

The general strategy here is this: the blood plasma is full of nutrients, proteins, ions, water, and other dissolved particles, some of which the body needs, some of which the body must remove. To remove the wastes, a large portion of the plasma is filtered from the blood and then the substances the body needs are put back into the blood. The substances that the body doesn't need are left behind and removed during urination.

The mammalian kidney consists of an outer **cortex** and an **inner medulla**. It is composed of units called nephrons. Each nephron consists of a glomerulus, situated in

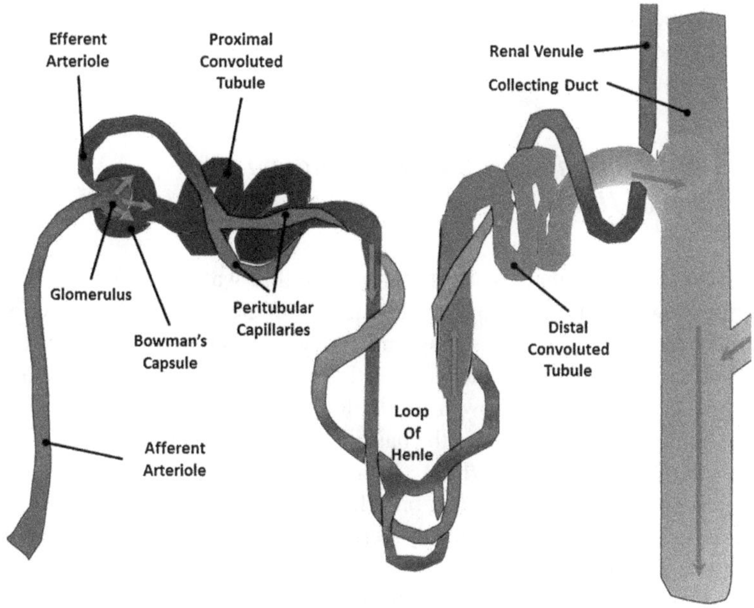

Fig. 14.6 Regulation of filtration in glomerulus

the cortex of the kidney, and a long U-shaped tubule (the loop of Henle) that extends into the medulla of the kidney and connects to collecting tubules which eventually merge to join the ureter. As one progresses inward from the cortex to the inner medulla, the concentration of solutes increases from about 300 to 1,200 mosm/L. The effects of this on water and solute balance are crucial physiological event.

The lumen wall of the epithelial cells of the proximal tubule is like the lumen wall of the small intestine—both are bordered with millions of microvilli to increase surface area. The role of these epithelial cells is to reabsorb ions, nutrients, and water and transport them to the blood vessels nearby.

Phase II: Reabsorption in the Proximal Tubule

1. A Na^+/K^+ ATPase located on the basolateral membrane of the epithelial cell (the side of the cell opposite the lumen) actively pump Na^+ out of the cell into the blood. This sets up a strong concentration gradient in the cell.
2. The gradient created by the Na^+/K^+ ATPase provides a potential to allow Na^+ cotransporters in the apical membrane to reabsorb nutrients and electrolytes. Water also flows in via osmosis.
3. Solutes exit the epithelial cells and enter the blood through channels.
4. Water flows from the epithelial cells into the blood via osmosis.

Note that because osmosis occurs, the osmolarity of the filtrate remains isotonic. The volume decreases.

Phase III: Creation of an Osmotic Gradient in the Loop of Henle

The loop of Henle has three distinct regions, the thin-walled descending limb, the thin-walled lower portion of the ascending limb and the thick-walled upper portion of the ascending limb. The descending limb of the loop is highly permeable to water but almost completely impermeable to solutes. As the filtrate travels down the descending loop, water will flow from the loop into the surrounding medium via osmosis. When the filtrate reaches the hairpin turn, it is isotonic with the surrounding medium (about 1,200 mosm/L).

The lower portion of the ascending branch of the loop of Henle, however, is highly permeable to Na^+ and Cl^-, moderately permeable to urea, and almost completely impermeable to water. As it travels up into the less-concentrated regions of the medulla, Na^+ and Cl^- will passively diffuse across the membrane. As the filtrate continues up the thick portion of the loop of Henle, Na^+ and Cl^- are actively pumped out of the filtrate into the surrounding medium. This requires energy, but helps to maintain the osmotic concentration gradient in the medulla. The water and solutes that flowed into the medulla can be reabsorbed by the vasa recta—a network of capillaries that surround the loop of Henle and reabsorb water and solutes filtered from the blood.

Phase IV: Regulating Water and Electrolyte Balance in the Distal Tubule and the Collecting Duct

The first three steps in urine formation, filtration, reabsorption, and establishment of an osmotic gradient, result in a fluid that is slightly hypotonic to blood. The major solutes still present in this fluid are urea and other wastes. Electrolytes and water are always absorbed by the distal tubule, the amount of Na^+, Cl^-, and water absorbed is variable. Regulation of these processes is under hormonal control (see Figs. 14.7 and 14.8).

If Na^+ levels in the blood are low, the hormone aldosterone is released, which leads to reabsorption of Na^+ and Cl^- in the distal tubule [1]. Water will also flow into the tubule via osmosis. However, if a person is dehydrated, the hormone ADH (antidiuretic hormone) is released. This causes *aquaporin channels* to be inserted in the membrane of the collecting duct so that large quantities of water can be reabsorbed.

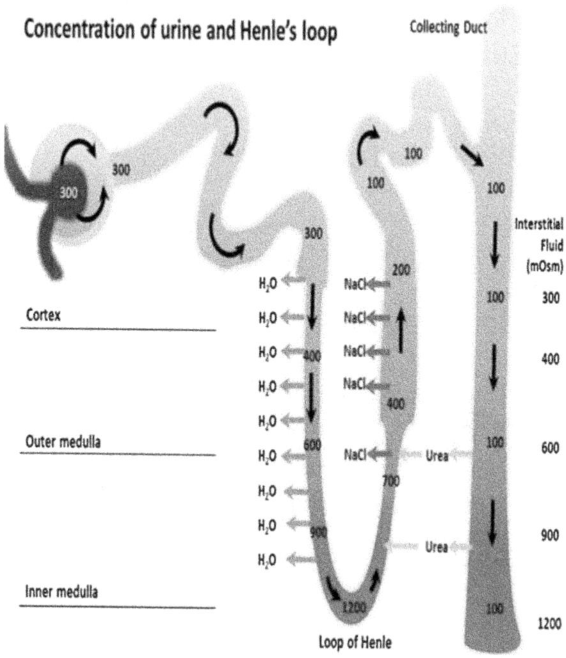

Fig. 14.7 Concentration gradient in cortex and medulla of kidney (*ADH present—collecting Duct is permeable to water and a small volume of urine is produced)

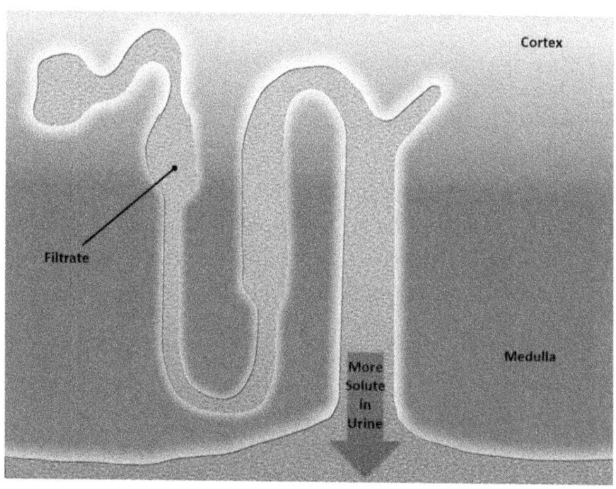

Fig. 14.8 Dilution of urine (No ADH present—collecting duct is NOT permeable to water and large volume of urine is produced)

The Concept of Clearance, Excretion in Urine, and Calculation for Different Solutes

Reabsorption and Secretion in the Tubules Through Transport Processes

Countercurrent Mechanism of Gradient Formation in the Kidney (Henle's Loop)

Peritubular Capillaries

The nephrons are surrounded by a fine network of capillaries called the peritubular capillaries [1, 2]. These perform an important role in direct secretion, selective reabsorption and the regulation of water [1].

Direct Secretion

In addition to glomerular filtration, some substances are secreted directly from the adjacent peritubular capillaries into the proximal tubule. These substances include potassium ions and some hormones.

Selective Reabsorption

Ultrafiltration is indiscriminate except for size of particle and useful substances are filtered from the blood as well as wastes. This situation is obviously unsatisfactory as the body would soon be depleted of amino acids, glucose, and sodium etc., which would need to be replenished from external sources. To resolve this problem, useful substances in the filtrate **are reabsorbed back** into the *peritubular capillaries* as the filtrate passes along the tubule, leaving only the wastes which are eliminated in the urine. This process is shown in Fig. 14.9.

Water Regulation by the Kidneys

The water content of the body can vary depending on various factors. Hot weather and physical activity such as exercise make us sweat and so lose body fluids. Drinking tends to be at irregular intervals when socially convenient. This means that sometimes the body has too little water and needs to conserve it and sometimes too much water and needs to get rid of it. Most of the control of water conservation

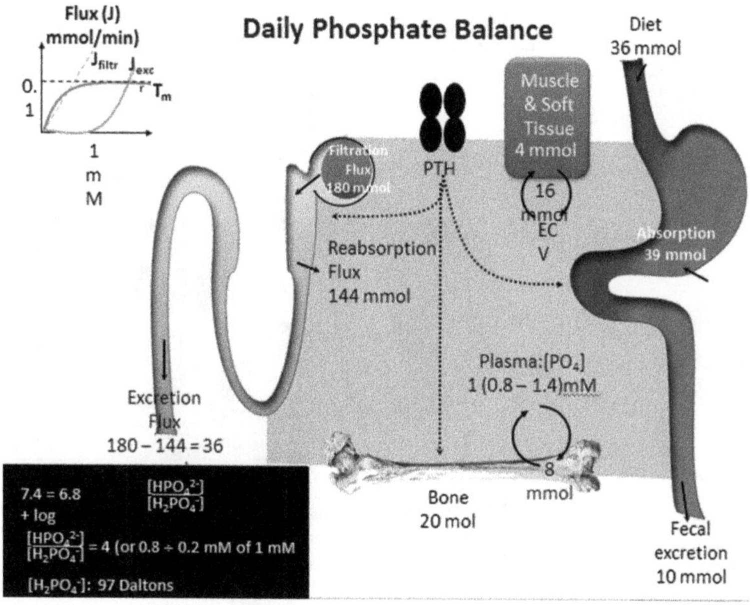

Fig. 14.9 Osmotic pressure as a driving force for water reabsorption in the tubules (Oncotic pressure is the pressure of proteins in the liquid)

takes place in the distal and collecting tubules of the nephrons under control of anti-diuretic hormone (ADH), sometimes called *vasopressin*. This hormone is released by the *posterior pituitary* under control of the *hypothalamus* in the mid-brain area. The hypothalamus monitors the water content of the blood [1]. If the blood contains too little water (indicating dehydration) then more ADH is released. If the blood contains too much water (indicating over-hydration) then less ADH is released into the blood stream [1].

Release of ADH from the Posterior Pituitary into the Bloodstream

ADH released from the pituitary travels in the blood stream to the peri-tubular capillaries of the nephron. ADH binds to receptors on the distal and collecting tubules of the nephrons which causes water channels to open in the tubule walls. This allows water to diffuse through the tubule walls into the interstitial fluid where it is collected by the peri-tubular capillaries. The more ADH present, the more water channels are open and the more water is reabsorbed (Fig. 14.10).

Over 99 % of the filtrate produced each day can be reabsorbed [1–3]. The amount of water reabsorbed from the filtrate back into the blood depends on the water situation in the body. When the body is dehydrated, most of the filtrate is reabsorbed but note that even in cases of extreme of water shortage, the kidneys will continue to produce around **500 ml of urine each day** in order to perform their excretory function.

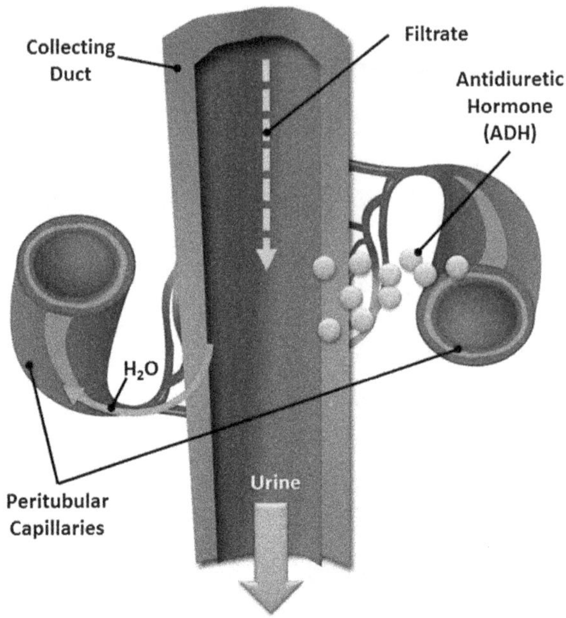

Fig. 14.10 Reabsorption of water from the filtrate

The Micturition Reflex

Micturition is another word for urination and in most animals it happens automatically. As the bladder fills with urine, stretch receptors in the wall of the bladder send signals to the parasympathetic nerves to relax the band of smooth muscle that forms the internal urethral sphincter [1]. As the muscle relaxes, the urethra opens and urine is voided to the outside environment. A second sphincter, the external urethral sphincter is skeletal muscle controlled by motor neurons (Fig. 14.11). These neurons are under conscious control and this means we are able to exercise control over when and where we urinate. This control is a learned response that is absent in the new-born infant.

Acid–Base Balance

The body controls the acidity of the blood very carefully because any deviation from the normal pH of around 7.4 can cause problems—especially with the nervous system. Deviations in pH can occur due to trauma or diseases such as diabetes, pneumonia and acute asthma. The mechanisms that resist and redress pH change are as follows:

1. Minor changes in pH are resisted by plasma proteins acting as buffers in the blood.
2. Adjustment to the rate and depth of breathing. An increase in acidity (decrease in pH) increases the rate and depth of breathing which gets rid of carbon dioxide from the blood and so reduces acidity.

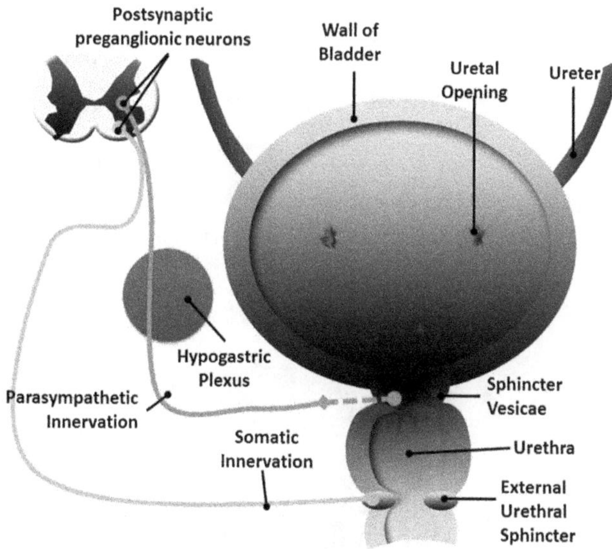

Fig. 14.11 The urinary bladder and urethra under the influence of ADH

3. The kidneys respond to changes in blood pH by altering the excretion of acidic or basic ions in the urine. If the body becomes more acidic, the kidneys excrete acidic hydrogen ions (H^+) and conserve basic bicarbonate ions (HCO_3^-). If the body becomes more basic, the kidneys excrete basic bicarbonate ions and conserve acidic hydrogen ions.

Together, these three mechanisms maintain tight control over the pH of the body.

Renin–Angiotensin–Aldosterone System

The long-term control of blood pressure is via the Renin–Angiotensin–Aldosterone (RAA) system. This system is also one of the body's compensatory mechanisms to a fall in blood pressure. The kidneys release renin into the bloodstream and this converts angiotensinogen to angiotensin I which in turn is converted to angiotensin II by angiotensin converting enzyme in the capillaries of the lungs. Under the influence of Angiotensin II, aldosterone levels increase. This increases blood sodium levels by decreasing the amount of salt excreted by the kidneys. Retaining salt instead of excreting it into urine increases the osmolarity of the blood and so the blood volume. As the volume increases, so does the blood pressure. Angiotensin II is also a potent vasoconstrictor which raises blood pressure by increasing vascular resistance (Fig. 14.12).

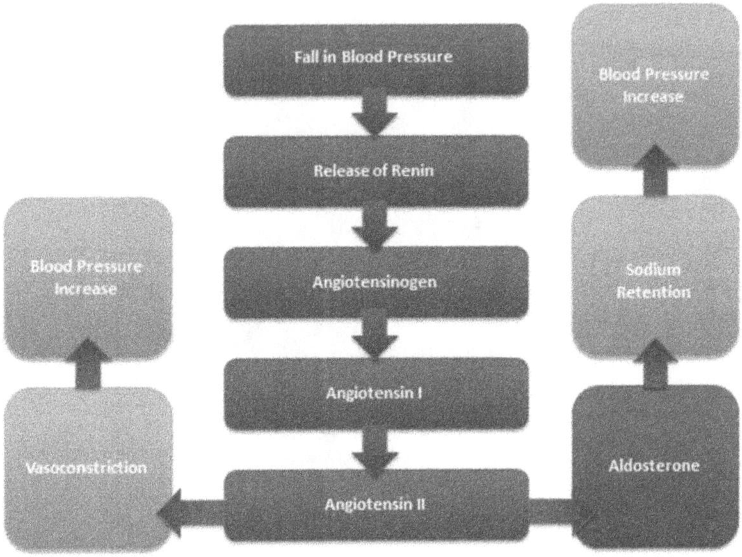

Fig. 14.12 The renin, angiotensin, aldosterone response to a fall in blood pressure

Emphasizing Bioengineering Aspect to Kidney Function

Dialysis is one of the most significant accomplishments that have been created in order to compensate for renal failed. It has had a very strong impact upon the mortality of kidney patients and their normal daily function despite the procedure.

Bladder (Transplantation Without a Donor) Atala

As a miraculous innovation must be mentioned Dr. Anthony Atala's 3-D printing of the patient's bladder cells into the mold and use of the memory of bladder cells to their old environment which created newly designed bladders of three sizes (children, female and male). The clinical trial is going on, while Dr. Atala is trying to create the entire kidney applying 3-D printing of the patient's cells (http://en.wikipedia.org/wiki/Artificial_urinary_bladder [Anthony Atala]). It is notoriously clear that to get a nephron is far more difficult than to get the bladder, due to complexity of this kidney—specific unit. However, this method known as "*Transplantation without a donor*" has a great potential since it uses patients own cells and avoids any immunological complications using relatively simple technique.

Liver: Detoxication and Bill Secretion

Biotransformation and Biliary Excretion

The liver is main organ for drug metabolism, but it is also an important excretory organ. Most of the venous blood from the intestine flows through the portal vein to the liver. This positioning is important for liver function: it appears from a review of the work of numerous authors that cells of many types, including mammalian and bacterial cells, can metabolize steroid hormones. Different cells of the mammal have quantitatively and qualitatively different capacities to perform steroid metabolism. Biosynthesis of steroids appears at present to be an exclusive property of cells derived from the genital ridge. The steroid hormones, once synthesized, have two fates. The first is that the substituted groups of the steroid nucleus are oxidized or reduced to form other steroid molecules. Some of these metabolites possess known functions. In the process of conversion, however, the transformed steroid usually loses, either quantitatively or qualitatively; it's original biological activity but may acquire a different function. The conversion of one steroid hormone to another by peripheral cells other than those capable of biosynthesis has been called biotransformation. A second fate is the conjugation of steroid metabolites which can then be excreted (Fig. 14.13).

Many peripheral cells perform the first type of metabolic transformation, but only the liver has been shown so far to be able to form tetra-hydrocorticosteroids, which is an essential step before conjugation. The kidney can conjugate corticosteroids

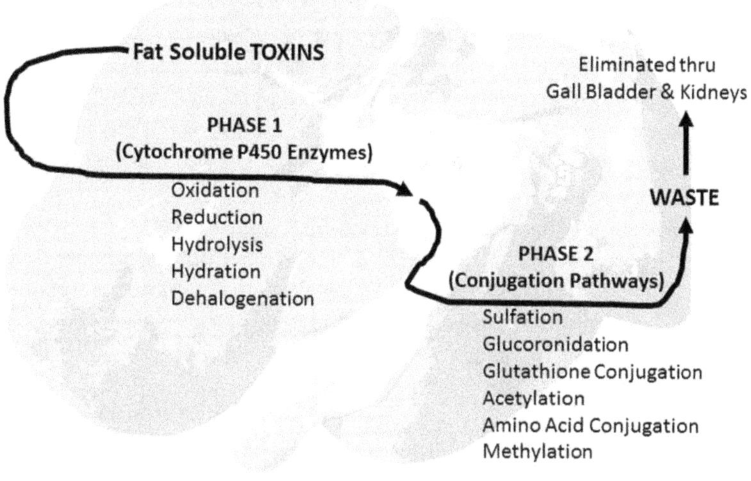

Fig. 14.13 Liver with basic detoxification and waste elimination processes

provided they are already in the tetra-hydro form. The oxidation and reduction of steroid molecules (biotransformation) may be reversible or irreversible. The degree of reversibility thus regulates partly the necessity for replenishing the supply of the steroid hormone.

The *concept of active and inactive steroid hormones* is used to indicate the presence or absence of a particular biological activity which is dependent upon the presence or absence of certain substituent groups on the steroid nucleus.

Prevention of oxidation or reduction of the steroid hormone keeps it in the active state, since the original structure–activity relation is maintained. It appears, although it has not been proved in each case, which increased potency of synthetic steroid hormones is due in part to the fact that essential positions for activity are protected from normal metabolism. The point of view presented here is that the actual mechanism of action for most if not all steroid hormones is still not understood and that the metabolism of these hormones is related not to mechanism of action but to their turnover, which imposes requirements for steroid renewal.

Non-steroidal hormones which influence steroid hormone biosynthesis, such as pituitary hormones, also directly influence steroid metabolism. Other hormones may influence steroid hormone metabolism by more indirect means increasing their biosynthesis, biotransformation and conjugation, and excretion. The major emphasis in this review has been upon the steroid metabolism of lymphocytes, fibroblasts, reticuloendothelial cells, hepatocytes, and their malignant counterparts. These cells appear to be among the most important target cells for corticosteroid activity.

Recent findings indicate that biological activity exists for metabolites which were thought to be inactive. Other metabolites may possess important functions which are at present unknown. The concept of biotransformation may have future practical and scientific importance because it may be possible to control the production of particular metabolites and thus control certain biological responses. For example, testosterone, an androgenic hormone, can he transformed to pyrogenic or hypo-cholesterolemic steroids. Estrogens seem to favor the formation of the metabolite which has hypocholesterolemic effect. It is possible that thyroid hormone might have a similar effect. The control of the preferential production of the pyrogenic steroid has not been investigated. In the opinions of the reviewers, the implications of this field of steroid hormone research for a further understanding of physiological and pathological processes are enormous.

The importance of abnormal metabolism by malignant cells may lead to new points of view concerning the fundamental processes of malignant growth and are already important in the diagnosis of some malignancies. Further research along this line offers rational avenues to the development of new steroid hormones for treatment of some types of cancer.

Emphasizing Bioengineering Aspect to Liver Function

Liver Transplantation

In 1985, Barnes-Jewish Hospital became the 16th hospital in the world with a dedicated liver transplant program and the first in Missouri to perform a successful liver transplant (http://www.barnesjewish.org/liver-transplant). Since then, the hospital has continued to be a world leader in liver transplantation and the management of end stage liver and hepatobiliary diseases including hepatocellular cancer and cholangiocarcinoma. With some of the best survival rates, Washington University transplant specialists have performed more than 1,200 adult liver transplants since the program's inception. The gastroenterology program ranks among the top programs in the nation by U.S. News & World Report (http://www.barnesjewish.org/liver-transplant). This is still far from building liver tissue from patient's own cells as it is proposed by some researchers. Such a successful approach would for sure eliminate timing and immunological complication that can follow liver transplantation.

References

1. Hall, G.E., Guyton, A.C.: Textbook of Medical Physiology. Sounders Elsevier, Philadelphia (2011)
2. Andjus, R.K.: General Physiology and Biophysics. Modules 1–7. Center for Multidisciplinary Studies. University of Belgrade, Belgrade, Serbia (2002)
3. Saltzman, M.: Biomedical Engineering. Cambridge University Press, New York (2009). ISBN: 978-0-521-840099-6 (hardback)

Chapter 15
Biomechanics: Principles

I do not know what I may appear to the world; but to myself I seem to have been only like a boy playing on the seashore, and diverting myself in now and then finding of a smoother pebble or a prettier shell than ordinary, whilst the great ocean of truth lay all undiscovered before me.

Sir Isaac Newton (1642–1727)

Scientists have discovered many laws of nature. They have learned some basic rules about how objects appear to interact, how light appears to travel and interact with other light and matter. They have studied gravity, magnetism and electricity and have learned laws which govern each. Using this knowledge, we can predict eclipses, build bridges, and put a man on the moon. But where did these laws come from? We know how the laws work, but what do we know about why they work? It has long been a tenet of science that it only deals with finding laws that describe how things work so that we can have a better world by applying them in inventions and technology.

Introduction: The Laws of Physics

Humans can hold their bodies erect, vertically above the earth, because their bodies are solid objects capable of supporting their own weight. The human skeletal system is the collection of 206 bones, connected by soft-tissue cartilage, ligaments, tendons, and muscles-that together provide a mechanical support system for the human body. **Scientists have discovered many laws of nature**. They have learned some basic rules about how objects appear to interact, how light appears to travel and interact with other light and matter.

M. Pavlovic, *Bioengineering*, DOI 10.1007/978-3-319-10798-1_15

Breakthrough the centuries: IsaakNewton (1642–1727), James Clerk Maxwell (1831-1879), Albert Einstein (1879-1955) and Erwin Schrödinger (1887-1961): from mechanics toward understanding the wave(s) and opening the door to bioengineering without knowing that.

They have studied gravity, magnetism and electricity and have learned laws which govern each. Using this knowledge, we can predict eclipses, build bridges, and put a man on the moon. But where did these laws come from? We know how the laws work, but what do we know about why they work? It has long been a tenet of science that it only deals with finding laws that describe how things work so that we can have a better world by applying them in inventions and technology. Science has no way to discover the ultimate cause behind the laws. It makes little difference to how your cell phone works to know whether the governing laws were written by a Creator, or whether they are self-existent truths. We mostly just want our high-tech devices to work, and the low-tech varieties also.

Thus, science has been **content to adopt the equations that came from some scientist's mind as explaining things**. Sir Isaac Newton discovered and articulated some of the best known laws of all time, including his **three laws of motion**, and **law of gravity** [1]. Maxwell later wrote the equations of **electricity and magnetism**, and Schrödinger defined an **equation for quantum wave mechanics** (**applied in NMR spectra determination**). It is rarely asked **why these equations work**, we are just glad that someone discovered that they do indeed work. When a scientist does ask "why" an equation works, the only answer allowed in science is to explain it with **an even more fundamental law**. For example, Johannes Kepler **defined three laws describing the motion of planets around the sun**. He gave no reason why they should work, but only claimed that they do. Then Sir Isaac Newton came up with his laws of motion and gravity, and was able to derive all of Kepler's laws from them. Thus, Newton's appear to be **more fundamental**, **while Kepler's are now an exercise for the student** [1].

Many laws about **how light travels and behaves have been discovered**. At first there was a major debate about whether light was **a particle** (like a little bullet) that could travel through a vacuum or a wave (like sound) that required a medium (like air) for transport. **Newton proposed that it was a particle**, because it didn't seem to go around corners like waves do. But then a host of wave properties of light were discovered, such as **interference** (when wave crests add), and **diffraction** (it does indeed go around corners) [1] (see Fig. 15.1).

When **Maxwell** showed that light could be explained as **an electromagnetic wave** that seemed to end the debate once and for all. Light was a wave [1].

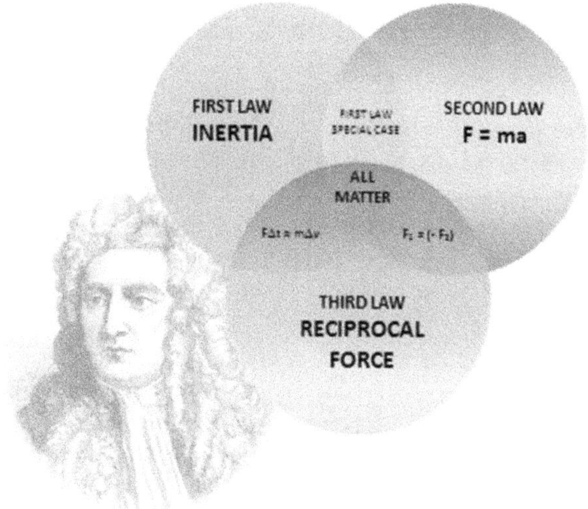

Fig. 15.1 Venn diagram of Sir Isaac Newton's primary principles

The medium through which light traveled was called the "*luminiferous ether*" or simply "*ether*." There were some famous, and now amusing, statements made at the end of the nineteenth century that physics had answered all the hard questions, and all that was left was to fill in some details. So, at that time, it looked like "light" as understood by science and religion had something in common. Now let's get back to science, to see how its story changed. One of those minor details yet to be determined at the end of the nineteenth century concerned just what was the precise speed of the earth moving through the ether. After all, the earth is believed to orbit the sun, the sun is moving relative to the nearby stars, and then our whole galaxy is rotating, so the earth's "absolute speed" must be fairly large. Then in 1887 the famous **Michelson–Morley** experiment failed to measure any speed at all relative to the ether [1].

The young Albert Einstein took that result and used it to rewrite much of physics. He proposed as one of his two fundamental postulates that **the speed of light is constant**, independent of the motion of either the emitter or observer. He declared that no experiment could be done to detect the motion through the ether because of that property of light. Then the big switch came, so watch the magician's hands closely. Because science only studies the observable, and because even in theory no experiment could detect the ether, then as far as science is concerned the ether does not exist. Thus, the ether has vanished. The vocabulary was amended, and it is now said that light travels through a *vacuum*. Light was then shown to have bullet-like properties, and **Einstein received the Nobel Prize for explaining the "photoelectric effect**." For a while, light seemed to behave both as a particle ("**photon**") and also as a wave [2]. Later it was shown that all particles, such as electrons, also have wave-like properties. In the minds of most physicists, the apparently wave-particle duality of light was resolved on the side of the photon, with all of the wave properties being explained by the amazing theory of quantum electrodynamics. It facilitated later on innovation of atomic force microscope (AFM) [2].

Mechanical Properties of Materials

In the realm of quantum mechanics, what holds an atom up has been a mystery. **For example, what keeps the electron from falling onto the proton in a hydrogen atom**? If the electron is orbiting the proton, then it should radiate energy and collapse. But it doesn't. This paradox was first explained by postulating **that there are certain rules it has to obey, and that light can only radiate away at certain energies which correspond to the electron jumping between "shells" in which the electrons supposedly reside**. But then, **quantum mechanics appeared** and proposed that the electrons were somehow just in probability clouds around the nucleus, and **that their state could not be known with certainty**. Now progress is being made to solve this enigma by the proposal that it may be the Zero Point Energy (**ZPE**) that is constantly **supplying energy to the atom to support it**. If so, then far from not being observable, the ZPE may actually be what allows matter even to exist (see Figs. 15.2 and 15.3).

Newton's First Law

According to Newton's first law, an object in motion continues in motion with the same speed and in the same direction unless acted upon by an unbalanced force. It is the natural tendency of objects to keep on doing what they're doing.

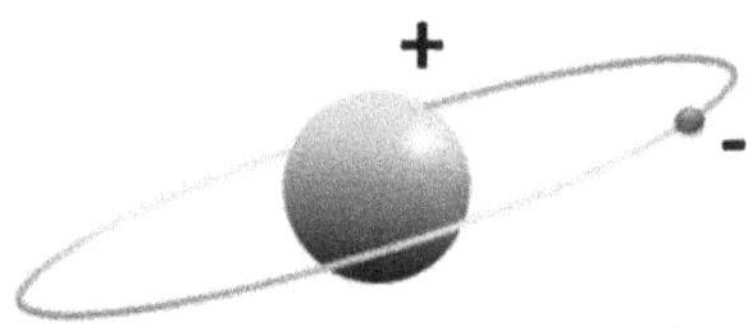

Fig. 15.2 Atom

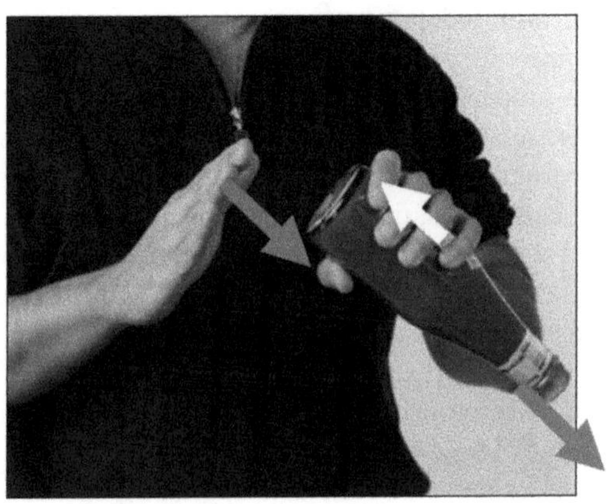

Fig. 15.3 Illustration of Newton's first law

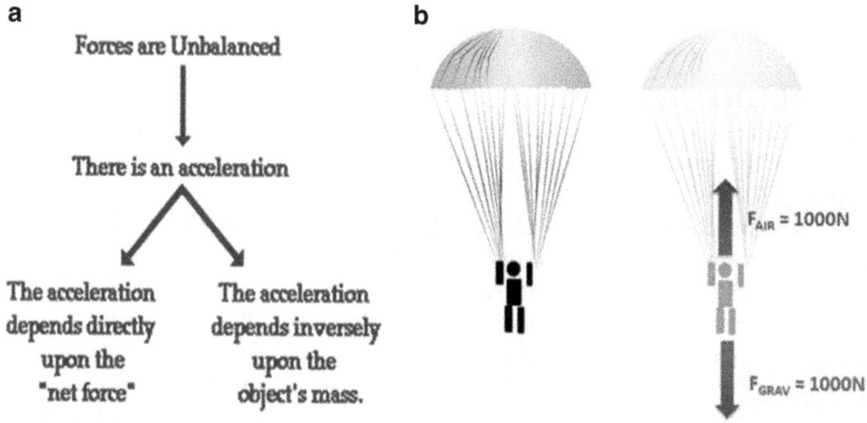

Fig. 15.4 **A** and **B** illustrating Newton's second (**A**) and third (**B**) laws

All objects resist changes in their state of motion. In the absence of an unbalanced force, an object in motion will maintain this state of motion. This is often called **the *law of inertia***. Since these two forces are of equal magnitude and in opposite directions, they balance each other. The book is said to be at equilibrium. There is no unbalanced force acting upon the book and thus the book **maintains its state of motion**.

The presence of an unbalanced force will accelerate an object—changing its speed, its direction, or both its speed and direction [1] (see Fig. 15.4).

Newton's Third Law

For every action, there is an equal and opposite reaction.

The statement means that in every interaction, there is a pair of forces acting on the two interacting objects. The size of the force on the first object equals the size of the force on the second object. The direction of the force on the first object is opposite to the direction of the force on the second object. Forces always come in pairs—equal and opposite action–reaction force pairs. Many of them are present in the cell and also used in bioengineering approaches.

Elasticity

In physics, **elasticity** is the physical property of a material that returns to its original shape after the stress (e.g. external forces) that made it deform is removed. The relative amount of deformation is called the strain [3]. The elastic regime is characterized by a linear relationship between stress and strain, denoted linear elasticity. The classic example is a metal **spring**. This idea was first stated by **Robert Hooke** in 1675 as **a Latin anagram** "ceiiinosssttuu" whose solution he published in 1678 as "***Ut tensio, sic vis***" which means "***As the extension, so the force***."

This linear relationship is called **Hooke's law**. The classic model of linear elasticity is the perfect spring [3]. Although the general proportionality constant between stress and strain in three dimensions is a 4th order **tensor**, when considering simple situations of higher symmetry such as a rod in one dimensional loading, the relationship may often be reduced to applications of **Hooke's law**.

Because most materials are elastic only under relatively small deformations, several assumptions are used to linearize the theory. Most importantly, higher order terms are generally discarded based on the small deformation assumption. In certain special cases, such as when considering a rubbery material, these assumptions may not be permissible. However, in general, elasticity refers to the linearized theory of the continuum stresses and strains (Fig. 15.5 A–C).

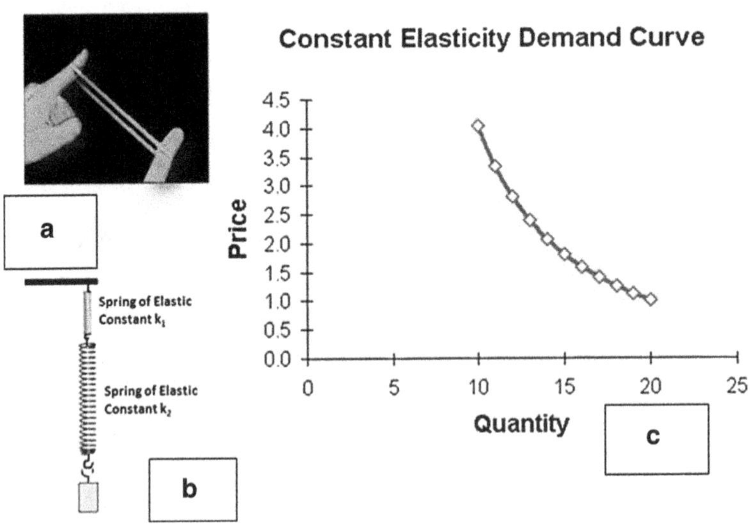

Fig. 15.5 (A–C) Principles of elasticity

Elastic Properties and Young's Modulus

For the description of the elastic properties of linear objects like wires, rods, columns, which are either stretched or compressed, a convenient parameter is the ratio of the stress to the strain, a parameter called the Young's modulus of the material. Young's modulus can be used to predict the elongation or compression of an object as long as the stress is less than the yield strength of the material. Note that the minimum energy (Fig. 15.5) is not necessarily at the dimension given the X-ray crystal structure. A deviation likely arises from inaccuracy in the DFT calculation [1]. The best model is obtained along the length of the carbon chains (x-direction). Here there is a clear minimum and Young's modulus could be calculated using these data. The deviation from experiment is only 1 % along the x dimension. The displacements are given in fractional coordinates along each unit cell direction.

Next we re-optimize the geometry for each deformed structure as shown in Fig. 15.6.

Energy conversion, storage and transport are rapidly emerging to be focal research topics in physical sciences due to socio-economic imperatives (see Figs. 15.7, 15.8, 15.9). An under-appreciated fact is that the so-called "energy problem" exists at multiple length scales: at the "global or macroscopic level", affecting individual cars to cities and at the "microscopic level" influencing next generation micro and nano-electronics (Figs. 15.10, 15.11, 15.12).

Viscosity

Viscosity is a measure of the resistance of a fluid which is being deformed by either shear stress or tensile stress. In everyday terms (and for fluids only), viscosity is "thickness". The word "viscosity" derives from the Latin word "viscum" for mistletoe.

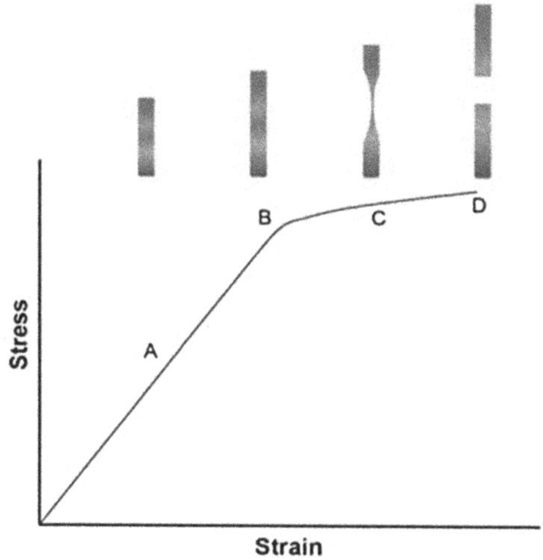

Fig. 15.6 Elastic deformation and Young's modulus

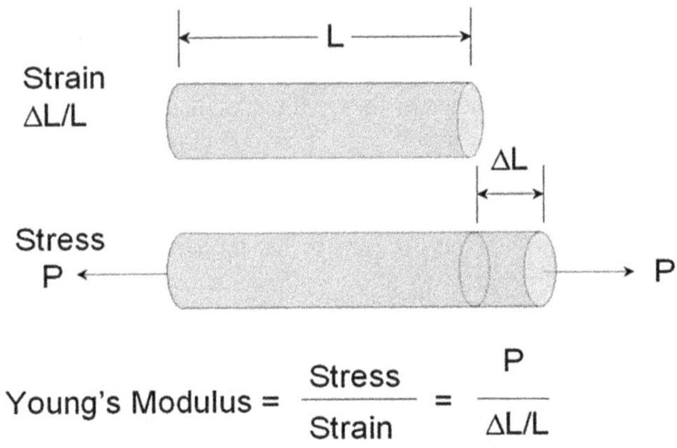

Fig. 15.7 Definition of terms used the calculation of elastic constants

Fig. 15.8 The binding energy of the high density (orthorhombic) unit cell with displacements along the given unit cell directions for a single point energy calculation

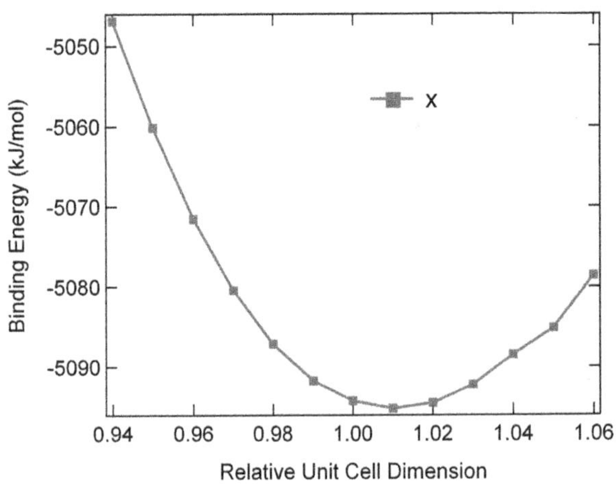

Fig. 15.9 The binding energy of the high density (orthorhombic) unit cell with displacements along the given unit cell directions for a single point energy calculation

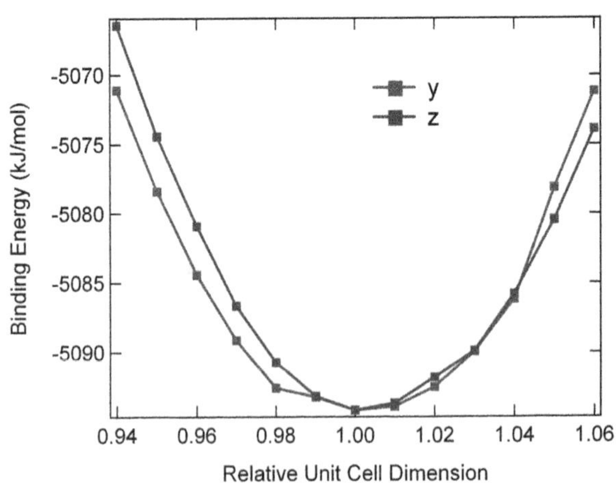

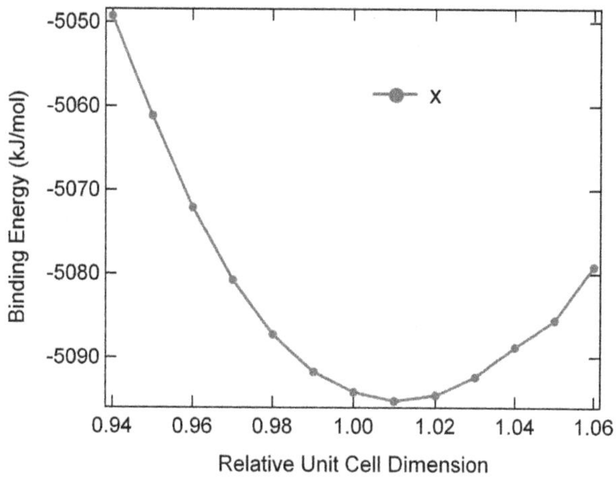

Fig. 15.10 The binding energy of the high density (orthorhombic) unit cell with displacements along the given unit cell directions for a re-optimized geometry at each deformation geometry. The displacements are given in fractional coordinates along each unit cell direction

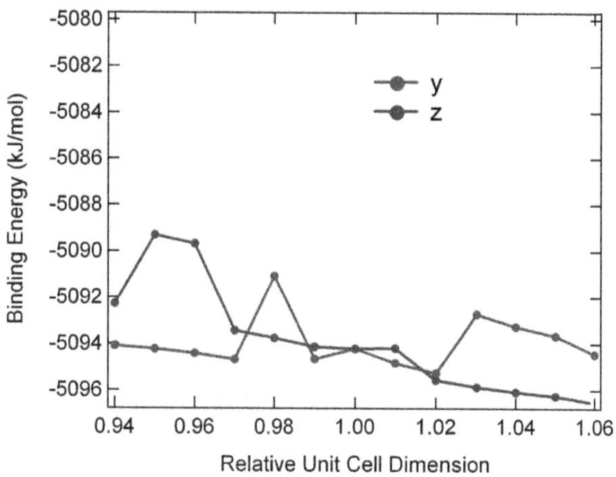

Fig. 15.11 The binding energy of the high density (orthorhombic) unit cell with displacements along the given unit cell directions for a re-optimized geometry at each deformation geometry. The displacements are given in fractional coordinates along each unit cell direction

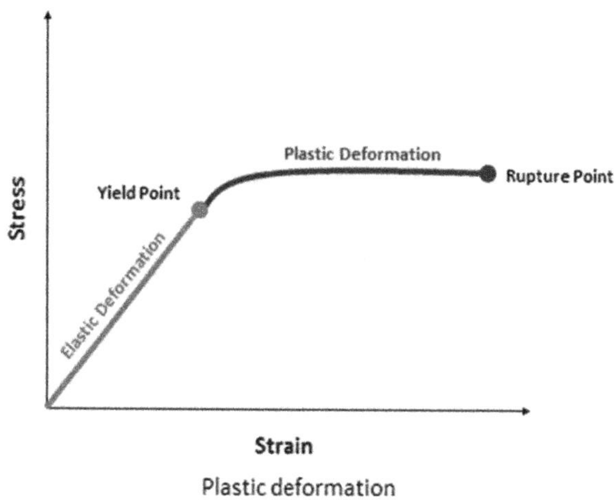

Fig. 15.12 Energy storage with deformation

A viscous glue was made from mistletoe berries and used for lime-twigs to catch birds. Thus, water is "thin", having a lower viscosity, while honey is "thick", having a higher viscosity. Viscosity describes a fluid's internal resistance to flow and may be thought of as a measure of fluid friction [3–5]. For example, high-viscosity magma will create a tall, steep stratovolcano, because it cannot flow far before it cools, while low-viscosity lava will create a wide, shallow-sloped shield volcano. Put simply, the less viscous something is, the greater its ease of movement (fluidity). All real fluids (except superfluids) have some resistance to stress, but a fluid which has no resistance to shear stress is known as an ideal fluid or in viscid fluid. The study of viscosity is known as **rheology**.

Viscosity Coefficients

Viscosity coefficients can be defined in two ways [1]:

- **Dynamic viscosity**, also **absolute viscosity**, the more usual one;
- **Kinematic viscosity** is the *dynamic viscosity* divided by the density.

Viscosity is a tensorial quantity that can be decomposed in different ways into two independent components. The most usual decomposition yields the following viscosity coefficients:

- **Shear viscosity**, the most important one, often referred to as simply **viscosity**, describing the reaction to applied shear stress; simply put, it is the ratio between the pressure exerted on the surface of a fluid, in the lateral or horizontal direction, to the change in velocity of the fluid as you move down in the fluid (this is what is referred to as a velocity gradient).
- **Volume viscosity** or **bulk viscosity**, describes the reaction to compression, essential for acoustics in fluids, see Stokes' law (sound attenuation).

Alternatively,

- **Extensional viscosity**, a linear combination of shear and bulk viscosity, describes the reaction to elongation, widely used for characterizing polymers.

For example, at room temperature, water has a dynamic shear viscosity of about 1.0×10^{-3} Pa s and motor oil of about 250×10^{-3} Pa s.

Newton's Theory

Laminar shear of fluid between two plates. Friction between the fluid and the moving boundaries causes the fluid to shear. The force required for this action is a measure of the fluid's viscosity. This type of flow is known as a Couette flow [1] (see Figs. 15.13 and 15.14).

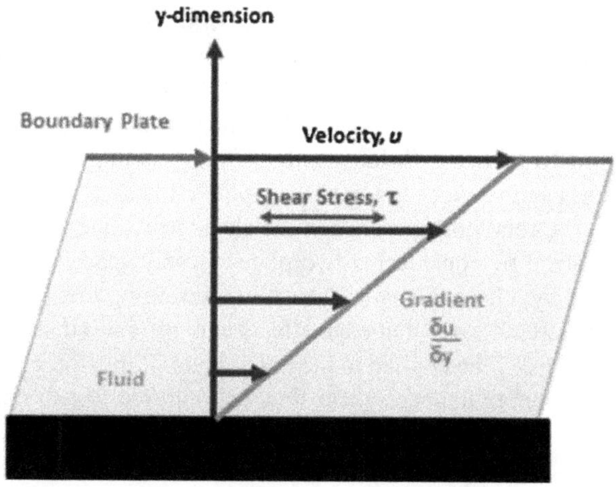

Fig. 15.13 Graphical representation of shear stress

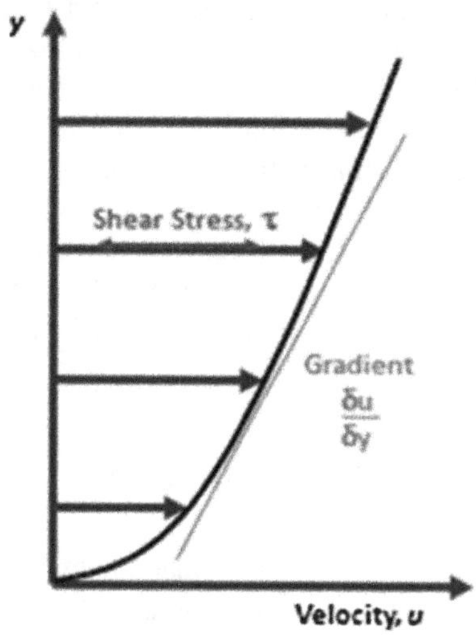

Fig. 15.14 Graphical representation of shear stress

Laminar shear, the non-constant gradient, is a result of the geometry the fluid is flowing through (e.g. a pipe). In general, in any flow, layers move at different velocities and the fluid's viscosity arises from the shear stress between the layers that ultimately oppose any applied force.

Isaac Newton postulated that, for straight, parallel and uniform flow, the shear stress, τ, between layers is proportional to the velocity gradient, $\partial u/\partial y$, in the direction perpendicular to the layers.

$$\tau = \mu \frac{\partial u}{\partial y}.$$

Here, the constant μ is known as the *coefficient of viscosity*, the *viscosity*, the *dynamic viscosity*, or the *Newtonian viscosity*.

This is a constitutive equation (**like Hooke's law, Fick's law, Ohm's law**). This means: it is not a fundamental law of nature, but a reasonable first approximation that holds in some materials and fails in others. Many fluids, such as water and most gases, satisfy Newton's criterion and are known as **Newtonian fluids. Non-Newtonian fluids** exhibit a more complicated relationship between shear stress and velocity gradient than simple linearity.

The relationship between the shear stress and the velocity gradient can also be obtained by considering two plates closely spaced apart at a distance y, and separated by a homogeneous substance. Assuming that the plates are very large, with a large area A, such that edge effects may be ignored, and that the lower plate is fixed, let a force F be applied to the upper plate. If this force causes the substance between the plates to undergo shear flow (as opposed to just **shearing elastically** until the shear stress in the substance balances the applied force), the substance is called a fluid. The applied force is proportional to the area and velocity of the plate and inversely proportional to the distance between the plates. Combining these three relations results in the equation $\boldsymbol{F=\mu\ (Au/y)}$, where $\boldsymbol{\mu}$ is the proportionality factor called the *dynamic viscosity* (also called *absolute viscosity*, or simply *viscosity*). The equation can be expressed in terms of shear stress; $\boldsymbol{\tau=F/A=\mu\ (u/y)}$. The rate of shear deformation is u/y and can be also written as a shear velocity, du/dy. Hence, through this method, the relation between the shear stress and the velocity gradient can be obtained.

James Clerk Maxwell called viscosity *fugitive elasticity* because of the analogy that elastic deformation opposes shear stress in solids, while in viscous fluids; shear stress is opposed by *rate* of deformation.

Viscosity Measurement

Dynamic viscosity is measured with various types **of rheometer**. Close temperature control of the fluid is essential to accurate measurements, particularly in materials like lubricants, whose viscosity can double with a change of only **5 °C**. For some fluids, it is a **constant** over a wide range of shear rates. These are **Newtonian fluids**.

The fluids without a constant viscosity are called **non-Newtonian fluids**. Their viscosity cannot be described by a single number. Non-Newtonian fluids exhibit a variety of different correlations between shear stress and shear rate. One of the most common instruments for measuring kinematic viscosity is the glass capillary viscometer.

In paint industries, viscosity is commonly measured with a Zahn cup, in which the **efflux time** is determined and given to customers. The efflux time can also be converted to kinematic viscosities (centistokes, cSt) through the conversion equations.

A **Ford viscosity cup** measures the rate of flow of a liquid. This, under ideal conditions, is proportional to the kinematic viscosity.

Also used in paint, a Stormer viscometer uses load-based rotation in order to determine viscosity. **The viscosity is reported in Krebs units** (**KU**), **which are unique to Stormer viscometers**.

Vibrating viscometers can also be used to measure viscosity. These models such as the *Dynatrol* use vibration rather than rotation to measure viscosity.

Extensional viscosity can be measured with various rheometers that apply extensional stress. Volume viscosity can be measured with **acoustic rheometer**.

Units

Dynamic Viscosity

The usual symbol for dynamic viscosity used by mechanical and chemical engineers—as well as fluid dynamics—is the Greek letter mu (μ). The symbol η is also used by chemists, physicists, and the IUPAC.

The SI physical unit of dynamic viscosity is the pascal-second (Pa s), which is identical to N m^{-2} s. If a fluid with a viscosity of one Pa s is placed between two plates, and one plate is pushed sideways with a shear stress of one pascal, it moves a distance equal to the thickness of the layer between the plates in one second.

The cgs physical unit for dynamic viscosity is the *poise* (P), named after Jean Louis Marie Poiseuille. It is more commonly expressed, particularly in ASTM standards, as *centipoise* (cP). Water at 20 °C has a viscosity of 1.0020 cP or 0.001002 kg/ ms.

$$1\mathrm{P} = 1\mathrm{g\,cm}^{-1} \cdot \mathrm{s}^{-1}.$$

The relation to the SI unit is

$$1\ \mathrm{P} = 0.1\ \mathrm{Pa} \cdot \mathrm{s},$$

$$1\ \mathrm{cP} = 1\ \mathrm{mPa} \cdot \mathrm{s} = 0.001\ \mathrm{Pa} \cdot \mathrm{s}.$$

Viscoelasticity

Viscoelasticity is the property of materials that exhibit both **viscous** and **elastic** characteristics when undergoing deformation. Viscous materials, like honey, resist shear flow and strain linearly with time when a stress is applied. Elastic materials strain instantaneously when stretched and just as quickly return to their original state once the stress is removed. **Viscoelastic materials have elements of both of these properties and, as such, exhibit time dependent strain**. Whereas elasticity is usually the result of bond stretching along crystallographic planes in an ordered solid, viscosity is the result of the diffusion of atoms or molecules inside an amorphous material.

Mechanical properties of body fluids, tissues, cells and organs (some topics to think about)

- Body Fluid mechanics (blood, plasma, interstitial fluid) (friction, lamination, microfluidics)
- Cellular mechanics (shear stress, elasticity, plasticity)
- Mechanical properties of cytoskeleton (complex)
- Tissue and organ mechanics (complex)
- Bones (brittle, elastic)
- Connective tissues (elastic, plastic)
- Lungs (elastic)

Emphasizing Bioengineering Aspects to Cell Biomechanics

Efforts from Kubo Laboratory

In order to create the world's leading-edge system and material technology for next-generation, the simulation and theoretical design of chemical reaction, structure, fluid, function, property, etc. is strongly required [6–20] (http://www.kubo.rift.mech.tohoku.ac.jp/eng/). Especially, the recent system, process, and material technologies are progressing toward the super-precision and super-miniaturization and constitute of the complicated multi-physics phenomena including chemical reaction, friction, impact, stress, fluid, photon, electron, heat, electric and magnetic fields, etc. [9–11]. Therefore, the individual and simple understanding of the chemical reaction, structure, and fluid on atomic-scale as well as the function and property on μm/cm/m-scale is insufficient for the development of the next-generation system and material, and then the multiple and deep understanding of the above complicated multi-physics phenomena are significantly essential. However, the traditional mechanical engineering is based on the macroscopic science and continuum approach such as the mechanics of machinery, mechanics of material, fluid mechanics, and thermodynamics and then it cannot solve the recent problems and not investigate the leading-edge research themes in a wide range of research fields because the multi-physics phenomena on electronic- and atomic-scale extremely affect the macro-scale function and performance in the state-of-the-art technologies [12–20] (http://www.kubo.rift.mech.tohoku.ac.jp/eng/).

Therefore, Kubo laboratory aims to pioneer and develop the multi-physics computational science simulation technology based on the first-principles molecular dynamics and SCF-tight-binding molecular dynamics simulation for clarifying the multi-physics phenomena including chemical reaction, friction, impact, stress, fluid, photon, electron, heat, electronic and magnetic fields on atomic- and electronic-scale (see Fig. 15.15).

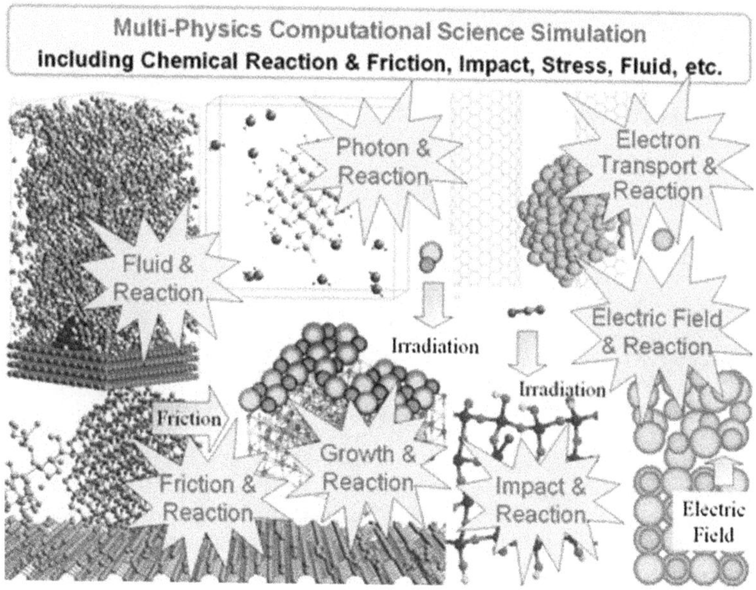

Fig. 15.15 Application fields for multiphysics computational science simulation

Efforts by Use of the AFM

Elasticity and responses of living cells to external forces have attracted tremendous attention in the modern research of tissue engineering, as well as in cell biology and cancer research [1–4]. During tissue development and wound healing, living cells respond to mechanical stimuli in their native environments with biological changes such as shape alternation of membranes and nuclei, cell-spreading, actin and microtubule reorganization or cross-linking under cell membrane, or cell bursting/motility [3–6]. These changes in turn may alter functional synergy as well as the mechanical behavior of cells. It is also known that tumor cells exhibit different elastic compliance compared to normal cells. The investigation of single-cell mechanics is essential for the characterization and control of the mechanical properties and functions of reconstituted tissues, an important task for the practical application of tissue engineering. Motivated by the interesting molecular mechanism of cell response to mechanical forces, as well as by the demand in applications, researchers have developed many elegant technologies, and methodologies have been developed during the past two decades in the metrology of cell mechanical properties and the investigation of the underlying biological and structural changes. The most recent and relevant techniques under this subject include (a) atomic force microscopy (AFM)-based imaging and force measurements and (b) micro device-based technology such as micropipette aspiration, microforce sensors, cell poker, and so forth [2]. Developed in 1986 *for high-resolution imaging*, AFM has also been used as a device for force measurements in the range of 10-5-10-11 N.20-23 In cell mechanics, AFM enables local imaging of cell membrane structures and forces at the nanometer scale [2].

References

1. Wang, Y.-L., Discher, DE: Cell Mechanics: 83 (Methods in Cell Biology) (July 5, 2007) [Kindle Edition]
2. Lulevich, V., Zink, T., Chen, H.-J., Liu, F.-T., Liu, G.-y.: Cell mechanics using atomic force microscopy-based single-cell. Langmuir **22**, 8151–8155 (2006)
3. Engler, A.J., Sen, S., Sweeney, H.L., Discher, D.E.: Matrix elasticity directs stem cell lineage specification. Cell **126**, 677–689 (2006)
4. Discher, D.E., Mooney, D.J., Zandstra, P.W.: Growth factors, matrices, and forces combine and control stem cells. Science **324**, 1673–1677 (2009)
5. Burdick, J.A., Vunjak-Novakovic, G.: Engineered microenvironments for controlled stem cell differentiation. Tissue Eng. Part A **15**, 205–219 (2009). PubMed Abstract
6. Vunjak-Novakovic, G., Scadden, D.T.: Biomimetic platforms for human stem cell research. Cell Stem Cell **8**, 252–261 (2011)
7. Yim, E.K., Sheetz, M.P.: Force-dependent cell signaling in stem cell differentiation. Stem Cell Res. Ther. **3**, 41 (2012). PubMed
8. Manivannan, S., Nelson, C.M.: Dynamics of branched tissue assembly. Stem Cell Res. Ther. **3**, 42 (2012)
9. Conway, A., Schaffer, D.V.: Biophysical regulation of stem cell behavior within the niche. Stem Cell Res. Ther. **3**, 50 (2012). PubMed Abstract
10. Fridley, K.M., Kinney, M.A., McDevitt, T.C.: Hydrodynamic modulation of pluripotent stem cells. Stem Cell Res. Ther. **3**, 45 (2012). PubMed Abstract | BioMed Central Full Text | PubMed Central Full Text OpenURL
11. Wan, L.Q., Ronaldson, K., Guirguis, M., Vunjak-Novakovic, G.: Micropatterning of cells reveals chiral morphogenesis. Stem Cell Res. Ther. **4**, 24 (2013)
12. Serbo, J.V., Gerecht, S.: Vascular tissue engineering: biodegradable scaffold platforms to promote angiogenesis. Stem Cell Res. Ther. **4**, 8 (2013)
13. Hung, B.P., Hutton, D.L., Grayson, W.L.: Mechanical control of tissue-engineered bone. Stem Cell Res. Ther. **4**, 10 (2013)
14. Thavandiran, N., Nunes, S.S., Xiao, Y., Radisic, M.: Topological and electrical control of cardiac differentiation and assembly. Stem Cell Res. Ther. **4**, 14 (2013)
15. de Peppo, G.M., Marolt, D.: Modulating the biochemical and biophysical culture environment to enhance osteogenic differentiation and maturation of human pluripotent stem cell-derived mesenchymal progenitors. Stem Cell Res. Ther. **4**, 106 (2013)
16. Chen, J.C., Jacobs, C.R.: Mechanically induced osteogenic lineage commitment of stem cells. Stem Cell Res. Ther. **4**, 107 (2013)
17. O'Conor, C.J., Case, N., Guilak, F.: Mechanical regulation of chondrogenesis. Stem Cell Res. Ther. **4**, 61 (2013)
18. Schiller, Z.A., Schiele, N.R., Sims, J.K., Lee, K., Kuo, C.K.: Adipogenesis of adipose-derived stem cells may be regulated *via* the cytoskeleton at physiological oxygen levels in vitro. Stem Cell Res. Ther. **4**, 79 (2013)
19. Zimmermann, W.H.: Biomechanical regulation of in vitro cardiogenesis for tissue-engineered heart repair. Stem Cell Res. Ther. **4**, 137 (2013). PubMed
20. Vunjak-Novakovic, G.: Physical influences on stem cells. Stem Cell Res. Ther. **4**(6), 153 (2013)

Chapter 16
Bioinstrumentation: Basic Information

Science and religion are not at odds. Science is simply too young to understand.

Dan Brown (1964–)

The discovery of reliable, sensitive, friendly and real time methods to detect the effect on humans and to the environment of chemical/physical exposure are of paramount importance. The great revolution in the clinical testing which is still time-consuming today will be replaced in the near future by Lab-on-a chip devices in which small volumes of blood or fluid are simultaneously subjected to multiple measurements. Microarray analysis chips will also "come into hospitals" and these new techniques will provide rapid information on the genes and proteins present in patient samples. Information from these microarrays may someday predict a patient's susceptibility to disease before the symptoms manifest, or an individual's response to drugs before they are taken. These instruments will increase "personalized" medical care by allowing physicians to ask and answer a spectrum of questions.

These instruments will increase "personalized" medical care by allowing physicians to ask and answer questions such as:

Will this drug be more effective for the person?

Will this treatment cause more dangerous side effects?

What is the optimal dose for this individual?

They will obviously include a spectrum of measurement systems in order to precisely identify individual disease profiles and achieve appropriate distinction between patients. They will also be made of material which is biocompatible, and in some cases even biodegradable and bioresorbable.

M. Pavlovic, *Bioengineering*, DOI 10.1007/978-3-319-10798-1_16

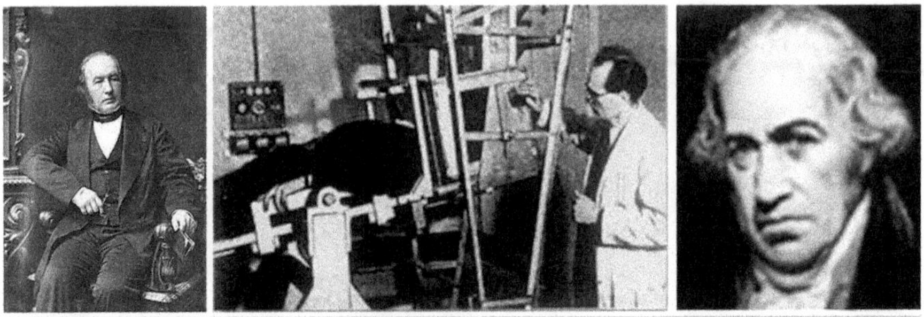

Claude Bernard (1813–1898) - inventor of heart catheter, Karl Theodor Dusik (1908–1968) - ultrasound imaging, Daniel Gabriel Fahrenheit (1686–1736) - mercury thermometer

Overview of the measurement systems: what are sensors, processors, amplifiers, receivers, etc.

The fundamental principles of each instrument are: **INPUT** and **OUTPUT**. The Part of Instrument that detects the input is called **the sensor** [1–3]. The sensor **converts** the input parameters into the **signal**, usually *electrical voltage* that allows for the reading of the signal. Many types of sensors use different mechanisms to convert an input variable into a measurable quantity. The input signal can be after detection *modified* by a **processor**, and that processing may include:

- *The amplification of the signal*
- *Filtering to remove unwanted information*
- *Comparison to signal to from previous measurement to control the signal* [2]

Receiver is the device that represents an *interpretable message* to humans, and signals can be (after processing) displayed, stored or communicated *via* that device [1].

Therefore, we can already understand that the use of computers in bioengineering in general and in this field especially, is of essential significance (Fig. 16.1).

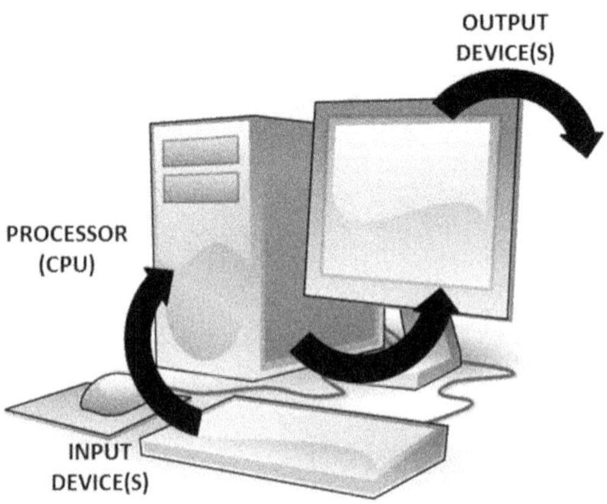

Fig. 16.1 Computer illustrates what is written and many other things

Table 16.1 Types of sensors and instruments used in clinical practice and research

Dependent on energy used	Used in clinical practice	Used in research lab	Lab-on-chip devices
Thermal	Thermometer	Thermocouple	Described at the end of the chapter
Mechanical	Microelectromechanical systems (MEMS)	Microelectromechanical systems (MEMS)	
Electrical	Silicon-nanowire field-effect devices for cancer detection	Electrodes-pH and ion sensitive, voltage sensitive, patch clamp	
Piezo-electrical	Piezoelectric force sensors	Piezoelectric force sensors	
Chemical	Glucose-sensitive glucometer	Clark-Oxygen electrode, Chance-Williams electrode, oxygen electrode	
Optical	Optical (photodiode, photomultiplier tubes)	Optical (photodiode, photomultiplier tubes)	
These are only essential examples given also as the pictorial samples, below. You can find it everywhere if you wish more knowledge and explanation	Spectrophotometer	Multiple plate reader	
		Fluorescent microscope	
		Flow-cytometer	
	Light Microscope	Transmission Electron Microscope (TEM)	
		Freeze–fracture electron microscope (SEM)	
		Atomic Force Microscope (AFM)	

Sensors have become integrated into our daily lives, for either chemical, biological, mechanical, or optical applications [1, 2]. Table 16.1 presents some types of sensors and instruments with certain sensors used in clinical practice and research as well. Of these different types of sensors, *chemical sensors have the widest application* in a multitude of areas. In particular, biosensors and electrochemical sensors have become increasingly important as biological and biochemical applications continue to emerge (Fig. 16.2) [4, 5].

- Electrical (electrodes-pH and ion sensitive, voltage sensitive, patch clamp)

One of the most prominent biological technologies that has found its application very quickly in vital medical situations is a Patch clump technique invented by Bert Sackman who has got a Nobel Prize for that in 1991. Everything is based on a microelectrode. The electrode is sealed to the patch of membrane, and the cell remains intact [6]. This allows for the recording of currents through single ion channels in that patch of membrane, without disrupting the interior of the cell. Our nerve cells convey the electric signal through ion channels of different kinds, and it is possible to be measured by this device. The microelectrodes will later on be used even on alive humans and gather a lot of information necessary for understanding of the work of nervous system.

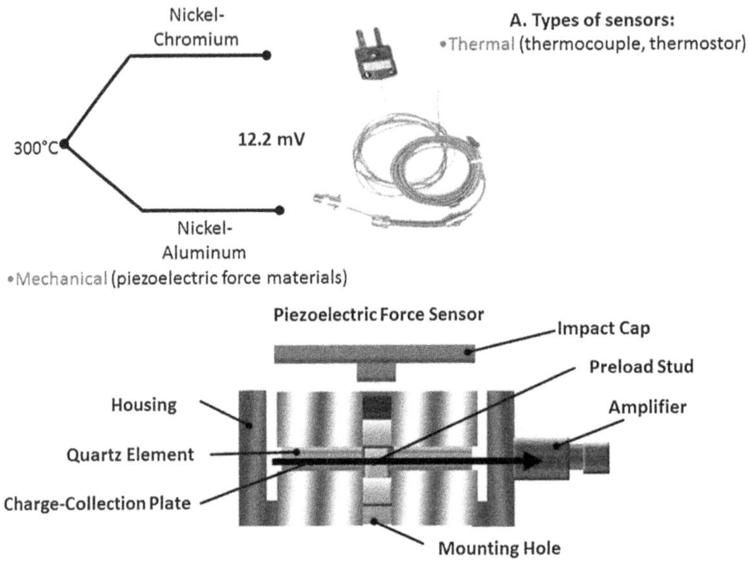

Fig. 16.2 Thermal, mechanical and piezoelectric sensors

Fig. 16.3 Patch–clamp technique with electrical type of sensor

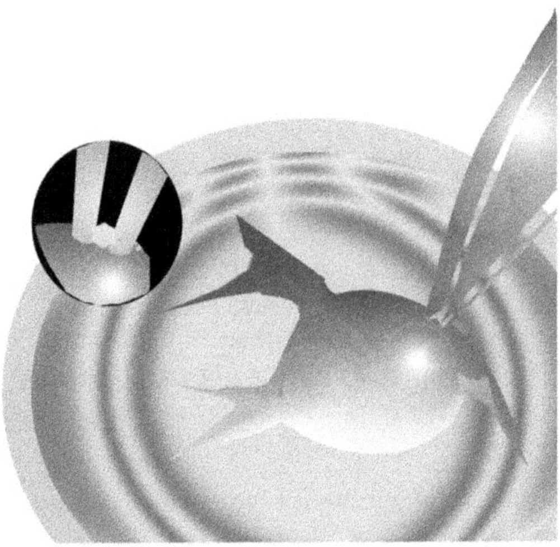

For ligand-gated ion channels or channels that are modulated by metabotropic receptors, the neurotransmitter or drug being studied is usually included in the pipette solution, where it can contact what had been the external surface of the membrane. While the resulting channel activity can be attributed to the drug being used, it is usually not possible to then change the drug concentration. The technique is thus limited to one point in a dose response curve per patch [6]. This technology is with different modifications elevated to much higher level than it was 20 years ago and used in animal models for better characterization of different neurological phenomenons including diseases and defects (Figs. 16.3, 16.4, 16.5, 16.6, and 16.7).

Fig. 16.4 Chemical (Clark-Oxygen electrode, Chance-Williams electrode)

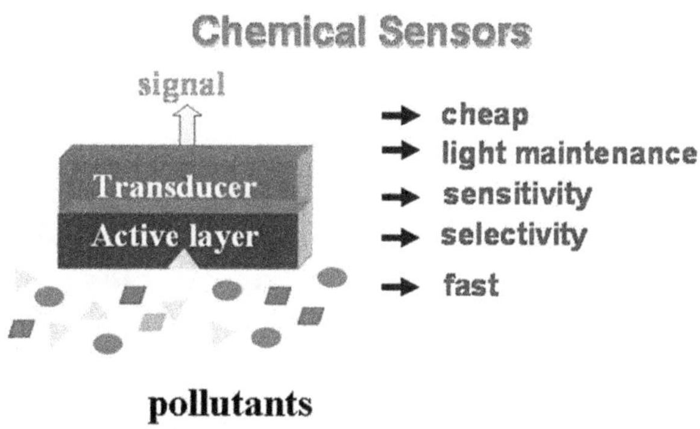

Fig. 16.5 Chemical sensors

Instruments in Medical Practice

Mostly picturesque presentations without descriptions will be given. The principles of work and their engineering architecture can be found in more specific books and articles [7].

- **Thermometer** (body temperature) (Fig. 16.8)

The volume of the **liquid in an expansion** thermometer depends on the temperature of the liquid. If 2 thermometers start out at the same temperature and then 1 is put in a hot environment and 1 in a cold, their volumes will change. A scale on the side of a thermometer shows how volume is related to temperature [2].

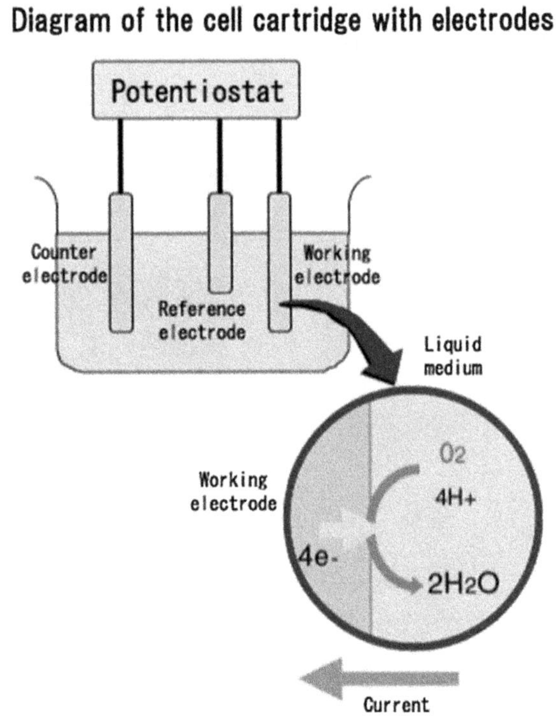

Fig. 16.6 Oxygen electrode

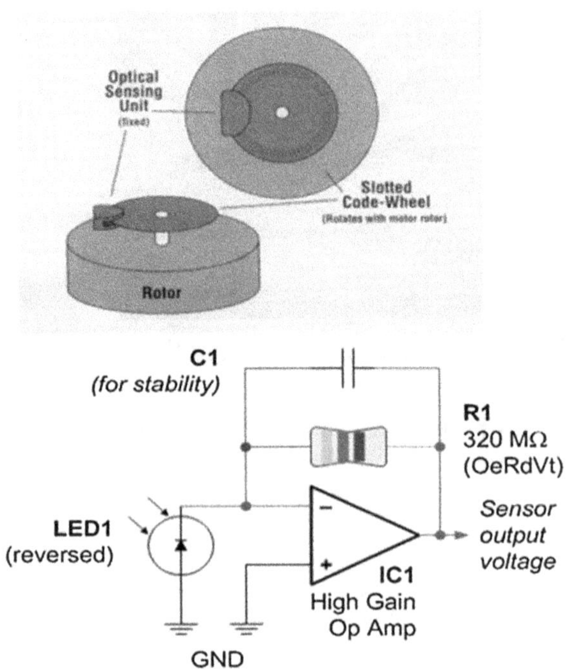

Fig. 16.7 Optical (photodiode, photomultiplier tubes)

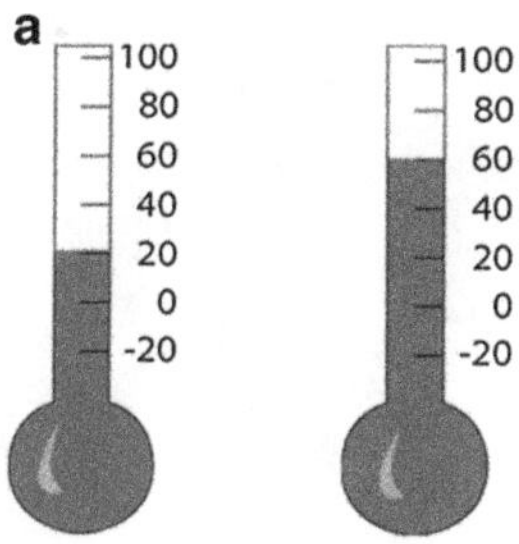

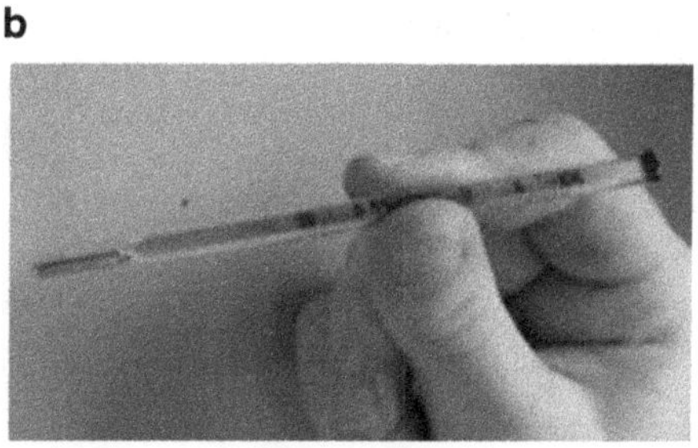

Fig. 16.8 (**a**, **b**). Different types of thermometers with different scales (filled with alcohol and with mercury)

Fig. 16.9 Double-lumen catheter used for hemodialysis

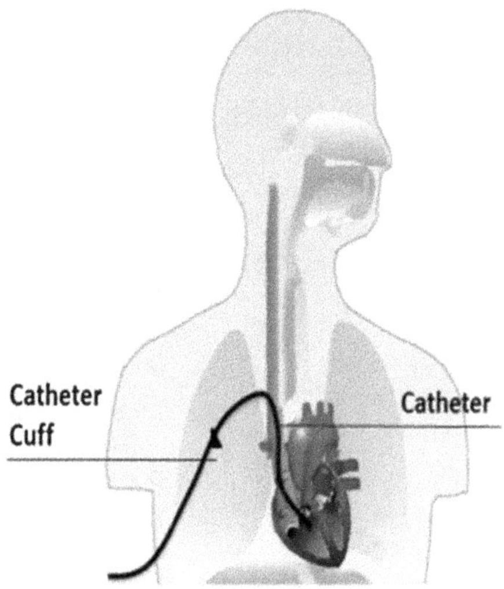

- **Catheter** (angioplasty, catheterization) (Figs. 16.9 and 16.10)
- **Oximeter** (Fig. 16.11)
- **Glucometer** (Fig. 16.12)
- **ECG** (Figs. 16.13 and 16.14)
- **Electrical stimulation devices: defibrillator** (Fig. 16.15)

Instruments in the Research Laboratory

- pH-meter (Fig. 16.16)
- **Spectrophotometers** (based on spectrophotometer-measurement of visible and invisible light: UV and infrared) (Fig. 16.17)

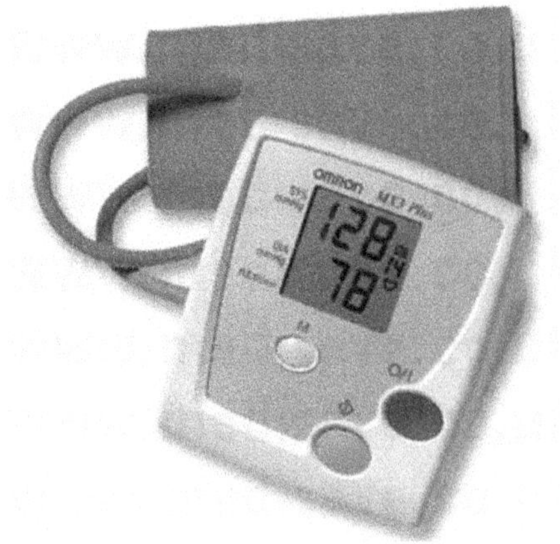

Fig. 16.10 Blood pressure apparatus

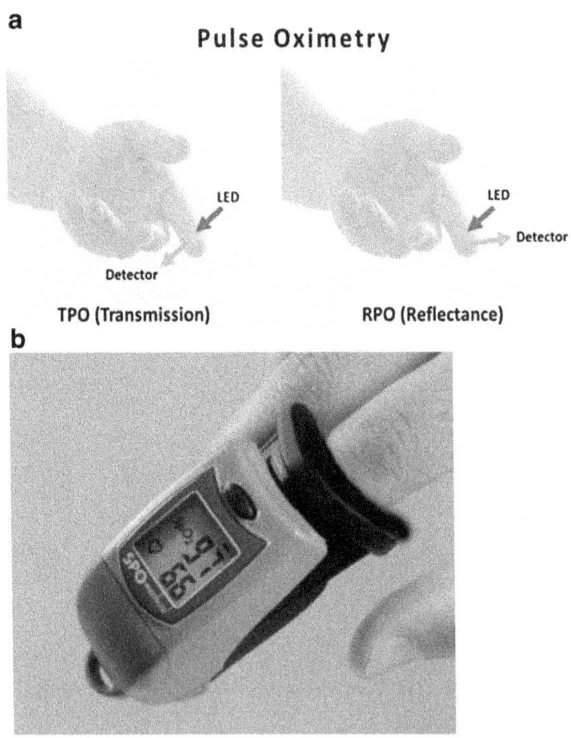

Fig. 16.11 (**a**, **b**): Principle (**a**) and apparatus (**b**) for pulse oximetry

Biosensors

Biosensors can be constructed by adding a biological sensing system-usually a molecule or a cell—that responds in a predictable and measurable way to the parameter of interest (Fig. 16.18).

Fig. 16.12 Glucometer (one of many versions)

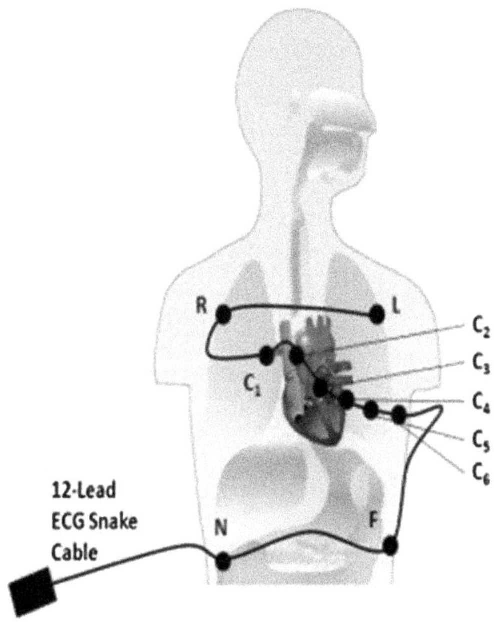

Fig. 16.13 The 12 lead ECG

Lab-on-Chip Devices

Agilent 2100 Bioanalyzer for molecular weight and antibody purity and concentration determination*: *Principles of Protein Analysis on a Chip

The Agilent electrophoretic assays are based on traditional gel electrophoresis principles that have been transferred to a chip format. The chip format dramatically reduces separation time and sample consumption. The system provides automated sizing and quantitation information in a digital format. On-chip gel electrophoresis is performed for the analysis of DNA, RNA, and proteins (Fig. 16.19) [9].

Charged biomolecules like DNA or RNA are electrophoretically driven by a voltage gradient—similar to slab gel electrophoresis. Because of a constant mass-to-charge ratio and the presence of a sieving polymer matrix, the molecules are separated by size.

Fig. 16.14 The 12 lead ECG

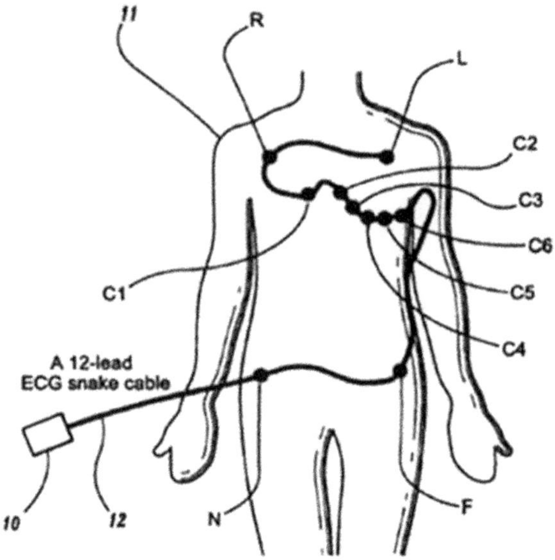

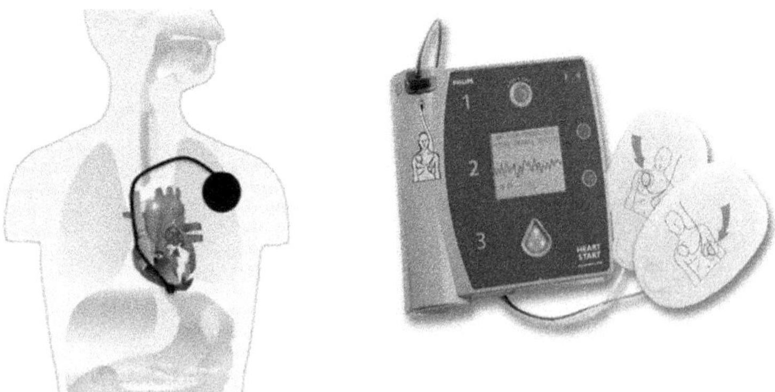

Fig. 16.15 Defibrillator (one of versions)

Fig. 16.16 Modern pH meter

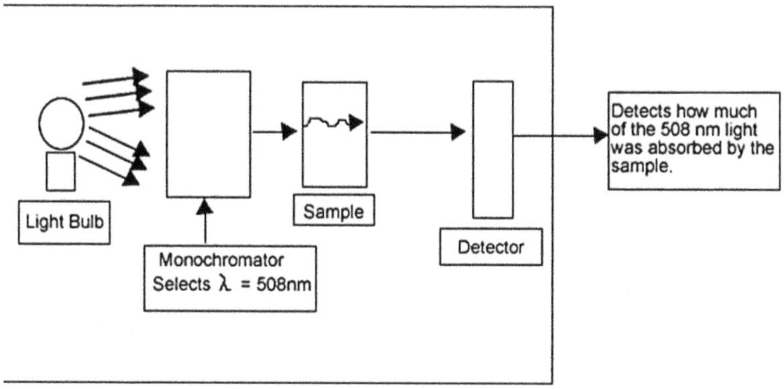

Fig. 16.17 Principle of work and illustration of classical spectrophotometer

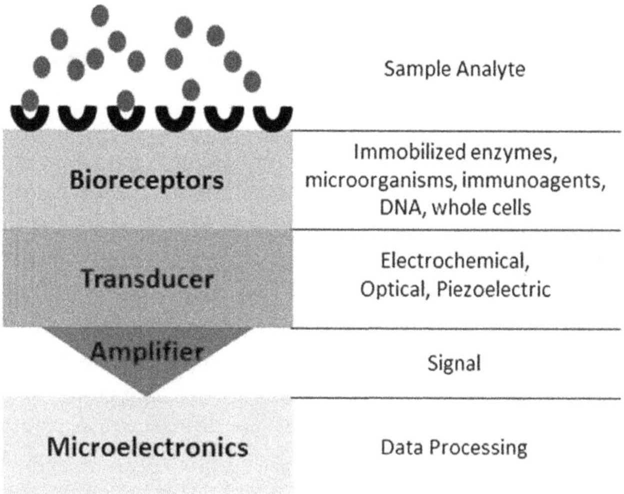

Fig. 16.18 Flow cytometric biosensor

Fig. 16.19 Lab-on-chip

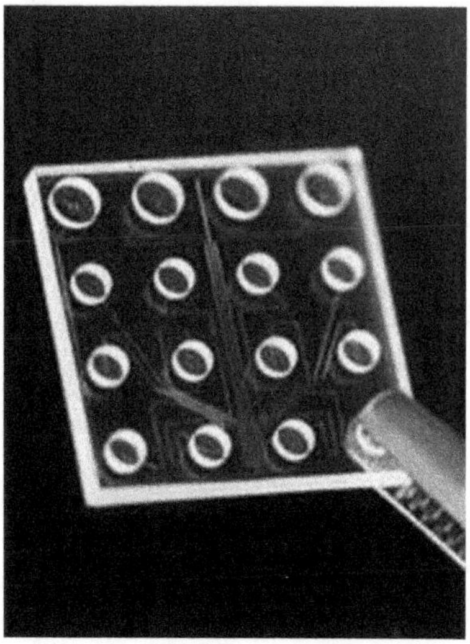

Smaller fragments are migrating faster than larger ones. Dye molecules intercalate into DNA or RNA strands or Protein-SDS micelles [9]. These complexes are detected by laser-induced fluorescence. Data is translated into **gel-like images (bands)** and **electropherograms (peaks)**. With the help of a ladder that contains fragments of known sizes and concentrations, a standard curve of migration time versus fragments size is plotted. From the migration times measured for each fragment in the sample, the size is calculated. Two marker fragments (for RNA only one marker fragment) are run with each of the samples bracketing the overall sizing range. The "lower" and "upper" markers are internal standards used to align the ladder data with data from the sample wells. This is necessary to compensate for drift effects that may occur during the course of a chip run. For DNA and protein assays, quantitation is done with the help of the upper marker. The area under the upper marker peak is compared with the sample peak areas. Because the concentration of the upper marker is known, the concentration for each sample can be calculated [9]. Besides this relative quantitation, an absolute quantitation is available for protein assays, using external standard proteins.

Emphasizing Bioengineering Aspect to Biosensors

The biologically sensitive elements can also be created by biological engineering, the transducer or the detector element (works in a physicochemical way; optical, piezoelectric, electrochemical, etc.) that transforms the signal resulting from the interaction of the analyte with the biological element into another signal (i.e., transduces) that can be more easily measured and quantified; biosensor reader device with the associated electronics or signal processors that are primarily responsible for the display of the results in a user-friendly way [8]. This sometimes accounts for the most expensive part of the sensor device; however it is possible to generate a user friendly display that includes transducer and sensitive element (Fig. 16.20).

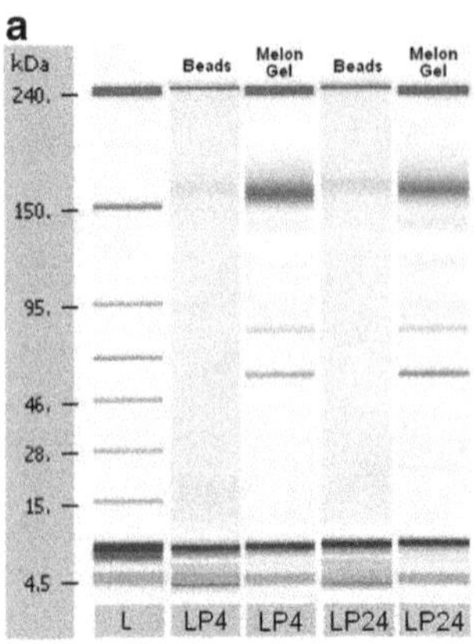

b

Agilent 2100 banding patterns of anti-ssDNA from lupus patient sera after applying two isolation methods. The superiority of magnetic beads over melon gel separation is observed through the absence of bands in the beads sample (ref.5, 8, 9). With permission of Dr Mirjana Pavlovic and Ms Ana Kats, MS/FAU, Boca Raton

Fig. 16.20 (**a, b**): Isolation of ssDNA lupus abzymes through two different methods (Pavlovic et al., 2012, Lab on chip) [9]

References

1. Ramsay, G.: Commercial Biosensors: Applications to Clinical, Bioprocess, and Environmental Samples. Wiley, New York (1998)
2. Harsányi, G.: Sensors in Biomedical Applications. Technomic, Lancaster, Basel (2000)
3. Pavlovic, M., Cavallo, F.M., Kats, A.M., Kotlarchyk, A., Zhuang, H., Shoenfeld, Y.: From Pauling's abzyme concept to the new era of hydrolytic anti-DNA autoantibodies: a link to rational vaccine design? – A review. Int. J. Bioinform. Res. Appl. **7**(3), 220–238 (2011)
4. Pavlovic, M., Kats, A., Cavallo, M., Shoenfeld, Y.: Clinical and molecular evidence for association of SLE with parvovirus B19. Lupus **19**, 783–792 (2010)
5. Cavallo, M.F., Kats, A.M., Chen, R., Hartmann, J.H., Pavlovic, M.: A novel method for real-time, continuous, fluorescence-based analysis of anti-DNA abzyme activity in systemic lupus. Autoimmune Dis (2012). Article ID 814048, 10 pages
6. Hamill, O.P., Marty, A., Neher, E., Sakmann, B., Sigworth, F.J.: Improved patch-clamp techniques for high-resolution current recording from cells and cell-free membrane patches. Pflugers Arch. **391**(2), 85–100 (1981). doi:10.1007/BF00656997. PMID 6270629
7. Hamburger, V.: The history of discovery of nerve growth factor. J. Neurobiol. **24**(7), 893–897 (1993)
8. Gerstel, I., Kats, A.M., Pavlovic, M.: A review of the applications of microfluidics in tissue engineering research. MD-Medical Data **4**(3), 273–279 (2012)
9. Kats, A.: Purification and analysis of autoimmune antibody reactive with single stranded DNA. MS thesis, Florida Atlantic University, Boca Raton (2008)

Chapter 17
Fundamentals of Bioimaging

Science is what you know. Philosophy is what you don't know.

Bertrand Russell (1872–1970)

Biologists and physicians have come to depend more and more on bioimaging as a tool to identify and locate specific proteins and molecules in their natural environment. With the recent development of probes and microscopes, biological processes can be monitored in real time in two or three dimensions. And observations can even be made non-invasively so a cellular process can be followed over an extended period of time.

The idea of bioimaging was born very fast after the first photograph has been taken. It advanced tremendously when scientists have realized that they can use different wavelengths of visible and invisible spectrum in order to get different images which will always improve on some particular and necessary parameter (X-rays, Magnetic resonance, fluorescence, ultrasound, etc.).

Modern Bioimaging mostly relies on *probes* that are attached to proteins or other molecules of interest [1]. These probes are usually fluorescent and as such emit light of a specific wavelength when they are excited by light of another—usually shorter—wavelength (laser for example).

M. Pavlovic, *Bioengineering*, DOI 10.1007/978-3-319-10798-1_17

Wilhelm Röntgen (1845-1923), Martin Chalfie (1947-), Max Knoll (1897-1969) and Ernst Ruska (1906-1988): the evolution of imaging from x-rays toward fluorescent and electron microscope

Cells and their components are mainly transparent and the myriad of molecules within them are indistinguishable from each other in a normal microscope [1]. But if the protein of interest lights up in bright green, it is easy to detect and distinguish from all other molecules in the cell. Many different imaging modalities—or types of imaging methods—fill different scientific and/or clinical niches. Every modality has limitations: a particular method can be low in quality, slow to acquire images, expensive, or not suitable for all patients. Bioimaging had its breakthrough in 1994 when **Martin Chalfie** from Colombia University in the US expressed a protein from a jellyfish in *E. coli* and the round worm *C. elegance* [1]. The protein was GFP (Green Fluorescent Protein) which is fluorescent and emits green light when excited by blue light. The result was overwhelming with bright green bacteria and worms, but the real power of bioimaging was demonstrated when scientists realized that GFP by means of genetic engineering can be fused to natural proteins in animals, plants or microorganisms. Now, not the whole organism but rather a particular protein will light up, so its distribution within the cell is visualized. Moreover, GFP was fused to promoters so it could tell where and when expression was directed.

The art of bioimaging is to develop probes that can be specifically attached to certain molecules and that emit light in distinct colors so that several probes can be used in the same sample in order to monitor more than one biological process at a time. GFP has been modified to emit cyan, blue and yellow light and other proteins add red to the palette. A classical method to label proteins or other molecules is to raise antibodies against them and couple these to a fluorescent probe. We have seen that already on one of the author's images from fluorescent microscope, where the macrophage's enzyme COX-2 is envisioned with fluorescently labeled antibody produced on this enzyme (**Fig. 5.3a**). Due to the size of the antibodies this method will, however, often render the molecule of interest biologically inactive and is accordingly not normally compatible with life processes. Also a large number of rather simple organic molecules have been developed as fluorescent probes [1–5]. These can be injected directly into cells or tissues and depending on their nature they will bind to certain proteins or molecules. One of these is DAPI which specifically binds DNA and emits blue light. Virtually, all imaging modalities are now **digital**: the images are acquired by a computer and are made up of individual picture elements, or **pixels**.

FC and FACS

Understanding of the fluorescence and fluorescent dyes has supported further development of knowledge on stem cells and cells in general since it was possible to label monoclonal antibodies for specific protein markers with characteristic dye and then detect it in the cell or on it by using fluorescent microscope or Flow cytometry. The classification of the cells in categories and their separation was facilitated by introduction of cell sorter (fluorescence acquisition cell sorter (FACS) [2, 5].

Today, beside fluorescence based, a spectrum of bioimage approaches based on different principles are used, in order to help clinical; and research studies. The most prominent are:

- X-rays and CT (conventional X-ray and CT)
- Ultrasound imaging
- Doppler imaging
- Nuclear medicine imaging methods (MRI structural and functional)
- PET
- SPECT
- Operation of a gamma camera
- Optical bioimaging (light microscope, electron microscope, atomic force microscope)

The discipline of bioimaging is taking on new dimensions as scientists develop new sensors to explore biological structure and function, and visualize/analyze this information in three and four dimensions [3]. Bioimaging research is fast becoming integrative in nature, both in terms of the type of sensor (e.g., NMR, X-ray, visible light for everything from microscopy to optical coherence tomography, ultrasound, etc.), scale (molecular to cellular to organ), and range of applications, from molecular crystallography to imaging the neuronal correlates of the mind [3]. The basic concepts of particularly significant images will be presented visually and roughly described, where necessary. One can always look for particular type of images in special books dedicated to that problem. In this chapter we just what to give a general idea on bioimaging.

Bioimaging on the Basis of Fluorescence

Certain molecules called fluorophores are fluorescent. They are used to label the molecule and measure their concentrations in different scenarios using appropriate readers and their scales (Fig. 17.1).

A new type of fluorescent probes is the quantum **dots** or **q-dots** [1–3]. They are **inorganic semiconductor nanocrystals**, and their fluorescent characteristics depend largely on their size. The general rule is that with increasing size of the crystal, the wavelength of the emitted light becomes longer, *i.e.* more reddish. By synthesizing them **in different colors it is possible to design a wide range of**

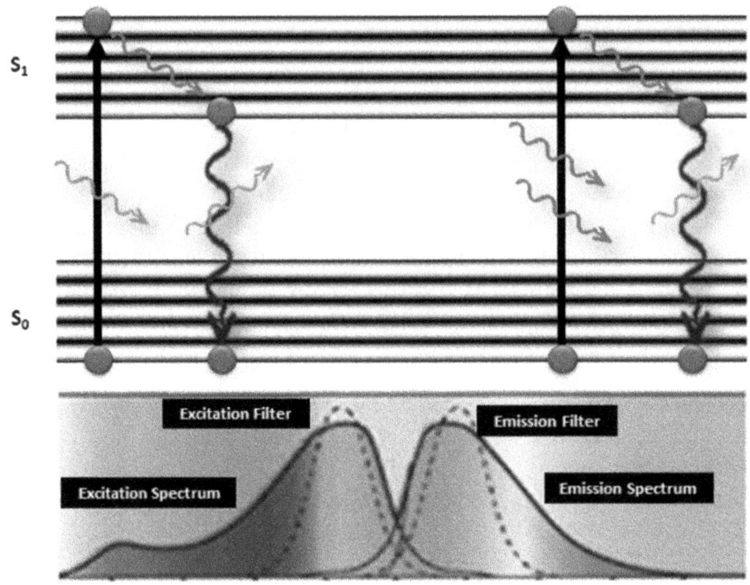

Fig. 17.1 Bioimaging on the basis of fluorescence

probes in different colors. Furthermore, it is relatively easy to design peptides where one end that binds the dots and the other has some kind of functionality, e.g., binding specifically to a molecule of interest. In this way it is possible to simultaneously study several processes within a cell or tissue (Fig. 17.2).

Irrespective of the probe used, it must be viewed through a fluorescence microscope in order to bee visualized. This is basically an ordinary microscope but it has a built-in *light source that produces light at a particular wavelength that will excite the probe being used*. The classical wide-field fluorescence microscope broadly illuminates the sample and emitted light from the whole field of sight is reflected to the microscopes eyepiece, where it can be detected by the naked eye or a digital camera. The main downside of this straight forward method is that the excitation light is difficult to focus so emission will occur from out-of-focus points, leading to somewhat blurred images with limited resolution. *The confocal laser scanning microscope uses a laser beam for excitation of a narrow point in the sample.*

Magnetic Resonance Imaging

Magnetic resonance imaging (**MRI**) is a procedure used mainly in hospitals to scan patients and determine the severity of certain injuries. It can also be used in research and include details such small as the molecule in the cell, or the whole animal.

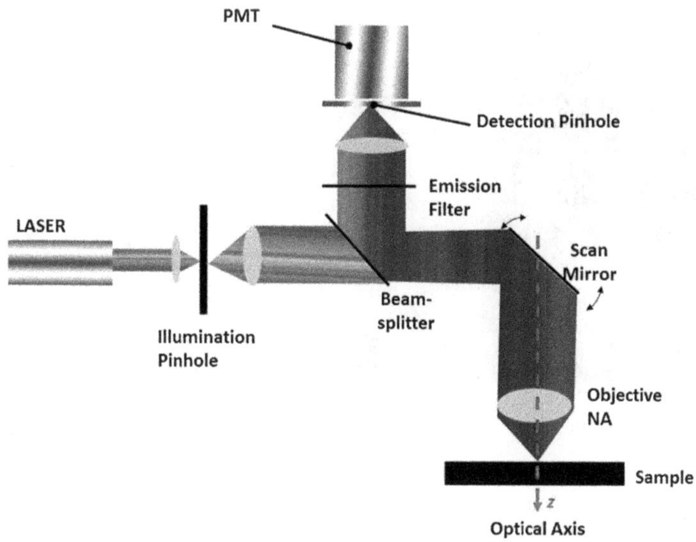

Fig. 17.2 Laser in its schematic presentation

It can also be structural and functional dependent on the purpose of investigation/examination. We shall give here only basic principles of the work of MRI in general.

An MRI machine uses a magnetic field and radio waves to create detailed images of the body [5]. Most MRI machines look like a long tube, with a large magnet present in the circular area. When beginning the process of taking an MRI, the patient is laid down on a table. Then depending on where the MRI needs to be taken, the technician slides a coil to the specific area being imaged. The coil is the part of the machine that receives the MR signal [5].

A strong magnetic field is created by passing an electric current through the wire loops. While this is happening, other coils in the magnet send and receive radio waves [5]. This triggers protons in the body to align them. Once aligned, radio waves are absorbed by the protons, which stimulate spinning [5]. Energy is released after "exciting" the molecules, which in turn emits energy signals that are picked up by the coil. This information is then sent to a computer which processes all the signals and generates it into an image. The final product is a 3-D image representation of the area being examined [1, 5]. Unlike CT scanning or general X-ray studies, no ionizing radiation is involved with an MRI.

Contrast: **gadolinium, T1 and T2 weighted are given in order to get better picture** (Fig. 17.3).

In MRI there are 3 kinds of magnetic fields:

1. B0—the main magnetic field
2. B1—an RF field that excites the spins
3. Gx, Gy, Gz—the gradient fields that provide localization

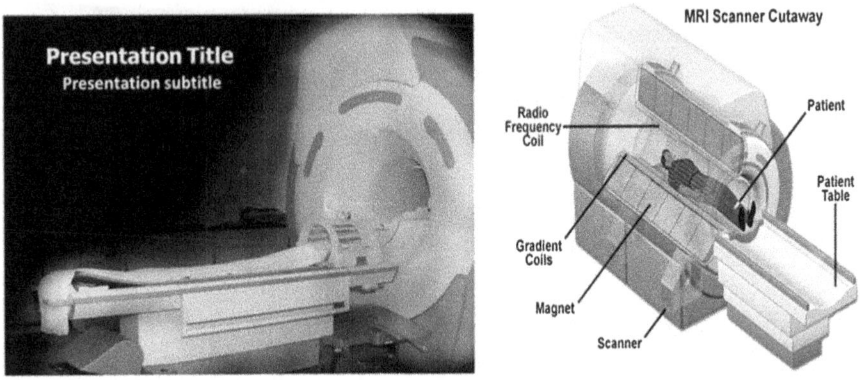

Fig. 17.3 MRI principles

What do we measure?
Two important time constants:

- **T1-time to recover longitudinal magnetization-realigning to B_0**
  Ranges from 200 ms to 2 s.
- **T2-time to diphase-lose horizontal**
  Ranges from 30 to 500 ms

Differences in T1 and T2 provide the basis in signal intensities and tissue contrast

Noll (2006) MRI Notes 1: page 31

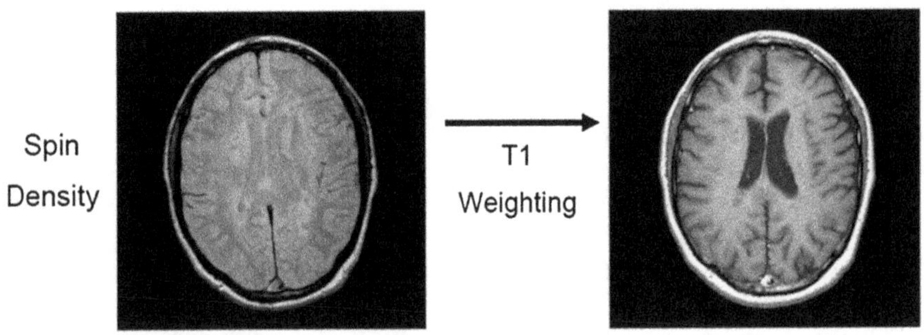

Finally, the signal intensity, for a particular tissue this thus a function of tissue parameters ρ, T1, and T2, and imaging parameters TE and TR:

$$\text{signal intensity} \propto \rho(1 - e^{-TR/T1})e^{-TE/T2}$$

Typical T1's, T2's and ρ's for Brain Tissues at 1.5 T

	T1	T2	Rel. density
Distilled water	3 s	3 s	1.0
Cerebro spinal fluid	3 s	300 ms	1.0
Gray matter	1.2 s	60–80 ms	0.98
White matter	800 ms	45 ms	0.80
Fat	150 ms	35 ms	1.0

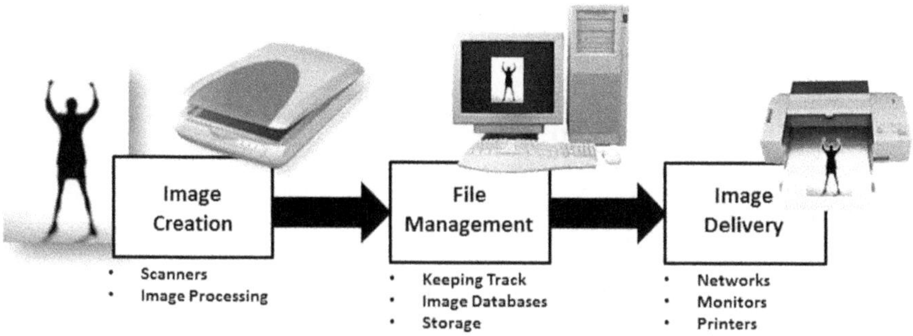

Fig. 17.4 Modern Leonardo and potential of today's imaging systems

Image Processing and Analysis

In order to get a good quality of any image, the image has to get through the processes of digitization, segmentation, registration and enhancement.

Digitization

Digitization is the process of converting information into a digital format. In this format, information is organized into discrete units of data (called **bits**) that can be separately addressed (usually in multiple-bit groups called **bytes**). This is the binary data that *computers* and many *devices with computing capacity* (such as digital camera s and digital hearing aid s) can process (Fig. 17.4).

Text and images can be digitized similarly: a scanner captures an image (which may be an image of text) and converts it to an image files, such as a **bitmap**. An optical character recognition (OCR) program analyzes a text image for *light* and *dark* areas in order to identify each alphabetic letter or numeric digit, and converts each character into an ASCII code. Audio and video digitization uses one of many analog-to-digital conversion processes in which a continuously variable (analog) signal is changed, without altering its essential content, into a multi-level (digital) signal. The **process of sampling** measures the amplitude (signal strength) of an analog waveform at evenly spaced time markers and represents the samples as numerical values for input as digital data. Digitizing information **makes it easier to preserve, access, and share**. For example, an original historical document may only be accessible to people who visit its physical location, but if the document content is digitized, it can be made available to people worldwide. There is a growing trend towards digitization of historically and culturally significant data.

According to an article in *The Guardian* in March 2007, if all spoken languages since the dawn of time were digitized, it would consume **five exabytes of storage space**. Total digital information, in 2006 was estimated at **161 billion exabytes**. Email alone made up six exabytes of that figure [1].

Registration and Segmentation

Segmentation

Medical images or volumes acquired using different modalities (like CT, MR, Angiography, Ultrasound, PET, and SPECT) are often not only analyzed visually by a surgeon but also by more or less **automated algorithms** trying to emulate human perception. **The partitioning of the original data into distinct regions is called "segmentation" and has a wide range of applications, but we shall consider only its use for registration and improvement in visualization**:

a) **Segmentation for Registration**: Feature-based registration relies on correspondence of anatomical features which thus have to be extracted beforehand. The registration of data sets by aligning the segmented vascular structure is one example [1].
b) **Segmentation for Visualization**: For gaining a 3-D understanding of CT or MR data, segmentation of the anatomical structure of interest can lead to much better visualizations. For virtual colonoscopy for example, a 3-D CT data set of the patient's abdomen is acquired and the colon's surface is reconstructed using segmentation. This segmentation now allows a virtual flight through the patient's colon by rendering its surface only [1]. The standard procedure of looking for polyps e.g., can now be done without an invasive colonoscopy. The term "augmented reality" denotes the overlay of medical image data onto the real (optical) view. Since it doesn't make sense to overlay the original, detailed data, one might want to add only specific anatomy which leads to the same segmentation task.
c) **Segmentation for Tracking**: Augmenting some data onto a real view requires a pose estimation for the current view. This is usually done by using **markers** or **features** which in either case must be detected using a fully automatic segmentation.
d) **Segmentation for Treatment Planning**: (Semi-) automatic segmentation of tumors can dramatically reduce a surgeons' workload when a complete labeling of some anatomical structures is needed. Instead of manually editing each slice of a 3-D data set, segmentation methods can be applied to segment the structure after a seed point is placed by the surgeon in the data set. This is especially useful for planning an operational intervention or radiation treatment.

Registration

This involves a spectrum of methods to register the image in the best possible way. Image registration can provide **enhanced information from different image modalities**. For that, the **modalities must be aligned such that anatomical features of one modality can be automatically detected in the other modality (or modalities)**. Modalities can differ in viewpoint (from where the images were taken), the sensor (e.g., X-ray, ultrasound, or laser scanner), and the time the images were recorded. Depending on which specific anatomy is imaged, or which device has been used for it respectively, **registration methods differ in their applicability**. Furthermore, if a real patient is to be aligned with preoperative data, one has to evaluate how to acquire data from the real patient that can be registered with the image data. For instance, laser scanners can compute **point clouds in space**, which are later registered with the preoperative image data. **For these entire different applications one has to determine the best registration method**. For instance, registering a region of the heart is hardly performed on rigid models since the beating heart deforms patient or modalities involved. Thus, non-rigid approaches are used in order to compensate for heart beat. Registration methods are split up into **feature-based** and **intensity-based approaches**, the former ones registering on anatomical features (landmarks, markers), and the latter ones only on image intensities (e.g., gray values). The above described segmentation methods are often used to **extract structure from medical images and register only them in feature-based approaches. Intensity-based registration, for instance, needs data comparison**, so filters or other image processing algorithms are applied to the modalities for suitable **sensor comparability**. Optimizing an initial guess of the registration parameters is crucial in almost any registration procedure. Again, the most suitable optimization algorithm (e.g., nonlinear methods like Gauss-Newton, Levenberg-Marquardt, or Best Neighbor iteration) must be chosen for a specific application [1].

Image Enhancement

The aim of image enhancement is **to improve the interpretability or perception of information in images for human viewers, or to provide "better" input for other automated image processing techniques**. Image enhancement techniques can be divided into two broad categories:

1. Spatial domain methods, which operate directly on pixels, and
2. Frequency domain methods, which operate on the Fourier transformation of an image.

There are no general criteria for determining what "good" image enhancement is when it comes to human perception. If it looks good, it is considered good. However, when image enhancement techniques are used as **pre-processing tools** for other image processing techniques, then **quantitative measures can determine which techniques are most appropriate**.

Emphasizing Bioengineering Aspect to Bioimaging

Spanish and Australian researchers have collaborated on a new imaging technique similar to magnetic resonance imaging (MRI), but offering resolution and sensitivity sufficient to scan individual cells. A paper describing their work explains how artificial atoms–diamond nanoparticles doped with nitrogen impurity–can probe very weak magnetic fields such as those generated in some biological molecules. To trap and manipulate these artificial atoms, the researchers use laser light. The laser works like tweezers, leading the atoms above the surface of the object to study and extract information from its tiny magnetic fields. This effort represents the first time nano diamonds have been optically trapped and manipulated in three dimensions [3].

Conventional MRI has a diagnostic resolution on a millimetric scale [6]. The new technique provides nanometer resolution (nearly one million times greater), making it possible to measure magnetic fields created by proteins, for instance. It could revolutionize the field of medical imaging, allowing for substantially higher sensitivity

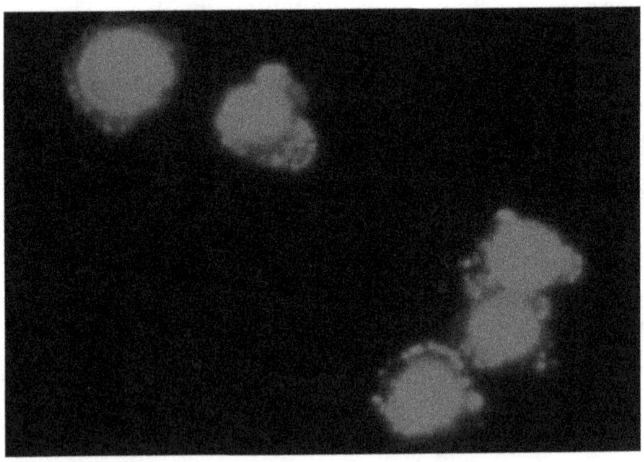

Fig. 17.5 Fluorescent microscope images (*two different magnifications*) of the cells stained with propidium iodide (PI) showing apoptotic blabbing of the membrane and apoptotic sequestration off the nucleus (apoptotic bodies) within mouse glioblastoma cells. From author's self-made collection (unpublished data). Year 1996-Penn University School of Medicine

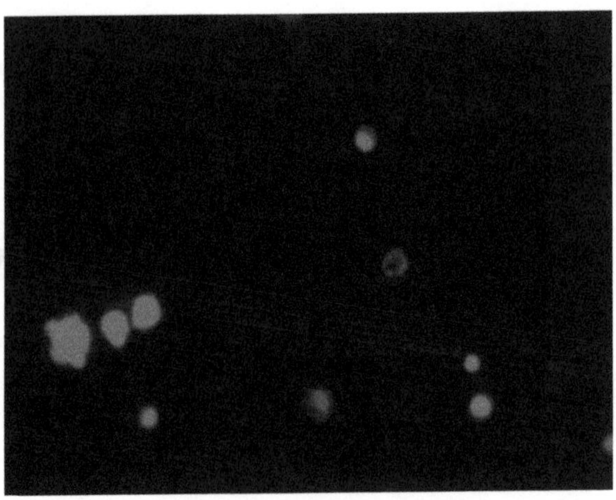

Fig. 17.6 Fluorescent microscope images (*two different magnifications*) of the cells stained with propidium iodide (PI) showing apoptotic blabbing of the membrane and apoptotic sequestration off the nucleus (apoptotic bodies) within mouse glioblastoma cells. From author's self-made collection (unpublished data). Year 1996-Penn University School of Medicine

in clinical analysis, an improved capacity for early detection of diseases, and thus a higher probability for successful treatment. The approach "will offer new sources of information and allow us to better understand the intracellular processes, enabling noninvasive diagnosis," explains Michael Geiselmann, who conducted the experiment. Geiselmann is with the Institute of Photonic Sciences (ICFO)—an associate institute of the Universitat Politècnica de Catalunya Barcelona Tech (UPC). The Spanish National Research Council (CSIC) and Macquarie University (Sydney, Australia) also contributed to the project (Figs. 17.5, 17.6, 17.7, and 17.8) [3].

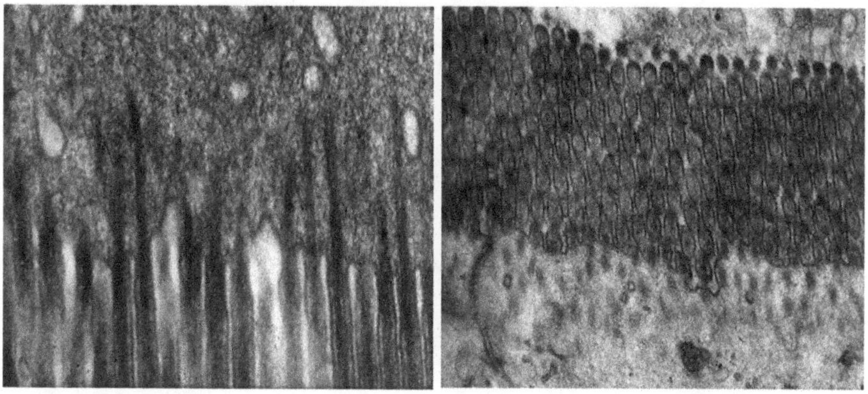

Fig. 17.7 Transmission Electron Microscopy (TEM) image of villous (absorptive) part of rat duodenal mucosa (longitudinal and vertical cut). From: Mirjana Pavlovic: Protein Nutrition and histologic-histochemical properties of rat duodenal mucosa. PhD Thesis, University of Belgrade, Serbia, 1984

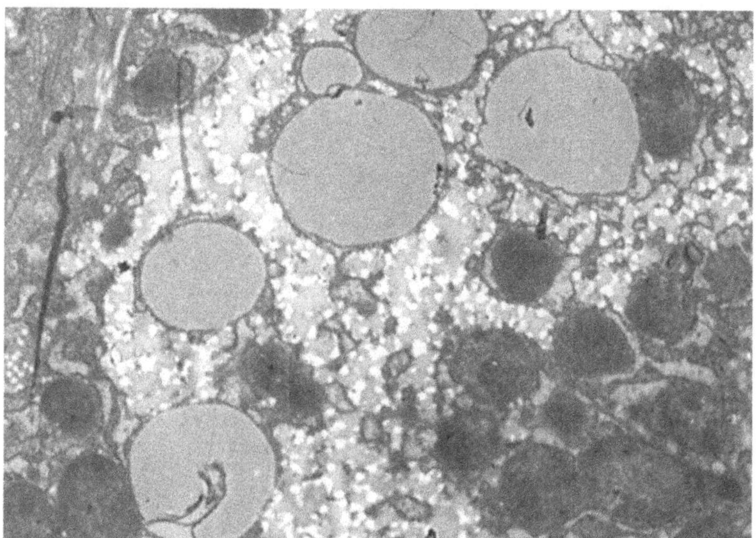

Fig. 17.8 TEM image of the part of mucous cell of rat duodenal mucosa identifying mucous granules: full (*dark*) and empty (*pale*). Same source as above [7]

References

1. Saltzman, M.W.: Biomedical Engineering. Cambridge University Press, New York (2009). hardback. ISBN 978-0-521-840099-6
2. http://bioimaging.yale.edu/index.aspx
3. http://www.biopticsworld.com/articles/print/volume-6/issue-2/departments/biooptics-breakthroughs/nanodiamonds-enable-magnetic-resonance-of-individual-cells-other.htm
4. http://lecb.physics.lsa.umich.edu/wl/carma/2004/fall-sat-morn-phys/20041211-umwlap001-01-dzakpasu/real/sld033.htm
5. http://www.medicalnewstoday.com/articles/146309.php
6. Noll 2006 MRI Notes
7. Pavlovic, M.: Protein Nutrition and histologic-histochemical properties of rat duodenal mucosa. PhD Thesis, University of Belgrade, Serbia (1984).

Chapter 18
What Are Biomaterials?

We think that the biggest breakthroughs in nanotechnology are going to be in the new materials that are developed.

Troy Kirkpatrick (1928–2011)

A biomaterial is any matter, surface, or construct that interacts with biological systems. Biomaterials can be found/derived in nature and they can be also synthesized for different purposes in bioengineering and especially Tissue Engineering in Regenerative Medicine. The application is very wide. Regardless of the origin, they have to be biocompatible, since they will be used in replacing living tissues such as heart valves, hips (replacement), heart electrical impulse generator (pace makers), prostheses, etc. Minimal or absent immune response is to be expected. However, in many cases biomaterial is also requested to be biodegradable or bioresorbable in order to disappear from organism after fulfilling their function. Biomaterials are extensively developed and used in drug delivery systems as capsules or nanoshells, or microbasket for carrying drugs toward the target, and in scaffold biofabrication for supportive growth of particular tissues used in regenerative therapy.

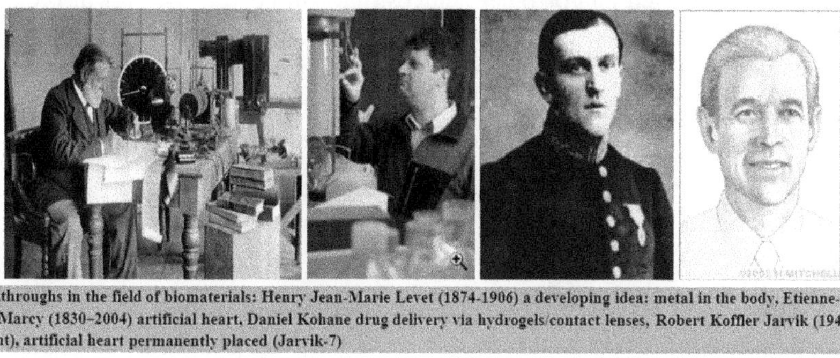

Breakthroughs in the field of biomaterials: Henry Jean-Marie Levet (1874-1906) a developing idea: metal in the body, Etienne-Jules Marcy (1830–2004) artificial heart, Daniel Kohane drug delivery via hydrogels/contact lenses, Robert Koffler Jarvik (1946-present), artificial heart permanently placed (Jarvik-7)

M. Pavlovic, *Bioengineering*, DOI 10.1007/978-3-319-10798-1_18

A **biomaterial** is any matter, surface, or construct that interacts with biological systems. As a science, biomaterials are very old [1]. Mayan culture used natural (shell) material in their primitive dentistry. Later on, it was metal in different parts of the body, including teeth (gold, silver, titanium) [2–6]. The study of biomaterials is called **biomaterials science** [7]. It has experienced steady and strong growth over its history, with many companies investing large amounts of money into the development of new products [8]. Biomaterials science encompasses elements of medicine, biology, chemistry, tissue engineering and materials science [5].

In modern era biomaterials are very popular in drug delivery systems. Dr. Daniel Kohane, has a long history of inventing new drug delivery methods: he has previously invented **hydrogels** that *can be inserted into the abdomen* after surgery in order to prevent tissues from sticking together and causing complications [9, 10]. He has also invented a hydrogel *with anti-fungal properties* that can coat medical devices and protect patients using those devices from fungal infections. The new *contact lenses* are also made from hydrogel, which includes a special polymer that holds the actual drug. The medication is slowly released at a rate controlled by the specific properties of the polymer and the hydrogel. Researchers have formulated a hydrogel and polymer combination capable of releasing drugs for a period of 100 days, well over the 30 day maximum imposed on single use contact lenses by the Food and Drug Administration. Materials scientists can develop substances with specific properties by manipulating the constituent elements and the way in which they are processed. Materials are characterized using various techniques from condensed-matter physics including electron microscopy, X-ray diffraction, neutron diffraction and atomic force microscope (AFM) [11–14]. Due to extremely wide range of biomaterials both natural and manufactured, it is impossible to write about each of them in this chapter. Therefore, the tabular presentation inspired recently by Parida et al., (2012) will replace extensive consideration [8]. For particular references one can look into PubMed and/or other sources that can be very helpful in providing specific details (Tables 18.1, 18.2, and 18.3).

Biomedical application of polymeric biomaterials is presented in Table 18.4.

Table 18.1 Where is biomaterial used?

Problem area	Examples
Replacement of diseased or damaged part	Artificial hip joint, kidney dialysis machine
Assist in healing	Sutures, bone plates, and screws
Improve function	Cardiac pacemaker, intraocular lens
Correct functional abnormality	Cardiac pacemaker
Correct cosmetic problem	Augmentation mammoplasty
Aid to diagnosis	Probes and catheters
Aid to treatment	Catheters, drains

*According to Parida et al., 2012, (IJAAS) 1, 3:31–35

Table 18.2 Biomaterials in organs

Organ	Examples
Hart	Cardiac pacemaker, artificial heart valve, total artificial heart, blood vessels
Lung	Oxygenator machine
Eye	Contact lens, intraocular lens
Ear	Artificial stapes, cochlea implant
Bone	Bone plate, intramedullary rod
Kidney	Catheters, stent, Kidney dialysis machine
Bladder	Catheter and stent

*According to Parida et al., 2012, (IJAAS) 1, 3:31-35

Table 18.3 Biomaterials in body systems

System	Examples
Skeletal	Bone plate, total joint replacements
Muscular	Sutures, muscle stimulator
Nervous	Hydrocephalus drain, cardiac pacemaker, nerve stimulator
Endocrine	Microencapsulated pancreatic islet cells
Reproductive	Augmentation mammoplasty, other cosmetic replacements

*According to Parida et al., 2012, (IJAAS) 1, 3:31-35

Table 18.4 Biomedical Application of Polymeric Biomaterials

Synthetic polymers	Applications
Polyvinylchloride	Blood and solution bag, surgical packaging, IV sets, dialysis devices, catheter bottles, connectors, and cannulae
Polyethylene	Pharmaceutical bottle, nonwoven fabric, catheter, pouch, flexible container, and orthopedic implants
Polypropylene	Disposable syringes, blood oxygenator membrane, suture, nonwoven fabric, and artificial vascular grafts
Polymethylmetacrylate	Blood pump and reservoirs, membrane for blood dialyzer, implantable ocular lens, and bone cement
Polystyrene	Tissue culture flasks, roller bottles, and filterwares
Polyethylenterephthalate	Implantable suture, mesh, artificial vascular grafts, and heart valve
Polytetrafluoroethylene	Catheter and artificial vascular grafts
Polyamide	Packaging film, catheters, sutures, and mold parts

*According to Parida et al., 2012, (IJAAS) 1, 3:31-35

Emphasizing Bioengineering Aspect to Biomaterials: Current Examples of Artificial Organs

This is rapidly developing filed, as well. It is impossible to touch all the aspects of the problem solutions in this research area [16, 17]. Therefore, we shall just present the efficient examples, mostly in the form of artificial organs, although the application is not so exclusive [18–22].

Artificial Organs

An **artificial organ** is a man-made device that is implanted into, or integrated onto, a human to replace a natural organ, for the purpose of restoring a specific function or a group of related functions so the patient may return to as normal a life as possible [6]. The replaced function doesn't necessarily have to be related to life support, but often is [18–20].

Implied by this definition is the fact that the device must not be continuously tethered to a stationary power supply, or other stationary resources, such as filters or chemical processing units. (Periodic rapid recharging of batteries, refilling of chemicals, and/or cleaning/replacing of filters, would exclude a device from being called an artificial organ.) Thus, a dialysis machine, while a very successful and critically important life support device that completely replaces the duties of a kidney, is not an artificial organ. At this time a successful portable self-contained artificial kidney has not become available. Reasons to construct and install an artificial organ, an extremely expensive process initially, which may entail many years of ongoing maintenance services not needed by a natural organ, might include:

- Life support to prevent imminent death while awaiting a transplant (e.g., artificial heart)
- Dramatic improvement of the patient's ability for self-care (e.g., artificial limb)
- Improvement of the patient's ability to interact socially (e.g., cochlear implant)
- Cosmetic restoration after cancer surgery or accident

The use of any artificial organ by humans is almost always preceded by extensive experiments with animals in order to determine their biocompatibility first of all [18]. Initial testing in humans is frequently limited to those either already facing death, or who have exhausted every other treatment possibility (Rarely testing may be done on healthy volunteers who are scheduled for execution pertaining to violent crimes.). Although not typically thought of as organs, one might also consider replacement bone, and joints thereof, such as hip replacements, in this context [4, 5]. Some examples based on representative biomaterials will be mentioned but their detailed architecture and physical work are not the goal of this chapter.

Artificial Heart

Artificial hearts can now completely, if temporarily, replace the ventricles and valves with a device made of plastic or other man-made materials, which does the job of pumping blood around [21]. While considered a success, the use of artificial hearts is limited to patients awaiting transplants whose death is imminent. The current state of the art devices are unable to reliably sustain life beyond about 18 months [23]. The artificial heart is really meant as a bridge to an actual transplant: The longest that someone has lived on one is just over 4 years (Fig. 18.1) [22].

Currently, researchers are working on producing the heart tissue from stem cells on particular scaffolds made from adequate material. One of the tricks is to find

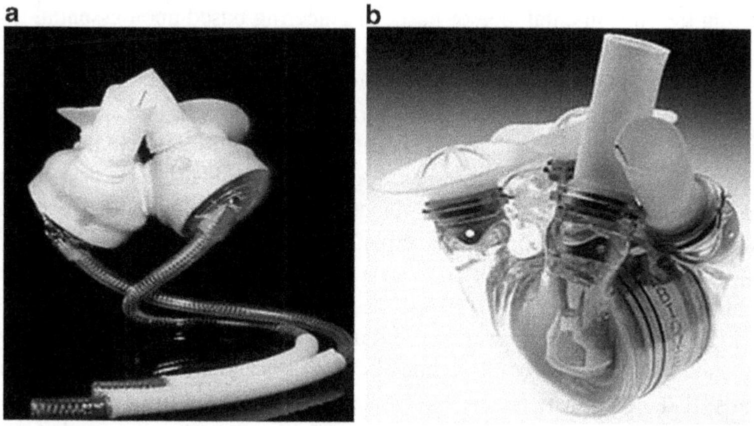

Fig. 18.1 (**a**) and (**b**) Artificial hearts mainly made of plastic

materials that, among other things, are nontoxic, won't get attacked by the body's immune system and allow for muscle cells to pass the electrical signals necessary for the heart to beat. Previous research has found that chitosan, which is obtained from shrimp and other crustacean shells, nearly fits the bill. In lab tests, scientists have used it as a scaffold for growing heart cells. But it doesn't transmit electrical signals well. Vunjak-Novakovic's team decided to build on the *chitosan development* and coax it to function more like a real heart. They added to the chitosan *carbon nanofibers*, which can conduct electricity, and grew neonatal rat heart cells on the resulting scaffold [15]. After two weeks, cells had filled all the pores and showed far better metabolic and electrical activity than with a chitosan scaffold alone. The cells on the chitosan/carbon scaffold also expressed cardiac genes at higher levels. Carbon nanofibers were used in this work as doping material to develop a highly conductive chitosan-based composite material [15]. Scaffolds based on chitosan only and chitosan/carbon composites were prepared by precipitation. Carbon nanofibers were homogeneously dispersed throughout the chitosan matrix, and the composite scaffold was highly porous with fully interconnected pores [15]. Chitosan/carbon scaffolds had elastic modulus of 28.1 ± 3.3 KPa, similar to that measured for rat myocardium, and excellent electrical properties, with conductivity of 0.25 ± 0.09 S/m [15]. The scaffolds were seeded with *neonatal rat heart cells* and cultured for up to 14 days, without electrical stimulation. After 14 days of culture, the scaffold pores throughout the construct volume were filled with cells. The metabolic activity of cells in chitosan/carbon constructs was significantly higher as compared to cells in chitosan scaffolds [15]. The incorporation of carbon nanofibers also led to increased expression of cardiac-specific genes involved in muscle contraction and electrical coupling. Thus, this study demonstrates that the incorporation of carbon nanofibers into porous chitosan scaffolds improved the properties of cardiac tissue constructs, presumably through enhanced transmission of electrical signals between the cells. This is an extraordinary example of what interaction between biomaterials and cells can do, supporting the fundamental idea of tissue engineering based upon essential triangle necessary for the efficient function of the 3-D growing tissue (cardiac) system.

Another trick-Doris Taylor regularly harvests organs such as hearts and lungs from the newly dead, re-engineers them starting from the cells and attempts to bring them back to life in the hope that they might beat or breathe again in the living [23]. Taylor is in the vanguard of researchers looking to engineer entire new organs, to enable transplants without the risk of rejection by the recipient's immune system. The strategy is simple enough in principle. First remove all the cells from a dead organ—it does not even have to be from a human—then take the protein scaffold left behind and repopulate it with stem cells immunologically matched to the patient in need [23]. Most researchers in the field use a mixture of two or more cell types, such as endothelial precursor cells to line blood vessels and muscle progenitors to seed the walls of the chambers. Some have been deriving these from iPS cells—adult cells reprogrammed to an embryonic-stem-cell-like state using growth factors—because these can be taken from a patient in need and used to make immunologically matched tissues. As they colonize the scaffold, some of the immature cells will take root and begin to grow. But urging them to become functional, beating cardiomyocytes requires more than just oxygenated media and growth factors. Cells sense their environment, including the growth factors, the stiffness and the mechanical stress, which in turn pushes the cells down their proper developmental path [23].

The heart must be placed into a **bioreactor** that mimics the sensation of beating. Some bioreactors use a combination of electrical signals—akin to a pacemaker—to help to synchronize the beating cardiomyocytes seeded on the scaffold, combined with physical beating motions induced by a pump. But researchers face a constant battle in trying to ape the conditions present in the human body, such as changes in heart rate and blood pressure, or the presence of drugs. The body reacts to things and changes the conditions so quickly it's probably impossible to mimic that in a bioreactor. Thus, a lot of work will be necessary to get the heart for permanent use in this way.

Artificial Pacemakers

These electronic devices, which can either intermittently augment (defibrillator mode), continuously augment, or completely bypass the natural living cardiac pacemaker as needed, are so successful that they have become common place. The TAH-t is a modern version of **the Jarvik-7 Artificial Heart** that was implanted in patient Barney Clark in 1982 (22). Scientists are at the moment busy with their experiments and we could possible see a fully reliable implantable artificial heart sometime in the future.

Oxygen Therapeutics Under Development (Artificial Blood)

Many researchers tried to use fluids such as beer, urine, milk, and animal blood as blood substitute (artificial blood) after William Harvey discovered blood pathways in 1616. The demand for more blood substitutes began during the Vietnam War as

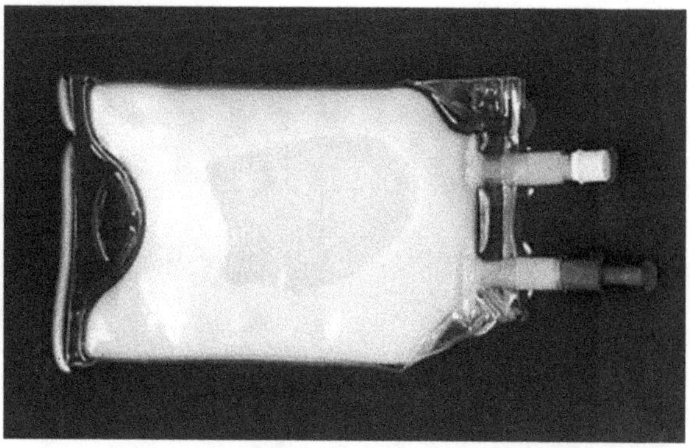

Fig. 18.2 Normal donor collected blood

Fig. 18.3 Artificial blood

wounded soldiers were unable to be treated at hospitals due to blood shortages (Figs. 18.2 and 18.3) [24].

Major worldwide blood shortages have led scientists to synthesize and test artificial blood [24]. Infected blood is a major problem for many countries. Each year 10–15 million units of blood are transfused without being tested first for HIV and hepatitis [25]. The second largest cause of new HIV infections in Nigeria comes from transfused blood. A major problem associated with donated blood is that after it is stored it loses nitric oxide and causes vasoconstriction for the recipient [26]. In the 1990s, because of the risk of undetected blood bank contamination from HIV, hepatitis C, and other emergent diseases such as Creutzfeldt-Jakob disease, there was additional motivation to pursue oxygen therapeutics.

Thus, the first approved oxygen-carrying blood substitute was a *perfluorocarbon-based product* called Fluosol-DA-20, manufactured by Green Cross of Japan. It was approved by the Food and Drug Administration (FDA) in 1989. Because of limited

Fig. 18.4 Earlier, Chris Mason, a Medical Research

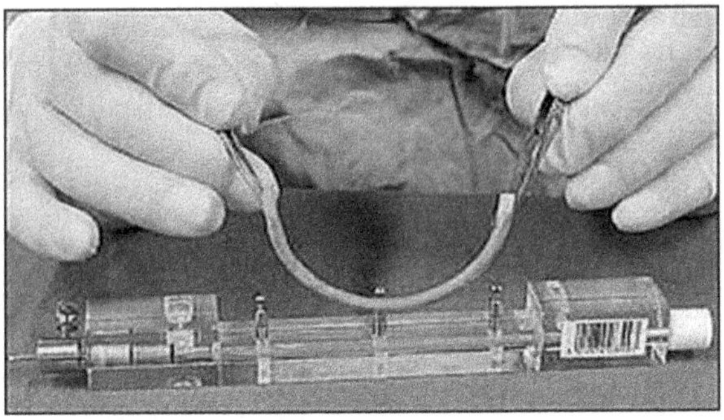

success, complexity of use and side effects, it was withdrawn in 1994. However, Fluosol-DA remains the only oxygen therapeutic ever fully approved by the FDA [24].

Significant progress was achieved, and a haemoglobin-based oxygen therapeutic called Hemopure which was approved for Phase III trial (in elective orthopedic surgery) in the U.S., and more widely approved for human use in South Africa. There are numerous controversies with respect to the use of artificial blood, but it is hoped that within the next few years, artificial blood may be widely used. The meaning of the products is under clinical trials. Another latest research in blood substitute's technology is being conducted by the *fluorinated water–oxygen carriers*.

Artificial Bones

A lot of work has been invested on experiments to develop artificial bones. The need for that is self-explanatory meaning how many different diseases and injuries make the spectrum of the needs for bone extremely wide (Figs. 18.4 and 18.5).

Artificial bone refers to bone-like material created in a laboratory that can be used in bone grafts, to replace human bone that was lost due to severe fractures, disease, *etc.* [27]. A lot of different material has been fabricated in order to be used instead of bones: wood, ceramics, stem cells, ink-jet printing combined material, *etc.* (Fig. 18.6) [28–31].

Status: Under clinical trials

Artificial Skin

Back in 1996, an artificial skin developed at method involved chemically bonding **collagen** derived from *animal tendons fibroblasts* in the lab only. These fibroblasts, which generate a protein called collagen, add to the strength and elasticity of the skin but as we age, the count of these cells keeps on declining. Therefore, when these cells will be injected into the wrinkles, it will lead to the regeneration of the collagen, encouraging the skin to regenerate again. Re-generating the human skin in itself is a

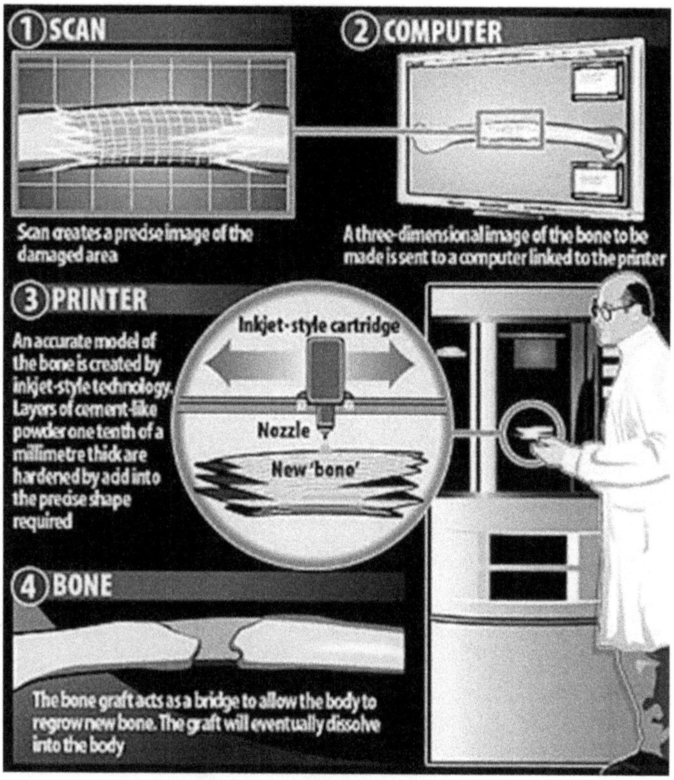

Fig. 18.5 **Research**: McGill University, Montreal

Fig. 18.6 Artificial skin

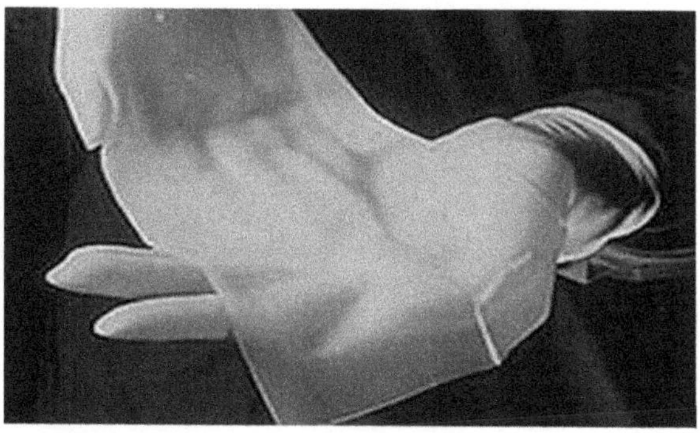

great achievement. Furthermore, researchers from Technion (Israel) have discovered how to make a new kind of flexible sensor that one day could be integrated into electronic skin. However, a lot of work is necessary in order to synchronize these inventions and get an integral, functional artificial skin. **Research**: MIT, Cambridge-based biotechnology firm. **Status**: Researches on the way for generating a real skin.

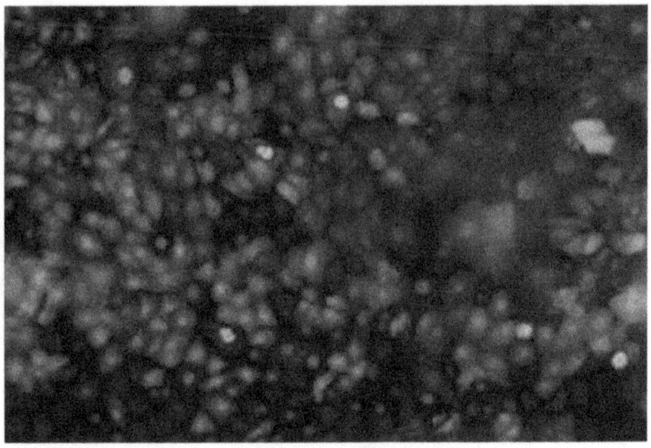

Fig. 18.7 Macular degeneration

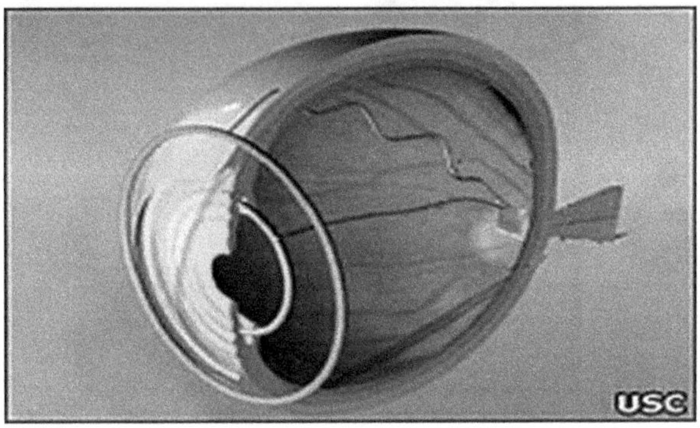

Fig. 18.8 Bionic eye

Macular Degeneration and Artificial Eye

There are two types of degenerative retinal disease: dry and wet macular degeneration where the sensorial cells of retina are loosing their shape and function. If diseased, patients tend to become completely blind until their 40s or 50s (Figs. 18.7 and 18.8) [32].

The cultivated retinal human embryonic cells (hESC) to the state of retinal sensorial cells (RP) have caused improvement in both dry and wet macular degeneration treated by group of scientists from MIT.

Research: U.S. scientists

Status: In the trials

Artificial Eye

The most successful function-replacing artificial eye so far is actually an *external miniature digital camera with a remote unidirectional electronic interface implanted on the retina*, *optic nerve*, or other related locations inside the brain. The present state of the art yields only very partial functionality, such as recognizing levels of brightness, swatches of color, and/or basic geometric shapes, proving the concept's potential. While the living eye is indeed a camera, it is also much more than that.

The final aim of the researchers in this field is to make people recognize faces and also tune electrodes to respond to light of different wavelengths and also allow the patients to see genuine color. As for retina, various researchers have demonstrated that the retina performs strategic image preprocessing for the brain. The problem of creating a 100 % functional artificial electronic eye is even more complex than what is already obvious. Steadily increasing complexity of the artificial connection to the retina, optic nerve or related brain areas advances, combined with ongoing advances in computer science, is expected to dramatically improve the performance of this technology [33].

For the person whose damaged or diseased living eye retains some function, other options superior to the electronic eye described above may be available. None of the current devices presents the cosmetic appearance of a living eye. For the nonfunctional cosmetic artificial eye, generically a "glass" eye, one can see instead Ocular prosthetic.

Limbs Regeneration

They have fruitfully grown extra arms on salamanders with the help of an extract of pig bladder (containing molecules of extracellular matrix). The research is related with re-growing a whole finger but growing enough of a finger that could be less than an inch, without a joint which is too complex to regenerate, yet [33]. However, if the results are optimistic, it could prove to be a stepping stone for further research to re-grow the whole fingers with the joints.

Artificial arms with semi-functional hands, some even fitted with working opposable "thumbs" plus 2 "fingers", and legs with shock absorbing feet capable of allowing a trained patient to even run, have become available. While the meaning of "full mobility" is debated, steady progress is made, as described in the main article Prosthesis.

There are now many artificial organs that have been implanted in humans, with varying degrees of success [33].

Brain Pacemaker

These devices, including deep brain stimulators, send electrical impulses to the brain in order to relieve depression, epilepsy, tremors of Parkinson's disease, and other conditions such as increased bladder secretions [33]. Rather than replacing existing neural networks to restore function, these devices often serve by disrupting the output of existing malfunctioning nerve centers to eliminate symptoms.

Artificial Cardiac

This pertains to gastric repairs, specifically of the valves at either end of the stomach. Artificial cardiac can be used to fight, between other diseases, esophageal cancer, achalasia and gastro esophageal reflux disease [33].

Artificial Corpora Cavernosa

To treat erectile dysfunction, both corpora cavernosa can be irreversibly surgically replaced with manually inflatable penile implants. This is a drastic therapeutic surgery meant only for men suffering from complete impotence that has resisted all other treatment approaches.

An implanted pump in the (groin) or (scrotum) can be manipulated by hand to fill these artificial cylinders, normally sized to be direct replacements for the natural corpus cavernosa, from an implanted reservoir in order to achieve an erection [33].

Artificial Ear

For external cosmetic repair one can look into plastic surgery.

For internal restoration of auditory function there is a *Cochlear implant*. While natural hearing, to the level of musical quality, is not typically achieved, most recipients are pleased, with some finding it useful enough to return to their surgeon with a request to do the other ear [33].

Artificial Liver

HepaLife is developing a bioartificial liver device intended for the treatment of liver failure using stem cells. The artificial liver, currently under development, is designed to serve as a supportive device, either allowing the liver to regenerate upon acute liver failure, or to bridge the patient's liver functions until a transplant is available.

It is only made possible by the fact that it uses real liver cells, and even then, it is not a permanent substitute for a liver.

On the other hand, researchers Dr. Colin McGucklin, Professor of Regenerative Medicine at Newcastle University, and Dr. Nico Forraz, Senior Research Associate and Clinical Sciences Business Manager at Newcastle University, say that pieces of artificial liver could be used to repair livers injured in the next 5 years. These artificial livers could also be used outside the body in a manner analogous to the dialysis process used to keep alive patients whose kidneys have failed [33].

Artificial Pancreas

For the treatment of **diabetes**, numerous promising techniques are currently being tested, including some that incorporate donated living tissue housed in special materials to prevent the patient's immune system from killing the foreign live components [33].

Artificial Urinary Bladder

This represents a **unique success in that these are autologous laboratory-grown living replacements, as opposed to most other artificial organs which depend upon electro-mechanical contrivances, and may or may not incorporate any living tissue** (**Antony Atala, 33**).

Artificial Ovary

Reproductive age patients who develop cancer often receive chemotherapy or radiation therapy which damages oocytes and leads to early menopause. An artificial human ovary has been developed at Brown University with self-assembled microtissues created using **novel 3-D petri dish technology**. The artificial ovary will be used for the purpose of in vitro maturation of immature oocytes and the development of a system to study the effect of environmental toxins on folliculogenesis [33].

Beyond Restoration

It is also possible to construct and install an artificial organ to give its possessor abilities which are not naturally occurring. Research is proceeding, particularly in areas of vision, memory, and information processing, however this idea is still in its infancy.

Some current research focuses on restoring inoperative short-term memory in accident victims and lost access to long-term memory in dementia patients. Success here would lead to widespread interest in applications for persons whose memory is considered healthy to dramatically enhance their memory of far beyond what can be achieved with mnemonic techniques. Given that our understanding of how living memory actually works is incomplete, it is unlikely this scenario will become reality in the near future [33].

One area of success was achieved in 2002 when a British scientist, **Kevin Warwick**, had an array of 100 electrodes fired into his nervous system in order to link his nervous system into the internet. With this in place he carried out a series of experiments including extending his nervous system over the internet to control a

robotic hand, a form of extended sensory input and the first direct electronic communication between the nervous systems of two humans.

Another idea with significant consequences is that of implanting a Language Translator for diplomatic and military applications [33]. While machine translation does exist, it is presently neither good nor small enough to fulfill its promise.

This might also include the existing (and controversial when applied to humans) practice of implanting subcutaneous "chips" (integrated circuits) for identification and location purposes. An example of this is the RFID tags made by VeriChip Corporation [33].

New technology is providing interesting opportunities to advance our knowledge of science and improve our lives. Indeed, technology is shrinking down to the nanoscale, putting objects and devices at up to one billionth of a meter in size. That is amazing, and it means that scientists are working on projects at a scale so small that they are working with molecules and even atoms in some cases.

Nanotechnology is providing the potential for a number of new discoveries and breakthroughs, including in materials. New materials are being discovered regularly, including biomaterials that have some amazing potential (Figs. 18.9 and 18.10).

Next chapter will present some of the top nanotech and biomaterial innovations around.

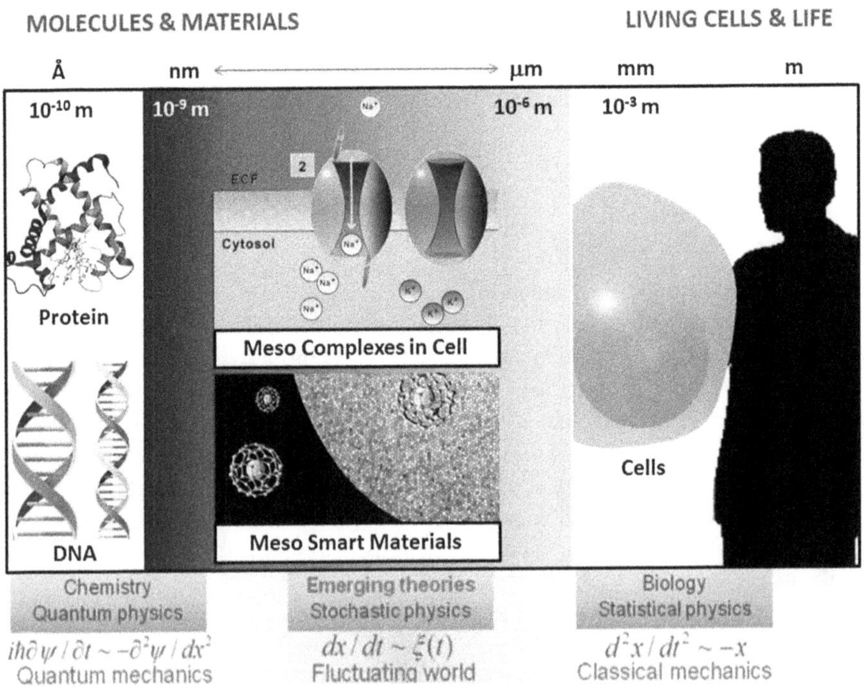

Fig. 18.9 An Editorial by Norio Nakatsuji (Director WPI-iCeMS) is now published. The Editorial explains that the mesoscopic domain exists between the nano-space and the bulk space, the scale between nanometers and micrometers. It goes on to say that understanding processes that occur in the mesoscopic domain will lead to breakthroughs in biomaterials science and biotechnology. Biomater. Sci., 2012, DOI: 10.1039/c2bm90001g, Advance article

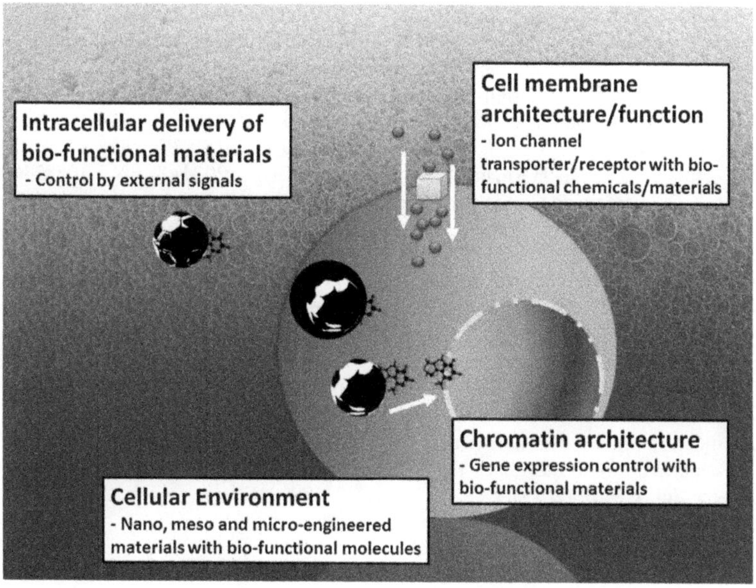

Fig. 18.10 Cell-biomaterial integration in the case of biocompatible materials

References

1. Helmus, M.N., Gibbons, D.F., Cebon, D.: Biocompatibility: meeting a Key functional requirement of next-generation medical devices. Toxicol. Pathol. **36**, 70–80 (2008). doi:10.1177/0192623307310949
2. Parida, P., Mishra, S.C.: Biomaterials In: Medicine. UGC Sponsored National Workshop on Innovative Experiments in Physics, 9–10 (2012)
3. Rourkela NM: http://hdl.handle.net/2080/1624.
4. Pramanik, S., Agarwal, A.K., Rai, K.N.: Chronology of total Hip joint replacement and materials development. Trends Biomater. Artif. Organs **19**(1), 15–26 (2012)
5. Chakraborty, G.V.: Biomaterials: Metallic Implant Materials, Technology, review essays. (2011)
6. Batista, G., Ibarra, M., Ortiz, J., Villegas, M.: Engineering Biomechanics of Knee Replacement, Applications of Engineering Mechanics in Medicine, GED-University of Puerto Rico, Mayagüez, pp. 1–12 (2004)
7. Bronzino, D.: The Biomedical Engineering Handbook, 2nd edition, Vol-1 (2000). ISBN: 3-540-66351-7.
8. Parida, P., Behera, A., Mishra, S.C.: Classification of biomaterials used in medicine. Int J Adv Appl Sci (IJAAS) **1**(3), 31–35 (2012)
9. Zhu, S.L., Wang, X.M., Qin, F.X., Inoue, A.: A new Ti-based bulk glassy alloy with potential for biomedical application. Mater. Sci. Eng. A **459**, 233–237 (2007). doi:10.1016/j.msea.2007.01.044
10. Zhang, E., Yin, D., Xu, L., Yang, L., Yang, K.: Microstructure, mechanical and corrosion properties and biocompatibility of Mg-Zn-Mn alloys for biomedical application. Mater. Sci. Eng. C **29**(3), 987–993 (2009)
11. Li, S.J., Niinomi, M., Akahori, T., Kasuga, T., Yang, R., Hao, Y.L.: Fatigue characteristics of bioactive glass-ceramic coated Ti-29Nb-13Ta-4.6Zr for biomedical application. Biomaterials **25**(17), 3369–3378 (2004)

12. Rajendran, V.: Bioactive glasses from implant applications. Defense Research and Development Organization, Ministry of Defense, New Delhi 110105, (2010). http://www.Drdo.gov.in/drdo/pub/monographs/Introduction bioactive glasses introduction
13. The education of the biomaterialist: Report of a survey. J. Biomed. Mater. Res. **16**(2), 159–167 (1982). doi:10.1002/jbm.820160208
14. Ratner, B.D., Hoffman, A.S., Schoen, F.J., Lemons, J.E. (eds.): Biomaterials Science: An Introduction to Materials in Medicine. Academic, San Diego (2004). ISBN 0-12582463-7
15. Spiller, K.L., Anfang, R.R., Spiller, K.J., Ng, J., Nakazawa, K.R., Daulton, J.W., Vunjak-Novakovic, G.: The role of macrophage phenotype in vascularization of tissue engineering scaffolds. Biomaterials **35**(15), 4477–4488 (2014)
16. ASM International, Introduction to Biomaterials in Orthopedic Surgery (2009). (#05233G).
17. Srivastav, A.: An overview of metallic biomaterials for bone support and replacement, biomedical engineering. Trends in Materials Science. pp. 153–168. http://cdn.intechweb.org/pdfs/12800.pdf
18. Atala, A.: Principles of Regenerative Medicine. (2011). ISBN 978-0-12-369410-2
19. http://thefutureofthings.com/4072-dispensing-drugs-through-contact lenses/#sthash.a71wqunB.dpuf
20. Nakatsuji, N. (ed.).: Mesoscopic science, where materials become life and life inspires materials. A great opportunity to push back the frontiers of life, materials, and biomaterials sciences. Biomater. Sci. **1**, 9–10 (2013). DOI: 10.1039/C2BM90001G
21. http://www.physics.org/article-questions.asp?id=74
22. http://web.mit.edu/invent/iow/jarvik.html
23. Ott, H.C., Matthiesen, T.S., Goh, S.K., Black, L.D., Kren, S.M., Netoff, T.I., Taylor, D.A.: Perfusion-decellularized matrix: using natur Schimmeyer, S. (2002, November 1). Illumin: article: The Search for a Blood Substitute. Retrieved December 2, 2010.
24. Schimmeyer, S.: Illumin: article: The Search for a Blood Substitute. (2002). Retrieved December 2, 2010
25. Brown University Division of Biology and Medicine. Home Page (2006). Retrieved December 3, 2010
26. Dow Jones & Co. (2010, November 12). Hard to Treat Diseases, Inc. (HTDS) Targets Merger with Biotechnology Company-MarketWatch. Retrieved December 4, 2010
27. http://en.wikipedia.org/wiki/Artificial_bone#cite_ref-1
28. Secrets of the Sea Yield Stronger Artificial Bone. Physorg. com. Retrieved 2013-11-16.
29. Injectable artificial bone developed. Cosmosmagazine com. Retrieved 2013-11-16.
30. Jaw bone created from stem cells. BBC News. October 10, 2009. Archived from the original on 11 October 2009. Retrieved 11 October 2009.
31. Turning wood into bone. BBC News. 8 January 2010. Archived from the original on 9 January 2010. Retrieved 8 January 2010.
32. Schwartz, S.D., Jean-Pierre, H., Heilwell, G., Franco-Cardenas, V., Pan, C.K., Ostrick, R.M., Mickunas, E., Gay, R., Klimanskaya, I., Lanza, R.: Embryonic stem cell trials for macular degeneration: a preliminary report. The Lancet **379**(9817), 713–720 (2012). www.thelancet.com, Published online January 23, 2012
33. Malchesky, S.P.: 38 (12 Issues in 2014), Print ISSN: 0160-564XOnline ISSN: 1525-1594

Chapter 19
Nanotechnology: Novel Emerging Concepts

Every industry that involves manufactured items will be impacted by nanotechnology research. Everything can be made in some way better–stronger, lighter, cheaper, easier to recycle—if it's engineered and manufactured at the nanometer scale.

Stan Williams (1951–)

Nanotechnology is a field of science that controls the manipulation of atomic properties so as to make the functional systems and other materials acquire exclusive capabilities. The amazing approach of nanotechnology has been developed within chemical engineering as an explosive response and inevitable consecutive inspiration caused by original development of fullerenes, designed and created by Sir Harry Kroto who was awarded Nobel Prize for his invention of the methodology for separation of members of fullerene family and engineering of fullerene molecules, in 1996. Ever since it is rapidly growing field of research with numerous applications in different branches of engineering with specific focus on biomedical and health solutions.

Nanotechnology is a field of science that controls the manipulation of atomic properties so as to make the functional systems and other materials acquire exclusive capabilities [1]. The amazing approach of nanotechnology has been developed within chemical engineering as an explosive response and inevitable inspiration caused by original development of fullerenes, isolated, designed and created by Sir Harry Kroto who was awarded Nobel Prize for his invention of the methodology for engineering and separation of members of fullerene family in 1996 [1].

M. Pavlovic, *Bioengineering*, DOI 10.1007/978-3-319-10798-1_19

Breakthrough in nanotechnology: Harry Kroto (NP 1996), for his discovery of fullerenes. Konstantin Novoselov, Andre Geim (NP 2010), for their work on graphene (two-dimensional material).

The term **"nanotechnology"** has become much more popular and is often used to describe any type of manufacturing or research that takes place at dimensions less than 1,000 nm [8–11]. Since in making the new molecules atoms have to collide with enough energy and under correct angle the idea is that we should soon be able to manipulate individual atoms and molecules to construct devices that are more powerful, more precise, lighter, smaller and despite all, stronger. Here are proposed three main aims of devices manufactured using nanotechnology which reflect the perspective of this fascinating developing field:

The prefix "nano" stems from the ancient Greek which stands for "dwarf". Nature (given in Figs. 19.1,19.2 and 19.3) clearly indicates differences in size between macro, micro and nano-words. These figures have emphasized first of all, the structures developed in nature of the nano-size that have also geometric character/microarchitecture, ranging from viruses and nano-bacteria to the wide spectrum of natural nanoparticles all over the nature surrounding our planet (especially ocean) of infinite structural forms [1–3]. The scale of things in the Fig. 19.3 is showing the shapes and architecture of many different entities from all levels, indicating that similar shapes can be formed not only at visible, but also at invisible parts of the scale.

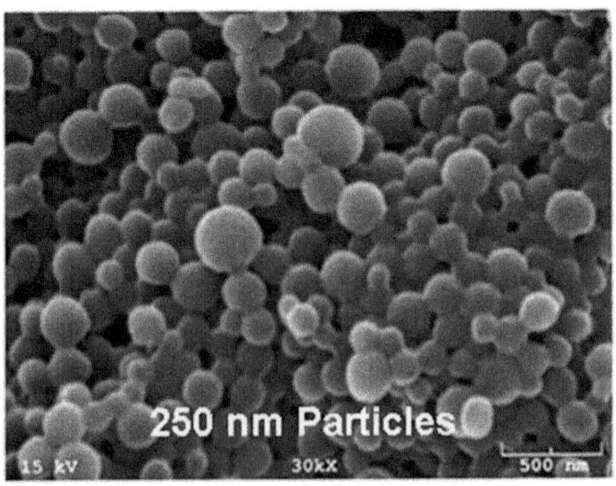

Fig. 19.1 The world of micrometer

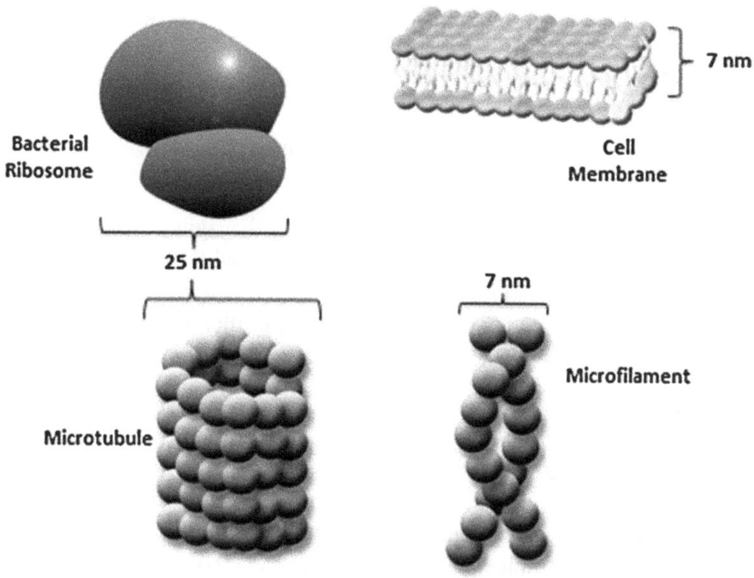

Fig. 19.2 The world of nanometer

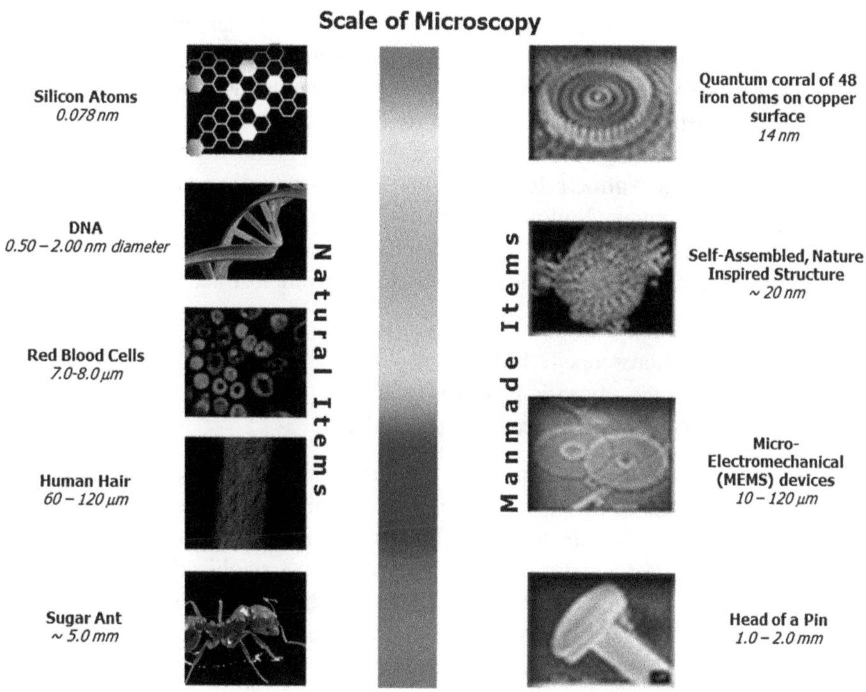

Fig. 19.3 The scale of things: architectonic parallelism

Nanotechnology is generally defined as “the engineering of functional systems at the molecular scale” [1–5]. In other words, it is manipulation of matter at the atomic and molecular scale to create materials with remarkably varied and new properties with huge potential in many sectors, ranging from healthcare to construction of electronics. In medicine, it promises to revolutionize drug delivery, gene therapy, diagnostics and many areas of research, development and clinical application.

1. We should be able to place every atom in exactly the right position.
2. We should be able to build anything that complies with the laws of physics, as long as we can understand it at a molecular or atomic level.
3. And perhaps, the one that drives most research in this field is the idea that the costs of manufacture should not be much higher than the costs of only the materials and energy required to put the product together [8–14].

Geometry and *minimization* are at the heart of nanotechnology. Look only into viruses! As of known nano-size, they have mostly stable icosahedron geometric shapes while in non-living state (out of the host) with their crystal DNA within the nano-particle, becoming alive during interaction with the host cells [2, 3]. The interaction involves their DNA/RNA replication which involves the “borrowing” of enzymes for that particular procedure within the host cell.

There are also certain names already given to the particular structures emerging from nano-technological approaches [1, 4–11] such as:

Nanotubes which are elongated carbon forms with one or more concentric polygonal cylinders. They are grown from carbon under high-temperature, electrically-charged conditions. Due to their geometric similarity to the geodesic domes designed by the greatest thinker of twentieth century, R. Buckminster Fuller, nanotubes are furthermore called “*Bucky tubes*”.

Nanohexagons are nanotubes with hexagonal ends. **Nanoctagons** are nanotubes with octagonal ends. **Nanocircles** are circular nanostructures and **nanospheres** are spherical nanostructures. **Nanoshells** are hollow nanospheres, and lately used in medical purposes [15–19].

Within last decade, nanotechnology has moved from abstraction to reality with the development of tools such as the:

- Atomic Force Microscope (AFM),
- Scanning Tunneling Microscope (STM), and the
- Virtual Surface Profiling Microscope (VSPM).

These microscopes do more than just let people view little materials. They also enable manipulation of matter on a perspective of nanometers in a vacuum, liquid or gas.

AFM has a probe that creates three-dimensional images of specific atoms and molecules at the nano-scale dimension as it moves across an object’s outside. STMs may etch surfaces and move particles on dimension of nanometers. VSPM digitizing the whole slide at high resolution, so that the viewer can zoom into areas and structures of interest on the slide. It creates what we call a **digital slide**, or a virtual slide, meaning that the user sees the same image in the screen as they would get if they projected the microscope image onto the screen, but it’s an electronic file. All of these methods help us to visualize invisible at the “nano” level.

How Did the "Adventure in Nano-space" Start and Who Are the Facilitators of This Event?

The adventure in nano-space and the synthesis and separation of Sir Harry Kroto's "fullerenes"

The discovery of fullerenes in 1985 led to a new field of study and a New Material Class of pure carbon that is significantly different from other forms of carbon, diamond and graphite [1]. Carbon fullerenes are the third allotropes of carbon beside graphite and diamond, spherical, caged molecules with carbon atoms located at the corner of the polyhedral structure consisting of pentagons and hexagons, much like the shape of a Soccer ball. Carbon fullerenes come in many forms. The most abundant form is Carbon 60 (which has a soccer ball shape), Carbon 70 (which has more of a rugby ball shape) and Carbon 84 (spherical). Fullerene get the name from the **geodesic dome shape** which was researched and promoted at macro-level by the most prominent thinker of twentieth century, **Buckminster Fuller** [1]. They need an carbon arc discharge (high pressure/temperature) in order to be created/synthesized.

Several interesting and important developments took place at the University of Sussex in UK between September 1985, when C_{60} was discovered by the Rice Group, and September 1990 when the brilliant paper on its extraction was submitted to *Nature* by Wolfgang Krätschmer, Lowell Lamb, Kostas Fostiropoulos and Donald Huffman [1]. During this period a parallel series of experiments to those of Krätschmer et al., was carried out at Sussex, by H. Kroto's group. A key reason for carrying out the experiment at Rice in the first place was an intriguing set of results obtained by Hintenberger and colleagues between 1958 and 1963 that showed, by mass spectrometry, that carbon species with as many as 33 carbon atoms were produced in a carbon arc discharge [1]. At Sussex, after the initial C_{60} discovery in 1985, Crotto had a hole drilled in an old carbon-arc evaporator they had, so that they could deposit carbon on a silica wafer at various argon pressures [1]. The idea was to follow up the Hintenberger et al. experiments by recreating roughly the same conditions, that the group had achieved with the Rice nozzle as cheaply, as simply as possible with an electric arc discharge. At this point Kroto conjectured that as the argon pressure was increased he might be able to use the electron microscope that was available at Sussex to see the formation of roundish carbon particles which they conjectured might provide some circumstantial evidence for C_{60} formation. The group thought that the assembly processes that created C_{60} might also lead to the formation of large spheroidal soot-like carbon particles. What they found was that the smooth carbon coating obtained under very low pressure changed, more-or-less suddenly, at 70-80 μm pressure of argon creating an undulating blistered rough surface of the kind they vaguely expected [1].

This observation was encouraging as it seemed to be some sort of confirmation that the idea might be valid and that C_{60} might be forming. Here Kroto says he made a fundamental mistake—and not for the first time! He assumed that C_{60} would only be formed in minuscule amounts and only detectable, if at all, by the most sensitive analytical technique available i.e. mass spectrometry. After all, how could C_{60} be easily made when it had avoided detection until nearly the end of the twentieth century, and then only fleetingly, when it's two more famous siblings, diamond and graphite had been known since time immemorial. It is now hard, more than twenty five years later, when C_{60} is in every school science textbooks to realize that C_{60} was, prior to 1990, considered by some to be highly suspicious character and indeed by some even an imposter [1].

These "maneuvers" have helped them to completely isolate C_{60} and show that it exists in pure form, by using NMR and analyzing the spectra. Since then, many fullerenes—based compounds have been synthesized, **displaying a range of biological activities potentially useful in anticancer or antimicrobial therapy, cytoprotection, enzyme inhibition, controlled drug delivery and contrast—or radioactivity-based diagnostic imaging** [15–26].

Discovery of the Structure of Graphene and Significance of its Impact on Nanotechnology: Gaim and Novoselov

What is graphene? Graphene is a single layer of carbon packed in a hexagonal (honeycomb)-lattice, with a carbon–carbon distance of 0.142 nm and the angle between the 2 of 120° [5]. It is the first truly natural two-dimensional crystalline material and it is representative of a whole class of 2D materials including for example single layers of Boron-Nitride (BN) and Molybdenum-disulphide (MoS_2), which have both been produced after 2004 [4, 5].

As mentioned, fullerenes, the large carbon cage molecules represent third carbon allotrope beside graphite and diamond [1]. The most abundant form of fullerenes is **buckminsterfullerene (C60)** with 60 carbon atoms arranged in a spherical structure. As indicated, the shape of the molecule, known as **truncated icosahedron**, resembles that of a soccer ball, containing 12 **C60** fullerene molecules. Carbon nanotubes, and graphite can all be thought of as being formed from graphene sheets, i.e. single layers of carbon atoms arranged in a honeycomb lattice. How is it obtained from the nature? It is interesting to consider that everyone who has used an ordinary pencil has probably produced graphene-like structures without knowing it. A pencil contains graphite, and when it is moved on a piece of paper, the graphite is cleaved into thin layers that end up on the paper and make up the text or drawing that we are trying to produce. A small fraction of these thin layers will contain only a few layers or even a single layer of graphite, i.e. graphene. Thus, the difficulty was not to **fabricate** the graphene structures, but to **isolate sufficiently large individual sheets in order to identify and characterize the graphene** and to verify its unique two-dimensional (2-D) properties. This is what Geim, Novoselov, and their collaborators from Manchester group succeeded in doing and got in 2010 Nobel Prize, for [4, 5, 12–14].

Essentials of technical approach are actually very simple. They used a simple but effective mechanical exfoliation method for extracting thin layers of graphite from a graphite crystal with Scotch tape and then transferred these layers to a silicon substrate. The development of this new material, opens new exiting possibilities.

It is the first crystalline 2-D-material with unique properties, which makes it interesting both for fundamental science and for future applications. The Manchester group succeeded by using an optical method, an Atomic Force Microscope (AFM) with which they were able to identify fragments made up of only a few layers, or even monolayer-that's how graphene was identified [4, 5].

Nanoparticles in the Nature and How Long the Adventure in the Nanospace Was?

It was Democritus of Abdera in ancient Greece who first told that the matter must have smallest, invisible particle retaining all properties of that matter [2]. Democritus' model was a small, invisible non-particulated sphere. Athomos in Greek's means: non-dividable. In his vision it contained no electrons or nucleus. This was the first atomic model ever designed.

He drew his model to show that all atoms are indestructible and unchangeable. Democritus also knew that atoms are different in size. He thought they were different in shape and temperature, as well. Much later on, in modern era it has been determined that radius of He atom for example is in the range of nano-values [2].

John Dalton in 1803 was the first to seriously revise Democritus's theory [2]. He stated that atom is particulated and gave the first taw model which was then modified with work of many scientists such as: Rutherford, Bohr, etc. [2].

However, Ernst Rutherford has proved that the nuclei of certain light elements, such as nitrogen, could be 'disintegrated' by the impact of particles coming from a radioactive source, and that during this process fast protons were emitted. It was the first artificially induced nuclear reaction and it would change the world forever. Along with the eventual founding of CERN in 1954, it would lead to nuclear power and the atomic bombs that devastated Hiroshima and Nagasaki in World War Two.

What did Bohr improve on atomic structure? Rutherford's atomic theory described an atomic model with all the mass concentrated in a nucleus with electrons circling the nucleus in a fixed orbit. This theory was shown incorrect by using Maxwell's equations, which states since the electrons are moving in a circular motion, they are accelerating. Accelerating electrons means they are emitting radiation and therefore losing energy and would eventually spiral in motion toward the nucleus and collapse [2]. Bohr's insight was that he declared an electron could orbit the nucleus but only in discrete orbits which didn't emit radiation. An electron moves to a higher orbit, with a larger radius, by absorbing radiation (a photon) and in contrast will emit a photon of energy when the electron moves to a lower orbit with a smaller radius. Each orbit corresponds to an angular momentum value relating to Planck's constant (h) divided by 2π. Insights regarding radiation and atoms were taken from Planck's Quantum Theory [2]. Bohr proposed that the outer orbits could accommodate more electrons than the inner orbits. **In total, the atomic structure theory that Bohr proposed included an atom which was 1/10,000 the size of the atoms proposed by other scientists.**

Somewhat before World War II Italian atomic scientist Enrico Fermi was invited to USA. What was Fermi's contribution to the development of atomic science? Within ten years of research he discovered that **when an element is bombarded by a slow moving neutron, it becomes radioactive and starts emitting radiations**. The result is one element changes into another element, very attractive idea suggesting the evolution of the Universe reflected in the periodic chart of elements as rather a dynamic system evolving with his time. In 1933 he discovered a neutral particle called neutrino [2]. He also produced 80 new artificial nuclei by neutron bombardment. Thus, Fermi opened the field of elemental particles and atomic manipulation through bombarding and radiation of elements (atoms of elements) and their changes from one into another, that will later on be elaborated by Gel Man who stated that this world is relying upon atoms that are based upon matter/antimatter co-existence of the particles within the atom, meaning that each particle has an anti-particle. Fermi was instrumental in developing of an atom bomb during the Second World War, **but his great contribution was also to the roots of nanotechnology. The roots of the idea of graphene's emergence from carbon atoms are in his work**. It has been shown that the two-dimensional nature of graphene leads to many interesting electronic, thermal and elastic properties that could be used in different fields of human life, including medicine.

Fermi Level and Graphene

Fermi showed by bombarding atoms with neutrons under optimal conditions that they can become **different elements**. He also defined **Fermi level**—a measure of the energy of the least tightly held electrons within a solid (i.e. the valence electrons, which are in the outermost orbit of the atoms of a solid). The value of the Fermi level at absolute zero (−273.15 °C) is called the **Fermi energy** and is a constant for each solid. The Fermi level changes as the solid is warmed and as electrons are added to or withdrawn from the solid.

Fermi-Dirac Statistics

The Fermi energy showed the temperature-dependent behavior similar to any other one-dimensional device in non-degenerate regime. However, in the degenerate regime, the normalized Fermi energy with respect to the band edge shown to be a function of carrier concentration [12, 13]. In other words, modeling of Graphene Nanoribbon Fermi Energy is possible and has been developed into alternative approach to carbon nanotube (CNT) in order to show that different banding of tube is possible under different conditions opening possibilities for semiconducting electronic properties as well as for the nanoshells capable of distributing targeted drug delivery [14–20] (Figs. 19.4 and 19.5).

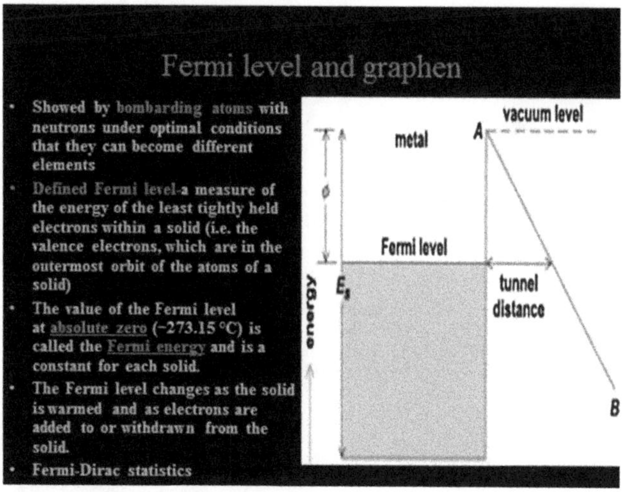

Fig. 19.4 Fermi level and graphene

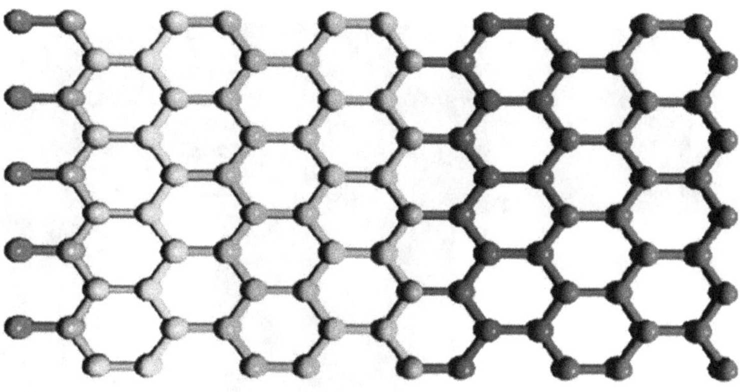

Fig. 19.5 Armchair of graphene nanoribbon (GNR) minimum energy band structure based on numerical solution of Fermi-Dirac integrals for nonparabolic region (Zaharah et al. 2010)

The breakthrough done by Geim, Novoselov and their co-workers with their paper from 2004 and 2007 which ignited the development of graphene-like materials brought them the Nobel Prize in Physics 2010 [4, 5].

Electrical and Thermal Conductivity of Graphene

Electrical Conductivity

Using the layer thickness we get a bulk conductivity of $0.96 \times 10^6\ \Omega^{-1}\ \text{cm}^{-1}$ for graphene. This is somewhat higher than the conductivity of copper which is $0.60 \times 10^6\ \Omega^{-1}\ \text{cm}^{-1}$ [14] (Figs. 19.6 and 19.7).

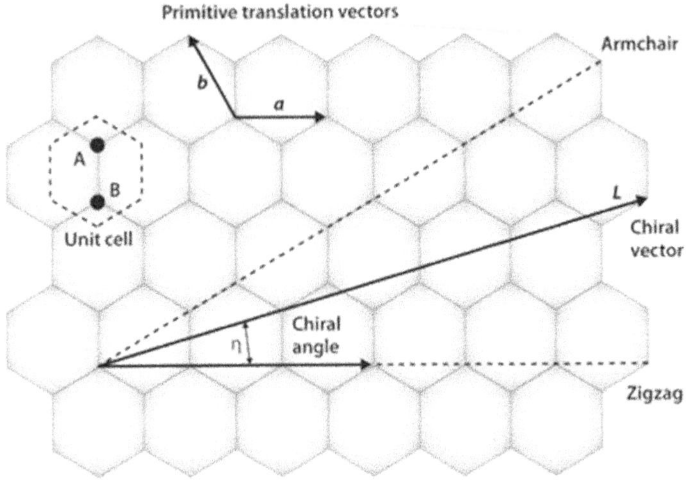

Fig. 19.6 Honeycomb lattice structure of graphene [21]

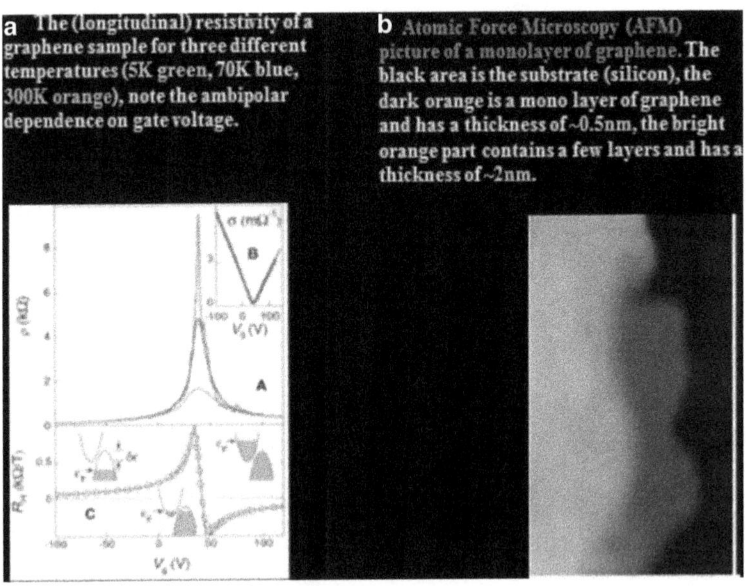

Fig. 19.7 Physicochemical aspects of isolated graphene

Physicochemical Aspects of Nanoparticles

Thermal Conductivity

The thermal conductivity of graphene is dominated by phonons and has been measured to be approximately 5,000 Wm^{-1} K^{-1}. Copper at room temperature has a thermal conductivity of 401 $Wm^{-1} K^{-1}$. Thus, graphene conducts heat ten times better than copper.

Graphene: Other Features

In contrast to low temperature 2-D systems based on semiconductors, graphene maintains its 2-D properties at room temperature. Graphene also has several other interesting properties, which it shares with carbon nanotubes. It is substantially stronger than steel, very stretchable and can be used as a flexible conductor. Its thermal conductivity is much higher than that of silver cooper [14, 21].

Nanotechnology Concepts

Here are two important concepts related to nanotechnology that are necessary if we are going to achieve the three aims mentioned before.

The first is known as "**positional assembly**", which is how we would get all of the molecules or atoms in the right place. This can be achieved through the use of tiny robots, molecular in size and produced using nanotechnology themselves [24, 27]. A second concept is "**massive parallelism**", which is a way to reduce the costs of manufacturing. Because one molecule-sized robot is going to take a very long time to build anything of substantial size, the idea is to have many robots that work in a production line, getting larger at each stage until the process is completed.

Emphasizing Bioengineering Aspects of Nanotechnology Important for Medicine

Although in essence nanotechnology has emerged as a revolution in the field of computer technology, it is no more confined to just computers. The success of nanotechnology in curing fatal diseases is reaching new horizons. The ability to manipulate structures and properties at the nanoscale in medicine is like having a sub-microscopic lab bench on which you can handle cell components, viruses or pieces of DNA using a range of tiny tools, robots and tubes. Scientists are working now to create nanostructures that serve as new kinds of drugs for treating cancer, to engineer nano-materials for use as artificial tissues that would replace diseased kidneys and livers, and even repair nerve damage, and to integrate nano-devices with the nervous system to create implants that restore vision and hearing, and build new prosthetic limbs [22–38].

The Current Use of Nanoparticles in Medicine

Considering the rapid progress in this field it is hoped that nanotechnology will soon be a huge force against diseases in the coming days and it will help in wiping away the most lethal diseases in the current era. **Nano medicine is the science of things**

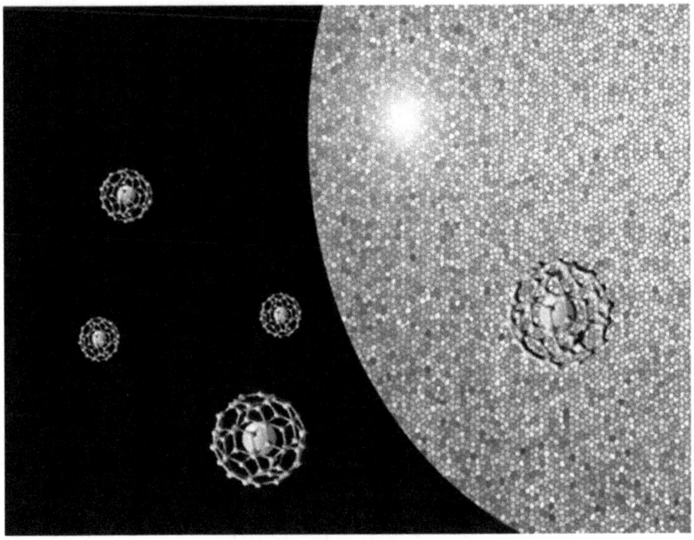

Fig. 19.8 The concept of drug delivery in bucky-ball

smaller than 1,000 nm in size, applied to the field of medicine. As we have seen from only potential of graphene, there are a wide variety of uses in this area, ranging from the use of **biosensors** for detection of anomalies in the body, such as high blood sugar concentration, which would suggest diabetes, to the concept of **nano-robots** [24]. Currently, nano medicine has applications in: drug delivery, cancer stem cell diagnosis and treatment, use in combination with laser beam as nanoshells in order to destroy cancerous tissue.

- *Concept of drug delivery*
  Pharmaceutical companies have started using nanotechnologies to develop genetically targeted drugs. This allows for much more precise drug development, and makes it faster to decide whether or not a substance is suitable for use in a drug delivery [28–38] (Fig. 19.8).

  These systems are already working in some diseases. Originally the idea is coming from Robert Langer who has synthesized drug-delivery systems based upon different materials, mostly polymers [28]. They are so far known as **controlled drug delivery systems**- a slow-releasing cancer (for example) medications that can be administered directly where the cancerous tumor had been removed. He also pioneered a variety of **remotely controlled drug delivery systems** that vary the amount of drug released through *electric impulse, ultrasound and magnetic field. Polymers* are incredibly versatile and one can use them to make them into all kinds of shapes and forms, including nanoparticles.

  How does the polymer work in developing a targeted drug delivery system? First, one has to inject the nanoparticle into the body-that can travel around the body for a long time, and ultimately find the target (diseased cells). A number of

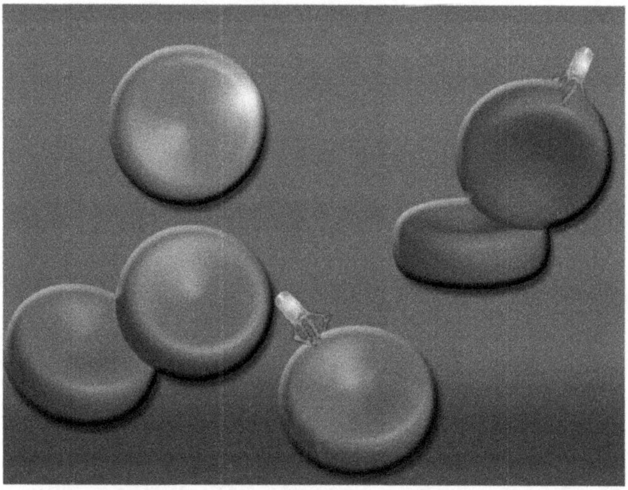

Fig. 19.9 Precursor of nanorobot-microscopic machine roaming throughout the body

breakthroughs have facilitated this control, coming from material science and life science, as well. The integral thinking of bioengineering principles has enabled that event. Polymer delivery systems have a great impact currently in making of transdermal patches, but they are also in the pills, different implants, stents, etc. [21, 23, 25–28]. They are expected to have a huge impact after clinical studies in next ten years. Some of them have antibodies for the targeted diseased cells surface molecules and in that way they are attached to it. The patents are numerous (about 300) and protected. So, controlled drug delivery has two important mechanisms: **one is the attachment of the drug exclusively for the target** (for example cancer cell and not the healthy one) while the other means that the **dosage is controlled and individualized/personalized which will avoid the side effects sometimes critically detrimental and fatal** [38].

- *Nanorobot concept*
  Nanotech medical robots ("**nano-medi-bots**") is the concept which might develop into the nano-device that might be able to: monitor body function; repair damaged tissue at the nanoscopic level; deconstruct pathologic or abnormal matter or tissue such as cancer or plaque; and refine human health and functioning. Although nanomedibots have not been developed, there are ongoing advances in nanofluidics and carbon nanotube flow sensors that could become their building blocks. As nanotechnology and biotechnology advance, nanomedibots and engineered useful microorganisms might be integrated (Fig. 19.9).

Carbon is the element the most suitable for medical nanorobots, probably as diamond or fullerene allotropic modification, due to extraordinary strength and chemical inertia of diamond. Most of other light elements such as: hydrogen, sulfur, oxygen, nitrogen, fluor, silicon, etc., will be used in special purposes and components.

Fig. 19.10 The idea of nanorobot and comparable nanoparticles in the cell

It is impossible to say exactly how would the typical nanorobot look like. Their purpose would have been to travel through the blood toward their target with precision of 500–3,000 nm. Nanorobot can be 500–3,000 nm in size and can be ingested with food or aerosol [24]. Each kind of Nanorobot would be designed to do a specific role.

Finally, maybe the most important: the nanorobot is being designed currently [24]. There is a lot of theoretical designs which look well on the paper, but they can be fundamentally changed after development and testing. However, it seems to be possible and this tendency to minimization will facilitate a lot of medical procedures and approaches (Fig. 19.10).

Beside nanorobot idea, where graphene could be potentially used, there are also other options like the one shown in Fig. 19.11. The antibacterial effect of the grapheme has been discovered as well, and could be potentially used in hospitals, sterile labs and other facilities that require it.

As technological advancement creates new opportunity in other realms of science, so too does it in the world of medicine. Two new treatment modalities are on the forefront of oncological intervention: nanoparticle therapy and alternating magnetic fields on replicating cancer stem cells. Although in the early phases of testing, the two show promise of accomplishing limited-to-no side-effects as well as being as non-invasive as possible. These novel treatments could be used in combination with chemotherapy or radiotherapy, as indicated.

The nanoparticle model has had difficulty from its inception with the use of quantum dots, later proven to be toxic. New concepts developed under strong influence of nanotechnology. However, recent models have proven dramatically more effective as anti-cancer agents through various methods of delivery [19]. The model proposes that nanoparticles that are composed of silicon and can be coated with

Fig. 19.11 Possible use of graphene as antibacterial bandage

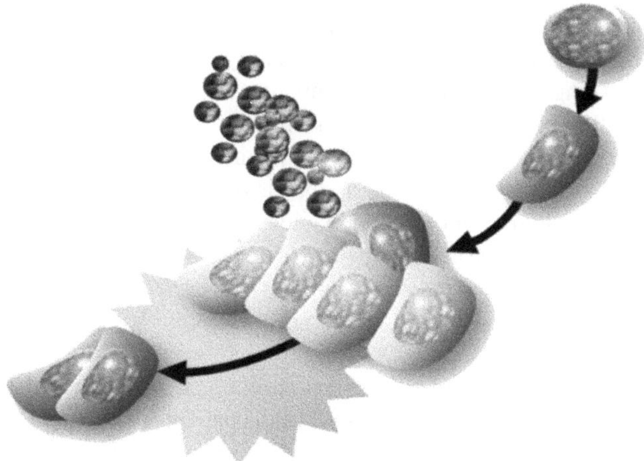

Fig. 19.12 Nanoparticles in drug-delivery systems in cancer treatment

antibodies specific for stem cell markers found on cancer stem cells, or similar molecular conjugates [38].

Through enhanced permeation retention, vascularized tumors would become infiltrated by these nanoparticles. Infrared radiation is then applied to the tumor, wherein the cancer stem cells that now contained these particles would increase in heat at a much greater rate than cells not containing the particles, thus killing the targeted cells. Studies have been carried out for a variety of cancer types including, breast, prostate and liver cancer [19] (Fig. 19.12).

Dynamic effects of therapeutic strategies directed against cancer stem cells (CSCs) are also a part of nanotechnology development. A tumor tissue is a complex mix of cancer cells at various stages of differentiation, from uncommitted CSCs through various stages of cancer progenitor cells to matured cancer cells, with a concomitant decrease in the levels of proliferative and/or metastatic potential. Both the CSC niche with supporting cell types and the matured cancer cell compartment create an intricate network of inter-dependency. Cancer therapy should ideally address both the CSCs and the matured cancer cells by slowing down proliferation and production of differentiated cancer cells and increasing apoptosis in both CSCs and matured cancer cells. In a fast-growing cancer, tumor therapy might come too late and/or be ineffective, or reduce tumor mass by killing matured cancer cells without targeting the CSC niche. The latter effect might stimulate CSC proliferation and increase the CSC pool, which would consequently result in a resurgence of even larger numbers of matured cancer cells. In another scenario, therapeutic intervention itself might provoke an enlargement of the CSC pool by selecting for **more radio- and chemoresistant CSC clones.**

These CSCs will have a superior ability to repair DNA damage upon radiotherapy and/or overexpress members of the ABC transmembrane pumps, resulting in the swift efflux of certain chemotherapeutics. Over time, this new generation of CSCs could also include new mutant CSCs with even more aggressive signatures. CSC therapy targets the CSC niche itself by attenuating the self-replicating potential of CSCs and disturbing cellular crosstalk within the CSC niche. Increased apoptosis of CSCs will result in a significantly smaller number of matured cancer cells, which can then be addressed successfully with common anticancer therapies [38]. Thus, anticancer therapy that only results in apoptosis of the matured cancer cells and/or only inhibits the proliferation of CSCs provides a potential window of opportunity for new and more aggressive CSC mutants to occur and might be unsuccessful if not dangerous. It is expected that the elimination of cancer should target the CSC pool, and successful treatment regimens would need to be the result of an orchestrated 'target and destroy" effect.

While the promise of reduced cross-effect oncological treatment via CSC targeting gives great hope to the future of oncology and the patients that suffer from cancer, a great deal of work still remains. CSCs are a moving target and exist in such a small population that effective use of treatment modalities, although more promising than some miRNA studies and the like, still does not, in its current state, exist as a viable treatment option for all cancers. Much as microbiologists have difficulty in the world of fighting an ever-adapting organism, so to shall the oncologist and researchers that pursue this path. In combination with other therapies, however, it does appear that a reduction in risk associated with current treatment modalities would be evident. In light of the difficulty of the manipulation of the CSC model, the research that has been done thus far is providing a solid framework upon which a new, improved paradigm of oncological treatment will be established [25, 26, 38] (Figs. 19.13 and 19.14).

Viruses, parasites and bacteria are continuously mutating causing new diseases of our natural immune system [3]. Theoretically, nanorobot could protect our body

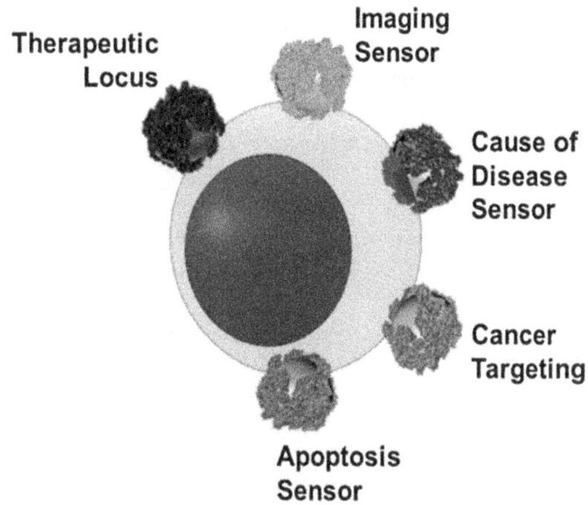

Fig. 19.13 Cellular nanosensors with diverse potentials: the future events?

Fig. 19.14 Nanoparticles permeating BB

from both current and future diseases. This would exclude the needed for vaccines in order to acquire immunity against certain diseases. Imagine the world without flu, or AIDS or other difficult viral or bacterial diseases! (Fig. 19.15)

The Use of Nanoparticles in Cancer Treatment

The nanoparticle model has had difficulty from its inception with the use of quantum dots, later proven to be toxic. However, recent models have proven dramatically more effective as anti-cancer agents through various methods of delivery. The model proposes that nanoparticles that are composed of silicon and can be coated with antibodies specific for stem cell markers found on cancer stem cells, or similar molecular conjugates. Through enhanced permeation retention, vascularized tumors

Fig. 19.15 Dendrimer as medically applied nanoparticle preventing the entrance of viruses into the cell

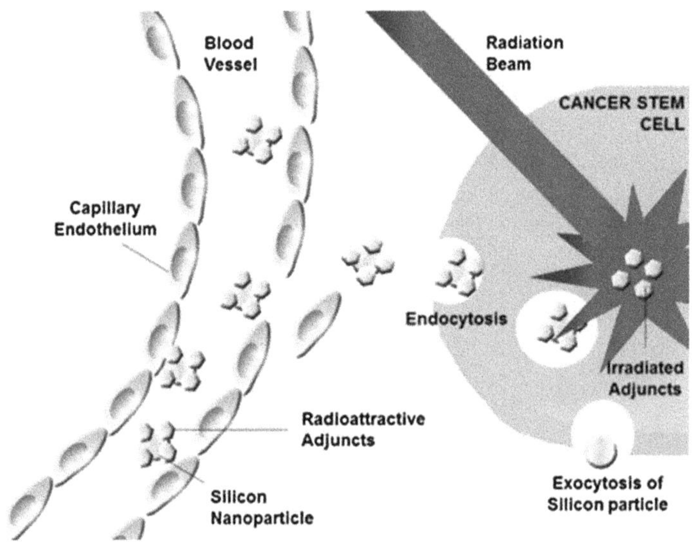

Fig. 19.16 Nanoparticle delivery of radioattractive adjuncts to cancer stem cells

would become infiltrated by these nanoparticles. Infrared radiation is then applied to the tumor, wherein the cancer stem cells that now contained these particles would increase in heat at a much greater rate than cells not containing the particles, thus killing the targeted cells. Studies have been carried out for a variety of cancer types including, breast, prostate and liver cancer [38] (Fig. 19.16).

- *The future of nanotechnology in medical arena*
  As drugs get smaller, they will be able to easily "sneak" past the body's defense mechanisms and will be able to reach places that the drugs available today, cannot. Because smaller compounds have a large surface area to volume ratio, these new drugs should also be more reactive. Tiny nanoparticles known as "**quantum dots**" can be made to give off different colors depending on their size, and so

used **for cellular disease detection** [34]. They can also be made to attach to different biological components, such as certain proteins in certain colors, making it much easier to analyze blood for specific components.

Nanotechnology used **for destruction of diseased tissues** makes use of Nano shells, microscopic balls of glass coated in gold. Nano shells can also be designed to bind to specific components in the body, and can then be heated by lasers to destroy damaged tissue without causing any more damage to skin or other close by tissue.

It has already been said that the successful development of a **medical nano-robot** would "change the world of medicine forever".

Conclusions and Future Directions

This chapter has taken in consideration the fundamental features of nanotechnology (geometry and minimization), the structure of the atom (as particulate entity), the synthesis of buckminsterfullerene's, discovery of graphene structure and its possible application in medical fields of drug-delivery (nanotubes, bucky balls) nano-robotics, anti-cancer therapy, anti-aging preparations, etc. [24, 35–38]. It does comprehend the most fundamental physicochemical aspects of this matter discovered and classified within last decade.

Nanotechnology for sure prepare us to looking in the future with optimism: In addition to innovation in other fields, nanotechnology will prompt the medicine to remove obstruction in the circulatory system, kill cancer cells, or take over the function of sub cellular organelles.

Killing viruses or bacteria, dissolving cholesterol or blood clots is very close to be done on nanotechnology basis. Discovering the earliest signs of the diseases, even before the actual symptoms appear, is not a dream anymore. We will be able to produce hearing aids that are actually a computer in each ear, artificial retinas to restore sight and many other medical wonders.

Nanotechnology would allow the gene combinations take place at the levels of molecules, rather than the larger gene. That would mean creating whole new organs for people, organs which their bodies won't reject as they do with transplanted human organs. Some of these will come within the next ten years, for sure. These persistent growing ideas will completely revolutionize the medical approach in sense of **diagnosis**, **treatment** and even **prevention**, minimizing equipment, space and errors that are inevitable today.

References

1. Curl Jr. RF, Kroto HE. Autobiography, nobelprize.org/ (2013)13 April
2. Brown, L., Holme, T.: Chemistry for Engineering Students. Mary Finch, Books/Cole, Cengage Learning, Belmont (2011). ISBN 13:078-1-4390-4791-0

3. Pavlovic, M.D.: Biomineralization and nanobacteria: at the door of the new concept? MD-Med Data Rev **1**(2), 21–24 (2009)
4. Geim, A.K., Novoselov, K.S.: The rise of graphene. Nat. Mater. **6**(3), 183–191 (2007)
5. Novoselov, K.S., Geim, A.K., Morozov, S.V., et al.: Electric field in atomically thin carbon films. Science **306**(5696), 666–669 (2004)
6. Berger, C., Song, Z., Li, X., et al.: Electronic confinement and coherence in patterned epitaxial graphene. Science **312**(5777), 1191–1196 (2006)
7. Zhang, X.Y., Tan, Y.-W., Stormer, H.L., Kim, P.: Experimental observation of the quantum Hall effect and Berry's phase in graphene. Nature **438**(7065), 201–204 (2005)
8. Novoselov, K.S., Geim, A.K., Morozov, S.V., et al.: Two dimensional gas of massless Dirac fermions in graphene. Nature **438**(7065), 197–200 (2005)
9. Berger, C., Song, Z., Li, X., et al.: Ultrathin epitaxial graphite: 2-D electron gas properties and a route toward graphene-based nanoelectronics. J. Phys. Chem. B **108**(52), 19912–19916 (2004)
10. Son, Y.-W., Cohen, M.L., Louie, S.G.: Energy gaps in graphene nanoribbons. Phys. Rev. Lett. **97**(21), Article ID 216803, 4 p (2006)
11. Xiang, H., Kan, E., Wei, S.-H., Whangbo, M.-H., Yang, J.: "Narrow" Graphene Nanoribbons Made Easier by Partial Hydrogenation. Nano Lett. **9**(12), 4025–4030 (2009)
12. Han, M.Y., Özyilmaz, B., Zhang, Y., Kim, P.: Energy bandgap engineering of graphene nanoribbons. Phys. Rev. Lett. **98**(20), Article ID 206805, 4 p (2007)
13. Areshkin, D.A., Gunlycke, D., White, C.T.: Ballistic transport in graphene nanostrips in the presence of disorder: importance of edge effects. Nano Lett. **7**(1), 204–210 (2007)
14. Zahra, J., Check, D.C.Y., Amin, N.A., Ismail, K.: Modelling of graphene nanorribon Fermi energy. J. Nanomater **90**, 9347 (2010)
15. Katre, N.V.: The conjugation of proteins with polyethylene glycol and other polymers altering properties of proteins to enhance their therapeutic potential. Adv. Drug Deliv. Rev. **10**, 91–114 (1993)
16. Kobayashi, H., Brechbiel, M.W.: Dendrimer-based nanosized MRI contrast agents. Curr. Pharm. Biotechnol. **5**, 539–549 (2004)
17. Kobayashi, H., Brechbiel, M.W.: Nano-sized MRI contrast agents with dendrimer cores. Adv. Drug Deliv. Rev. **57**, 2271–2286 (2005)
18. Anderson, J.M., Shive, M.S.: Biodegradation and biocompatibility of PLA and PLGA microspheres. Adv. Drug Deliv. Rev. **28**, 5–24 (1997)
19. Davis, M.E.: Nanoparticle therapeutics: an emerging treatment modality for cancer. Nat. Rev. Drug Discov. **7**(9), 771–782 (2008)
20. Stupp, R., et al.: NovoTTF-100A versus physician's choice chemotherapy in recurrent glioblastoma: a randomised phase III trial of a novel treatment modality. Eur. J. Cancer **48**(14), 2192–2202 (2012)
21. Brannon-Peppas, L., Blanchette, J.O.: Nanoparticle and targeted systems for cancer therapy. Adv. Drug Deliv. Rev. **56**(11), 1649–1659 (2004)
22. Gagliardi, G., et al.: Long-term cardiac mortality after radiotherapy of breast cancer–application of the relative seriality model. Br. J. Radiol. **69**(825), 839–846 (1996)
23. Gaitanis, A., Staal, S.: Liposomal doxorubicin and nab-paclitaxel: nanoparticle cancer chemotherapy in current clinical use. Methods Mol. Biol. **624**, 385–392 (2010)
24. http://www.bionano.neu.edu/NanoRobotics_in_Medicine.pdf
25. Popovic, J., Petrajcic, J., Pocac, M., Popovic, J.K., Popovic, D.J., Tepavcevic, V.: Kinetics of molecular and nanostructural drugs. MD-Medical Data **5**(2), 125–133 (2013)
26. Pavlovic, M., Mayfield, J., Balint, B.: Nanotechology and its Application in Medicine. In: Furth, B., Agarwal, A. (eds.) Handbook of Medical and Healthcare Technologies, pp. 181–205. Springer, New York (2013)
27. Fahmy, T.M., Fong, P.M., Park, J., Constable, T., Saltzman, W.M.: Nanosystems for simultaneous imaging and drug delivery to T cells. AAPS J. **9**(2), 19 (2007)
28. Langer, R., Folkman, J.: Polymers for the sustained release of proteins and other macromolecules. Nature **263**, 797–800 (1976)

29. Naked, K., Fujita, M., Dresselhaus, G., Dresselhaus, M.S.: Edge state in graphene ribbons: nanometer size effect and edge shape dependence. Physical Review B **54**(24), 17954–17961 (1996)
30. Barone, V., Hod, O., Scuseria, G.E.: Electronic structure and stability of semiconducting graphene nanoribbons. Nano Lett. **6**(12), 2748–2754 (2006)
31. Liang, G., Neophytou, N., Nikonov, D.E., Lundstrom, M.S.: Performance projections for ballistic graphene nanoribbon field-effect transistors. IEEE Trans Electron Devices **54**(4), 677–682 (2007)
32. Nakada, K., Fujita, M., Dresselhaus, G., Dresselhaus, M.S.: Edge state in graphene ribbons: nanometer size effect and edge shape dependence. Physical Review B **54**(24), 17954–17961 (1996)
33. Wakabayashi, K.: Electronic and magnetic properties of nano graphite. In: Makarova, T., Palacio, F. (eds.) Carbon–Based Magnetism—An Overview of the Magnetism of Metal Free Carbon-Based Compounds and Materials, pp. 279–304. Elsevier, Amsterdam (2006)
34. Yan, Q., Huang, B., Yu, J., et al.: Intrinsic current-voltage characteristics of grapheme nanoribbon transistors and effect of edge doping. Nano Lett. **7**(6), 1469–1473 (2007)
35. Wildoer, J.W.G., Venema, L.C., Rinzler, A.G., Smalley, R.E., Dekker, C.: Electronic structure of atomically resolved carbon nanotubes. Nature **391**(6662), 59–62 (1998)
36. Jishi, R.A., Inomata, D., Nakao, K., Dresselhaus, M.S., Dresselhaus, G.M.W.: Electronic and lattice properties of carbon nanotubes. J. Physical Soc. Japan **63**(6), 2252–2290 (1994)
37. Satoh, S., Nakamura, Y., Kanade, T.: Name-it: naming and detecting faces in news videos. IEEE. MultiMedia **6**(1), 22–35 (1999)
38. Gil, J., Stenbalska, A., Pesz, K.A., Sasiadek, M.M.: Cancer stem cells: the theory and perspectives in cancer therapy. J. Appl. Genet. **49**(2), 193–199 (2008)

Chapter 20
Tissue Engineering Breakthroughs

Stem cell research can revolutionize medicine, more than anything since antibiotics.

Ron Reagan (1911–2004)

Tissue engineering (TE) is an emerging multidisciplinary field involving biology, medicine, and engineering that is likely to revolutionize the ways we improve the health and quality of life for millions of people worldwide by restoring, maintaining, or enhancing tissue and organ function. Three essentials components of TE are: cells, scaffolds and molecules of extracellular matrix, designed to repair tissue defect. In addition to having a therapeutic application, where the tissue is either grown in a patient or outside the patient and transplanted, tissue engineering can have diagnostic applications where the tissue is made in vitro and used for testing drug metabolism and uptake, toxicity, and pathogenicity. The foundation of tissue engineering for either therapeutic or diagnostic applications is the ability to exploit living cells in a variety of ways. Tissue engineering research includes: biomaterials, cells, biomolecules, engineering design aspects, biomechanics, bio-informatics in order to help interventions not only at organ but also at cellular and molecular level. Therefore, gene therapy, manipulations with abzymes and rational vaccine design are also parts of TE. This Chapter will present the crucial breakthroughs in this field.

The fundamental breakthroughs in this century in the field of Bioengineering and Tissue engineering: A. Atala: transplantation without a donor, Steven Badylak: regeneration, Robert Langer: Drug delivery systems, Gordana Vunjak Novakovic: artificial bones and regenerative therapy for Cardial Tissues

M. Pavlovic, *Bioengineering*, DOI 10.1007/978-3-319-10798-1_20

Introduction

Tissue engineering crosses numerous medical and technical specialties: cell biologists, molecular biologists, biomaterial engineers, computer-assisted designers, microscopic imaging specialists, robotics engineers, and developers of equipment such as bioreactors, where tissues are grown and nurtured.

Tissue engineering thus involves a combination of disciplines to achieve new therapies and in some cases, entirely new approaches to therapy. Transplantability of engineered tissue and artificial organs in combination with immunosuppressive therapy had become a daily reality. The biggest problem is the huge number of patients requiring that kind of interventions and new technology is needed to reduce this deficit.

What is TE? There are two official definitions accepted in scientific community:

NIH Definition of Tissue Engineering (TE)

Tissue engineering is an emerging multidisciplinary field involving biology, medicine, and engineering that is likely to revolutionize the ways we improve the health and quality of life for millions of people worldwide by

- **restoring,**
- **maintaining, or**
- **enhancing**

tissue and organ function. In addition to having a **therapeutic application**, where the **tissue is either grown in a patient or outside the patient and transplanted, tissue engineering can have diagnostic applications where the tissue is made in vitro and used for testing drug metabolism and uptake, toxicity, and pathogenicity.** The foundation of tissue engineering for either **therapeutic** or **diagnostic** applications is the ability to exploit living cells in a variety of ways. Tissue engineering research includes:

- **biomaterials,**
- **cells,**
- **biomolecules,**
- **engineering design aspects,**
- **biomechanics,**
- **informatics**

to support tissue engineering and stem cell research.

The Pittsburgh Tissue Engineering Initiative Definition

Tissue engineering **is the development and manipulation of laboratory-grown**

- **molecules,**
- **cells,**
- **tissues, or**
- **organs**

to replace or support the function of defective or injured body parts. Although cells have been cultured, or grown, outside the body for many years, the possibility of growing complex, three-dimensional tissues—literally replicating the design and function of human tissue—is **a recent development** [1]. The intricacies of this process require input from many types of scientists, including **the problem solving expertise of engineers,** hence the name tissue engineering.

Laboratory techniques for inducing tissue formation are frequently inspired by natural mechanisms. **Central among these new strategies is the idea that synthetic materials can serve as degradable templates for tissue regeneration.** Blood cells circulate and can be injected, while solid organ cells need a mechanical foundation when transplanted into the organ which are localized at a particular site (liver, skin, heart, brain), and chances for cell-cell communication [2, 3]. In tissue engineering, **degradable polymers** are often used to provide **scaffolds** for mechanical support.

A strong limitation with stem cell research is the ever-present question of embryo ethics but we can envisage that what medical breakthroughs we could achieve if scientists were not so constrained by federal regulations. The field of regenerative medicine (RM), encompassing stem cell (SC) technologies, therapeutics, tissue engineering (TE), biomaterials, scaffolds and other enabling technologies provides a wide gamut of tools and tracks to combat, manage and hopefully cure serious human and animal injuries, dysfunctions and diseases [5–7]. The trends that are becoming the major platforms in this field within the last ten years in itself are including: multi-track directives of adult stem cell translational technologies, tissue and organ engineering protocols, iPS cell applications, and understanding of the role of cancer stem cells to develop effective, targeted anti-cancer regimens. With the rapid advances of RM translational research, further advances are expected to be implemented for personalized repair and curative outcomes. RM future is bright although laden with challenges of global fragmentation which needs coherent consolidation, stringent cost and time effective regulation and long-term funding mechanisms, so clinical and diagnostic solutions are realized and recognized to combat unmet medical needs [8–10]. As a result, there has been an exponential increase in the expectations for translation of these technologies, both on a national and global scale. This is because RM offers potential solutions to many unmet clinical needs and challenges, and provides a variety of innovative CT/TE tools to repair, regenerate and rebuild tissues and organs [2].

Development and Examples of Tissue Engineering

Bioreactor refers to a device or system meant to grow cells or tissues in the context of cell culture. These devices are being developed for use in TE [1]. Bioreactor is mimicking the situation in the body with regard to temperature, humidity and ratio of CO_2/O_2.

Medical Technology Breakthroughs

Development of Cell Patterns for Possible TE Use and Technology of Multiple Staining-Flow Cytometry

The first use of stem cells in humans was done by E. Donald Thomas whose work was later recognized with a Nobel Prize and Physiology or Medicine. His work showed that bone marrow cells infused intravenously could repopulate the bone marrow and produce new blood cells. His work also reduced the likelihood of developing a life-threatening complication known as graft-versus-host disease (GVHD). With the availability of the stem cell growth factors most hematopoietic stem cell transplantation procedures are now performed using stem cells collected from the peripheral blood rather than from the bone marrow [10]. Collecting peripheral blood stem cells provides a bigger graft, does not require that the donor be subjected to general anesthesia collect the graft, results in a shorter time to engraftment, and may provide for a lower long-term relapse rate [10]. The first recorded attempt at cellular therapy occurred in 1912 when German physicians attempted to treat hypothyroid children with thyroid cells [10]. Cellular therapy, as practiced today, was developed in the early 1930s by Paul Niehans, MD (1882–1971), a Swiss physician who became known as "the father of cell therapy." It soon became popular with celebrities as a means of rejuvenation. A 1990 article in In Health magazine described Niehans as a "public relations genius" and stated that the Clinic La Preire, which he had founded in Clarens-Montreaux, Switzerland, had attracted 65,000 patients. Its (1999) one-week "revitalization program" costed about $8,000. However, Niehans made a great breakthrough in the last century by applying cell based therapy, which was unimaginable before him.

Generally, the Stem Cell (SC)—compartment is divided into embryonic and tissue specific or adult SCs. Embryonic SCs (ES or ESC) are by definition the "master cells" with the largest spectrum of differentiation potential, e.g. capable of differentiating into every type of cells either in vitro or in vivo [10–14]. Thanks to the presence of embryonic body, these cells have ability to develop into three primary layers: endoderm, ectoderm and mesoderm. The discovery of SCs inside cell mass of embryos and in adult tissue has revolutionized the medical field by introducing new therapeutic dimensions into previously untreatable diseases and injuries [14]. Several experimental or preclinical studies have suggested that application of embryonic SC could be promising in the treatment of various diseases.

However, recognition of appropriate ethical aspects, regulatory acts and standardization in embryonic SC mediated regenerative medicine is needed as it is still the matter of controversy. Besides, permanent, persistent and accurate updating of the facts regarding their immunologic characteristics is an essential requirement for safe clinical application of SCs. Some authors stand that the initial theory that embryonic SCs are ignored by immunocompetent hosts was overlooked. On the contrary, it is even more evident that embryonic SCs could protect themselves actively by several immunomodulatory mechanisms against T lymphocytes and natural killer cells of host, and actively participate in immune-mediated events [13].

Recent isolation of fetal SCs from several sources either at the early stages of development or during the later trimesters of gestation, sharing similar growth kinetics and expressing pluripotency markers, provides strong support to the statement that these cells may be biologically closer to embryonic SCs . In fact, they represent intermediates between embryonic and adult mesenchymal SCs with regards to proliferation rates and plasticity features, thus being able to confer an advantage over postnatal mesenchymal SCs derived from conventional adult sources, including mesenchymal cells [12].

Historically, bone marrow was the primary source of SCs for transplant. However, peripheral blood and umbilical (cord) blood are also currently used as sources. SCs derived from these sources may have therapeutic potential (without severe adverse effects) only when given to the individual from whom they were derived (autologous transplants) or from an immunologically matched donor (allogeneic transplants) [4–6].

Despite the fact that the ideal type and source of cells have not yet been defined, immature SCs are capable of colonizing different tissues due to ability of homing and trans-differentiation or lineage-plasticity, in the settings of regenerative medicine [5–9]. Furthermore, there are several facts suggesting that adult SCs and even differentiated somatic cells, under appropriate microenvironmental cues or signals, are able to be "reprogrammed" and contribute to a much wider spectrum of differentiated progeny than previously anticipated. This has been demonstrated by using tissue-specific SCs—which like embryonic SCs—do not express CD45 as an exclusive hematopoietic marker [11]. Consequently, adult mesenchymal SCs and endothelial precursors are clinically applicable for cell-mediated, regenerative therapy of patients with myocardial, brain, vascular, liver, pancreas and some other tissue damages [4, 5, 15–22].

Allogeneic transplants are still the most efficient treatment for patients with liver failure. However, there is a lack of donors and some alternative therapeutic approaches are needed. Transplantation of mature hepatocytes has been evaluated, but the long-term efficacy remains unclear and the paucity of donor cells limits this strategy. The use of SC-therapy transplantation is perhaps a more promising alternative approach.

The intensification of myeloablative radio-chemotherapy, enlarged use of SC transplants, as well as the introduction of cell-mediated therapeutic approaches in regenerative medicine resulted in increased needs for both specific blood-derived progenitor/cells and practical operating procedures inducing minimized cellular damages during their collection or processing and storage in frozen state. Therefore, successful performance of SC transplantats or cell-mediated therapy requires efficient collection, processing, and (cryo) preservation procedures for obtaining an acceptable cell yield and post-thawing recovery, as well as advantageous clinical outcome.

In South Korea, Woo Suk Hwang of Seoul National University was known as the cloning king and his lab enjoyed strong support—and funding—from the South Korean government. It helps that therapeutic cloning was (and still is) much less controversial in South Korea than in the West. And unlike most of the rest of the world, the Korean public offers almost unequivocal support. Hwang says that a poll showed that more than 70 % of South Koreans agree with therapeutic cloning, whereas a recent poll in the US suggests 75 % are opposed to it. Two factors in particular seemed to be critical to Hwang's success. First, an incredible 1,200 nuclear transfers (on cow and sheep cells) would take place every day in his lab. It really has been a case of practice makes almost perfect. Second, was the supply of eggs. South Korean law allows Hwang to use fresh eggs from young women who are prepared to donate their eggs by undergoing ovarian stimulation, which can be a risky and painful procedure. Yet, the results of Hwang have been withdrawn at the certain point as false. Was it political issue? Or was it true? It seems that this drama continues and we have to wait for the final scenario.

In the UK, by contrast, scientists are only allowed to use eggs rejected or left over from IVF treatment. A group in Newcastle, UK (2005), announced that they had cloned a very early stage embryo, but this took 36 leftover eggs and the team failed to isolate any stem cells from the embryo. Hwang's team found that when the eggs used come from donors under 30, an average of only 14 eggs were needed to generate a cell line, a 16-fold improvement in efficiency from only a year ago. He has also reduced the use of animal "feeder cells" to nourish the developing embryos, making it less likely that the stem cells will become infected with viruses or prion diseases. When it comes to cloning techniques, Hwang's laboratory is now way ahead of the field, says stem cell biologist Stephen Minger of King's College London, who recently visited Korea. He said that there is a good chance that the US will be left behind. And really, the wind came from the other side when after Dr. Hwang withdraw his discovery as not having meritory values. Dr. Stojkovic [9] from New Castle (now in Serbia, Leskovac) has shown that analysis of arrested embryos demonstrated that these embryos express pluripotency marker genes such OCT4, NANOG, and REX1. Derived hESC lines also expressed specific pluripotency markers (TRA-1-60, TRA-1-81, SSEA4, alkaline phosphatase, OCT4, NANOG, TERT, and REX1) and differentiated under in vitro and in vivo conditions into derivatives of all three germ layers [9]. All of the new lines, including lines derived from late arrested embryos, have had normal karyotypes [9]. These results demonstrate that **arrested embryos are additional valuable resources to surplus and donated developing embryos, and should be used to study early human development or derive pluripotent hESC** [9]. The line of work of this researcher and his groups is actually the first one which has clearly and with no doubts with regards to reproducibility, demonstrated the possibility of establishing embryonic

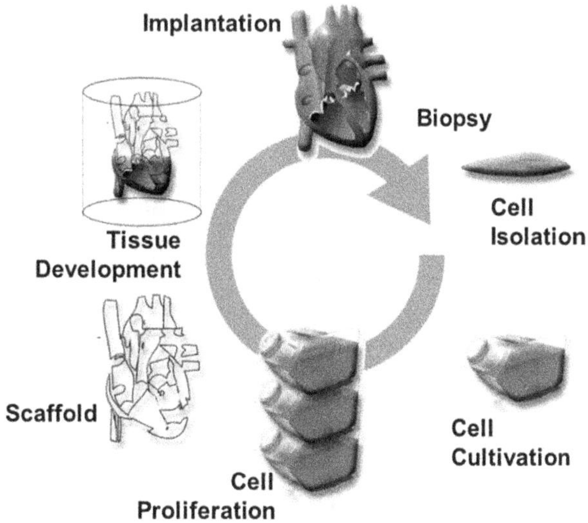

Fig. 20.1 Principles of TE

human cell line. The use of ESC in therapy of different diseases is more problematic then the story that preceeds their establishment. They are actually alogeneic transplants which need immunosupression or in vitro intervention called stem cell genetic (therapeutic) cloning, where embryonal cell cytoplasm is fused into hybrid with the nucleus of chosen adult cell in order to continue self-renewal and repair the desired tissue line or organ from the patient's proteins orchestrated by nuclear coding from patient's genetic material (nuclear gene transfer technique) (Fig. 20.1).

This is a critical issue from both scientific and clinical point of view, given that the concept of VSELs (very small embryonic like adult, non-hematopoietic stem cells) have been recently discovered, and proved by Dr. Ratjczak's group [23]. This emerging new concept raises the questions whether these cells should be used at all, if VSELs can already replace them, successfully? Not only that, but quite a distinct CD34- population of mesenchymal stem cells (MSCs) has been isolated from adult tissues, including bone marrow, and defined phenotypicaly, and functionaly as a distinguished category of adult stem cells with respect to hematopoietic stem cells and VSELs.

Summary and Conlusions on the Role of Stem Cells in TE

Hematopoietic stem cell transplantation remains a risky procedure with many possible complications. It has traditionally been reserved for patients with life-threatening diseases, such as malignancies. While occasionally used experimentally in nonmalignant and nonhematologic indications such as severe disabling autoimmune and cardiovascular diseases, the risk of fatal complications appears too high to gain wider acceptance [24–33]. Yet, this is the most-well known and the most developed stem-cell regenerative approach, given that if successfully engrafted, it repopulates and later on recruits the new, healthy bone marrow cells in circulation.

Embryonic stem-cell research is still the matter if controversies at a very stratified levels, although many researchers agree that it might be the source of stem cells with the highest differentiation potential.

Apparently, basic adult stem cell research is still evolving, and is the matter of everchangeing issues. Due to our extensive studies, but yet limited knowledge on their behavior and potentials, it is not yet easy to determine how to act in clinical arena. It is obvious that each approach to any particular disease or damage has to be optimized within team work and by bridgeing the gap between fundamental and clinical studies. Knowing molecular level in depth, will help clinicians to orcestrate the team work and overcome critical obstacles in each particular scenario. There is no doubt that adult stem cell therapy (and probably embryonic as well) belong to the future, but we have to act as that we shall belong to the future, as well. Continuous efforts in both molecular and clinical directions will lead to the unique and optimal plan for each particular regenerative treatment. Are we too far away from that goal or not, it will be shown very soon.

Microfabrication of Scaffolds and 3-D Growth of Tissues

As the research field of particular stem cell patterns and their optimization for the cellular therapy and TE purposes is still evolving, one of the most significant/fundamental achievements in the TE filed is definitely the development of the scaffolds as the cell supporters and strong motivation for 3-D growth of tissue. So far, cell culture done on different surfaces enabled two-dimensional/planar growth of cells which could not reproduce all of the tissue dimensions neither their complex architecture. Thus, scaffolds represent an important component in TE. What is also important it is to know the guide- lines for selecting scaffolds and the major scaffolding approaches as the part of specific TE design [9, 11, 20]. As the role of TE is to restore, replace or regenerate defective tissue, it is apparently multitask area providing a complex evolution of the new-growing components of defective tissue. We have seen that one of the possibilities is transplantation of stem cells, however it does function in particular occasions and is not the answer to all tissue damages that require reparation. Therefore, development of scaffolds with consecutive 3-D growth of the cells was one the most striking technological breakthroughs in this field.

Scaffolds have the origin in biomaterials that have to be biocompatible, biodegradable and bio-resorbable in order to fully satisfy its role in supporting the 3-D growth of the cells chosen for tissue damage repair [9]. It is not easy neither to find (natural) nor to manufacture (fabricated) such material. Therefore, the technology of scaffolds is still developing as a separate scientific approach. The technologies for micro-fabrication are numerous and not the matter of this chapter. The key of the structure are holes in the architecture of the scaffolds on different levels (Fig. 20.1) that are no bigger than 50 μm and represent a physical challenge for cells to go through.

This is causing three-dimensional growth and differentiation at the same time which results in a piece of tissue either in the organism or in the dish, where it will be taken from in order to be implanted into right spot/locality. Thus, using the scaffolds and Growth factors from ECM (regulatory cytokines) a suitable biochemical and biomechanical microenvironment is created and cell multiplication fills the scaffold with the tissue and allows the cells to grow into the correct shape. When implanted into the body, the seeded scaffold becomes integrated concomitantly supporting and directing cell proliferation. As the cells proliferate and differentiate, the scaffolds slowly biodegrade, gradually allowing blood vessels and host cytokines to make contact with the cells [34–41]. Through this process, the scaffold furthure biodegrades while the cells proliferate and differentiate into desired tissue. Finally, the scaffold completely dissolves and the formed tissue starts functioning in its new surrounding.

CATE (Computer Aided Tissue Engineering) as a Leading Concept

The inevitable consequence of the recent revolution in the biological sciences and bioengineering has brought about the new field of computer-aided tissue engineering (CATE) [18, 19]. It really highlights the interdisciplinary nature of this technology breakthrough. Particular focus in this field is placed on rapid prototyping and direct digital fabrication for cell and organs, construction of tissue analogues and precursors of 3-D scaffolds [41, 42]. This emerging field encompasses computer-aided design (CAD), image processing, manufacturing and solid-free-form fabrication (SFF) for modeling, designing simulation and manufacturing of biological tissue and organ substitutes [41–46]. This involves imaging based 3D model reconstruction, computer aided-tissue informatics, with a wide array of image modalities, DNA microarrays, etc., computer-aided cell analysis (cell counting, geometry, chromosomal counting interpreting fluorescence data, etc.), computer-aided tissue identification and analysis, computer-aided tissue scaffold design and manufacturing.

Ink-jet Printing of the Cells and Liquid Scaffolds

Organ printing is today possible thanks to application of brilliant idea that cells can be printed onto scaffolds as it is the ink [19, 45–49]. This in essence, very simple technique provides a broad spectrum of TE maneuvers including the very fast recovery from large burns, the event that is extremely useful and highly necessary in those situations. Today many cell types can be printed as bio-ink using ink-jet printers: the cells survive, maintain their phenotype, differentiate and show function. They can be printed uniformly and homogeneously into confluent layers. They can be printed into 3-D structures. However, there are considerable technical barriers in the development of this emerging inkjet printing technology, such as the ability of the modified printers to deliver viable cells and the capability of the inkjet printing to fabricate functional, viable and functionally vascularized 3-D configurations. [49]. This is the future problem which needs to be gradually solved.

Transplantation Without a Donor

For more than century, medical doctors were looking for alternative solutions to help address the shortage of organs for those needing transplantation. The whole concept of this greatest probably technological breakthrough in TE is that patient can use his own tissue in order to replace damaged organ. Dr. Anthony Atala is doing that now with the bladder [11, 20]. After a small biopsy, patient's own cells are shaped, grown on three-dimensional polymer mold in the shape of a bladder and eventually the temporary scaffold deteriorates inside the body, leaving behind, a healthy, functional organ. Most important is that since the cells are taken from the patient, the body will not reject the organ grown in that manner.

Other Techniques of Great Relevance for Advance of TE

Tissue engineering is a very complex and multidisciplinary field. Many discoveries are in tight relationship with the development of the idea. Looking into the roots, probably the discovery of microscope was the most important for establishment of the theory of the cell. Compact light microscope is later on upgraded into microscope with phase contrast which has enabled scientists to see through the medium the cells that they were growing, including stem cells. Electron microscopy will give the ultrastructure and inspire biochemistry to develop in order to explain molecular conversions and other processes in the cells. Atomic force microscope (AFM) will help understanding of chemical content of the cells. The CO_2 incubator will later on enable the growth of the cells in culture and afterwards in tissue culture. Development of fluorescent labeling and Flow Cytometer in combination with fluorescent microscope will enable molecular detection and phenotypic distinction between the cells. This had incredible impact on development of stem cell theories and concepts. The explosion and evolution of the field is infinite. The future world will due to that have more comfortable and much higher quality of life and probably the life itself will be somewhat longer although humans will never fly [51–55].

Skin grafts (optimized tissue culture techniques by culture with feeder layer of irradiated 3 T3 cells and medium with growth factors)

Enhancing vascularization (microsphere that slowly release vascular growth factor) mixed with endothelial cells will form the vessel system in a bioreactor).

Examples of TE/Cell Therapy Treatments in Development with Help of Neuralstem Inc (Overview)

- ALS
  FDA-approved Phase II NSI-566/ALS clinical trial, expanded to two centers with significant increase in dosing, commenced in September 2013. In Mexico City, ALS Phase I/II trial expected to commence in 2014.
- Ischemic Stroke
  NSI-566 Phase I/II trial for ischemic stroke commenced in December 2013, in collaboration with BaYi Brain Hospital in Beijing.
- Spinal Cord Injury
  FDA approved Phase I safety trial transplanting NSI-566 into the T2-T12 lumbar regions of chronic SCI patients. Phase I commencement expected in 2014. Acute SCI trial expected to commence in Seoul, South Korea in 2014.
- Multiple Sclerosis (MS)
  Preclinical phase with NSI-566. Further clinical development on multiple sclerosis application pending NSI-566 ALS ongoing trial.
- Optic Neuritis
  Preclinical phase with NSI-566. Further clinical development on optic neuritis application pending NSI-566 ALS ongoing trial.
- Alzheimer's Disease
  Preclinical phase with NSI-566. Further clinical development on Alzheimer's disease application pending NSI-566 ALS ongoing trial.
- Traumatic Brain Injury
  Preclinical phase with NSI-566. Further clinical development on traumatic brain injury application pending NSI-566 ALS ongoing trial.
- Peripheral Nerve Injury
  Preclinical phase with NSI-566. Further clinical development on peripheral nerve injury application pending NSI-566 ALS ongoing trial.
- Diabetic Neuropathy
  Preclinical phase with NSI-566. Further clinical development on diabetic neuropathy application pending NSI-566 ALS ongoing trial.
- Lysosomal Diseases
  Preclinical phase with NSI-566. Further clinical development on lysosomal diseases application pending NSI-566 ALS ongoing trial.
- Parkinson's Disease
  Preclinical phase with NSI-566. Further clinical development on Parkinson's disease application pending NSI-566 ALS ongoing trial.
- Huntington's Disease
  Preclinical phase with NSI-566. Further clinical development on Huntington's disease application pending NSI-566 ALS ongoing trial.
- Cerebral Palsy
  Preclinical phase with NSI-566. Further clinical development on cerebral palsy application pending NSI-566 ALS ongoing trial.

- Ischemic Spastic Paraplegia
  Preclinical phase with the University of California at San Diego. Lead collaborator, Dr. Martin Marsala, PhD authored a peer-reviewed paper on study results.

See more at: http://www.neuralstem.com/patient-info-treatments-in-development#celltherapy

References

1. Temple, J.P., Yeager, K., Bhumiratana, S., Vunjak-Novakovic, G., Grayson, W.L.: Bioreactor cultivation of anatomically shaped human bone grafts. Methods Mol. Biol. **1202**, 57–78 (2014)
2. HepaLife—Artificial Liver
3. British scientists grow human liver in a laboratory. Mail Online
4. World's First Artificial Human Liver Grown In Lab. LiveScience
5. Van de Kamp, J., Jahnen-Dechent, W., Rath, B., Knuechel, R., Neuss, S.: Hepatocyte growth factor-loaded biomaterial for mesenchymal stem cell recruitment. Stem Cells Int. Article ID **892065**, 9 (2013)
6. Krotz S, Robins J, Moore R, Steinhoff MM, Morgan J, Carson S. Model Artificial Human Ovary by Pre-Fabricated Cellular Self-Assembly. (2008) 64th Annual Meeting of the American Society for Reproductive Medicine, San Francisco, CA
7. Warwick, K., Gasson, M., Hutt, B., Goodhew, I., Kyberd, P., Schulzrinne, H., Wu, X.: Thought Communication and Control: A First Step using Radiotelegraphy. IEE Proc. Commun. **151**(3), 185–189 (2004)
8. Taylor, D.A.: Cell therapy-a 21st century hope for treating cardiovascular disease-a five-year retrospective and predictive view. Am. Heart Hosp. J. **9**(1), E24–E27 (2011)
9. Stojkovic, M., Stojkovic, P., Leary, C., Hall, V.J., Armstrong, L., Herbert, M., Nesbitt, M., Lako, M., Murdoch, A.: Derivation of a human blastocyst after heterologous nuclear transfer to donated oocytes. Reprod. Biomed. Online **11**(2), 226–231 (2005)
10. Badylak, S.F., Taylor, D., Uygun, K.: Whole-organ tissue engineering; decellularization and recelullarization of three-dimensional matrix scaffolds. Annu. Rev. Biomed. Eng. **13**, 27–53 (2011)
11. Mhashilkar, A., Atala, A.: Advent and maturation of regenerative medicine. Curr. Stem Cell Res. Ther. **7**, 430–445 (2012)
12. Boland, T., Xu, T., Damon, B., Cui, X.: Application of inkjet printing tissue engineering. Biotechnol. J. **1**, 910–917 (2006)
13. Schneider, R.K., et al.: The role of biomaterials in the direction of mesenchymal stem cell properties and extracellular matrix re-modelling in dermal tissue engineering. Biomaterials **31**, 7948–7959 (2010)
14. Chen, B.P., Leong, K.W.: Scafolding in tissue engineering: general approaches and tissue-specific considerations. Eur. Spine J. **17**(Suppl 4), 467–479 (2008)
15. Nemeno-Guanzon, J.G., et al.: Trends in tissue engineering. J. Biomed. Biotechnol. **2012**, 956345 (2012). doi:10.1155/2012/956345. 14p
16. Lee, J.I., Sato, M., Kim, H.W., Mochida, J.: Transplantatation of scaffold-free spheroids composed of synovium-derived cells and chondrocytes for the treatment of cartilage defects of the knee. Eur. Cells Mater. J. **22**, 275–290 (2011)
17. Michael, A., Liebchner, K. (eds.): Computer Aided Tissue Engineering. Humana, New York (2012). ISBN 978-1-61779-764-4
18. Sun, W., Darkling, A., Starly, B., Nam, J.: Computer-aided tissue engineering: overview, scope and challenges. Biotecnol. Appl. Biochem. **39**(Pt1), 29–47 (2004)

19. Boland, T., Xu, T., Damon, B., Cui, X.: Application of inkjet printing to tissue engineering. Biotechnol. J. **9**, 910–917 (2006)
20. Atala, A., Bauer, S., Soker, S., Yoo, J.J., Retik, A.: Tissue-engineered bladders for patients needing cytoplasty. Lancet **367**(9518), 1241–1246 (2006)
21. Vunjak-Novakovic, G.: Tissue Engineering. J. Serbian Soc. Comput. Mech. **5**(2), 29–36 (2011)
22. Rees, H.W., Panzer, S., Deter, M., Gabriel, C., Lambermont, M., Deneys, V., Sondag, D., Dickmeiss, E., Fischer-Nielsen, A., Korhonen, M., Krusius, T., Ali, A., Tiberghien, P., Schrezenmeier, H., Tonn, T., Seifried, E., Klüter, H., Politis, C., Stavropoulou-Gioka, A., Parara, M., Flesland, Ø., Nascimento, F., Balint, B., Fernández, P.M., Bart, T., Chen, F.E., Pamphilon, D.H.: International forum: new cellular therapies: is there a role for transfusion services? Vox Sang. **97**, 77–90 (2009)
23. Kucia, M., Zuba-Surma, E., Wysoczynski, M., Dobrowolska, H., Reca, R., Ratajczak, J., Ratajczak, M.Z.: Physiological and pathological consequences of identification of very small embryonic like (VSEL) stem cells in adult bone marrow. J. Physiol. Pharmacol. **57**(Suppl 5), 5–18 (2006)
24. Huang, G.T.-J., Gronthos, S., Shi, S.: Mesenchymal stem cells derived from dental tissues vs. those from other sources: their biology and role in regenerative medicine. J. Dent. Res. **88**, 792 (2009)
25. Balint, B., Ivanović, Z., Petakov, M., Taseski, J., Jovčić, G., Stojanović, N., et al.: The cryopreservation protocol optimal for progenitor recovery is not optimal for preservation of marrow repopulating ability. Bone Marrow Transplant. **23**(6), 613–619 (1999)
26. Balint, B., Ljubenov, M., Stamatović, D., Todorović, M., Pavlović, M., Ostojić, G., et al.: Stem cell harvesting protocol research in autologous transplantation setting: large volume vs. conventional cytapheresis. Vojnosanit. Pregl. **65**(7), 545–551 (2008)
27. Balint, B.: Stem and progenitor cell harvesting, extracorporeal "graft engineering" and clinical use—initial expansion vs. current dillemas. Clin. Appl. Immunol. **5**(1), 518–527 (2006)
28. Balint, B., Stamatovic, D., Todorovic, M., Jevtic, M., Ostojic, G., Pavlovic, M., et al.: Stem cells in the arrangement of bone marrow repopulation and regenerative medicine. Vojnosanit. Pregl. **64**(7), 481–484 (2007)
29. Balint, B., Todorovic, M., Jevtic, M., Ostojic, G., Ristanovic, E., Vojvodic, D., et al.: The use of stem cells for marrow repopulation and in the field of regenerative medicine. Mak. Med. Pregl. **63**(Suppl 75), 12–17 (2009)
30. Alvarez, A., Unda, F., Cañavate, M.L., Hilario, E.: Stem cell and regenerative medicine. Curr. Stem Cell Res. Ther. **4**(4), 287–297 (2009)
31. Mayorga, M., Finan, A., Penn, M.: Pre-transplantation specification of stem cells to cardiac lineage for regeneration of cardiac tissue. Stem Cell Rev. Rep. **5**(1), 51–60 (2009)
32. Pavlovic, M., Balint, B.: The use of stem cells to repair cardiac tissue. Anest. Reanim. Transfuziol. **34**, 129–150 (2006)
33. Sell, S. (ed.): Stem cells. In: Stem Cell Handbook, pp. 1–18. Humana, Totowa (2004)
34. Weissman, I.: Stem cell research: paths to cancer therapies and regenerative medicine. JAMA **294**(11), 1359–1366 (2005)
35. Körbling, M., Estrov, Z.: Adult stem cells for tissue repair—a new therapeutic concept? N. Engl. J. Med. **349**, 570–582 (2003)
36. Tsai, R.Y.L.: A molecular view of stem cell and cancer cell self-renewal. Int. J. Biochem. Cell Biol. **36**, 684–694 (2004)
37. Cai, J., Weiss, M.L., Rao, M.S.: In search of "stemness". Exp. Hematol. **32**, 585–598 (2004)
38. Doyonnas, R., Blau, H.M.: What is the future of stem cell research? In: Sell, S. (ed.) Stem Cell Handbook, pp. 491–499. Humana, Totowa (2004)
39. Fuchs, S., Satler, L., Kornowski, R., Okubagzi, P., Weisz, G., Baffour, R., et al.: Catheter-based autologous bone marrow myocardial injection in no-option patients with advanced coronary artery disease. J. Am. Coll. Cardiol. **41**, 1721–1724 (2003)
40. Perin, E.C., Dohmann, H.F.R., Borojevic, R., Silva, S.A., Sousa, A.L.S., Mesquita, C.T., et al.: Transendocardial, autologous bone marrow cell transplantation for severe, chronic ischemic heart failure. Circulation **107**, 2294–2302 (2003)

41. Strauer, B.E., Brehm, M., Zeus, T., Köstering, M., Hernandez, A., Sorg, R.V., et al.: Repair of infarcted myocardium by autologous intracoronary mononuclear bone marrow cell transplantation in humans. Circulation **106**, 1913–1918 (2002)
42. Aicher, A., et al.: Assessment of the tissue distribution of transplanted human endothelial progenitor cells by radioactive labeling. Circulation **107**, 2134–2139 (2003)
43. Beauchamp, J.R., Morgan, J.E., Pagel, C.N., Partridge, T.A.: Dynamics of myoblast transplantation reveal a discrete minority of precursors with stem cell-like properties as the myogenic source. J. Cell Biol. **144**, 1113–1122 (1999)
44. Taylor, D.A., et al.: Regenerating functional myocardium: improved performance after skeletal myoblast transplantation. Nat. Med. **4**, 929–933 (1998)
45. Vassilopoulos, G., Wang, P.R., Russel, D.V.: Transplanted bone marrow regenerates liver by cell fusion. Nature **422**(6934), 901–904 (2003)
46. Wang, X., Willenbring, h., Akkari, Y., et al.: Cell fusion is the principal source of bone-marrow derived hepatocytes. Nature **422**(6934), 897–901 (2003)
47. Chen, J., Zhang, Z.G., Li, Y., Wang, L., et al.: Intravenous administration of human bone marrow stromal cells induces angiogenesis in the ischemic boundary zone after stroke in rats. Circ. Res. **92**(6), 692–699 (2003)
48. Chen, J., Leong, K.W.: Scafolding in tissue engineering: general approaches and tissue-specific considerations. Eur. Spine J. **17**(Suppl 4), 467–479 (2008)
49. Nemeno-Guanzon, J.G., et al.: Trends in Tissue engineering. J. Biomed. Biotechnol. **956345**, 14 (2012). doi:10.1155/2012/956345
50. Lee, J.I., Sato, M., Kim, H.W., Mochida, J.: Transplantatation of scaffold-free spheroids composed of synovium-derived cells and chondrocytes for the treatment of cartilage defects of the knee. Eur. Cells Mater. J. **22**, 275–290 (2011)
51. Liebchner, M.A.K. (ed.): Computer Aided Tissue Engineering. Humana, Totowa (2012). ISBN 978-1-61779-764-4
52. Sun, W., Darkling, A., Starly, B., Nam, J.: Computer-aided tissue engineering: overview, scope and shallenges. Biotecnol. Appl. Biochem. **39**(Pt1), 29–47 (2004)
53. Boland, T., Xu, T., Damon, B., Cui, X.: Application of inkjet printing to tissue engineering. Biotechnol. J. **9**, 910–917 (2006)
54. Atala, A., Bauer, S., Soker, S., Yoo, J.J., Retik, A.B.: Tissue-engineered bladders for patients needing cytoplasty. Lancet **367**(9518), 1241–1246 (2006)
55. http://www.neuralstem.com/patient-info-treatments-in-development#celltherapy

External Links

American Society for Artificial Internal Organs (ASAIO)
Artificial organ experts
Hardware Models of Hippocampus: Toward Brain Implants as Neural Prostheses for Memory Loss
Adaptive Optoelectronic Eye: Hybrid Sensor/Processor Architecture
Information on extracorporeal liver support
Retrieved from "http://en.wikipedia.org/wiki/Artificial_organ"
Categories: Implants | Prosthetics | Artificial organs

Chapter 21
Cell Culture in Bioengineering-Working on 3-Dimensional Culture and Ink-Jet Printing: Regenerative Medicine (RM)

It requires a very unusual mind to undertake the analysis of the obvious.

Alfred North Whitehead (1861–1947)

Cell culture in TE is three-dimensional (3-D) while classical seeding of the cells constituted monolayer. This was limitation that was bridged by inventing tissue scaffolds and using the knowledge on the cell growth parameters and influences. Scaffolds are supportive materials, biocompatible and usually biodegradable and bioresorbable after its function once is finalized. They do have different surfaces and holes which enable cells to move through and establish contact activation and contact inhibition stimulating growth and differentiation supported by molecules of extracellular matrix that are in cell media. Combining cells with scaffolding materials to generate functional tissue constructs describes tissue engineering at its most basic level. However, understanding and manipulating the complex relationship between the cells and the scaffolding materials, represents the great challenge for tissue engineers (Figs. 21.1 and 21.2).

Chester Carlson (1906–1968) inventor of Xerography: inspiration for improved Ink-jet printing, Charles Hull (1939–present)-father of 3-D printing (stereolitography)

M. Pavlovic, *Bioengineering*, DOI 10.1007/978-3-319-10798-1_21

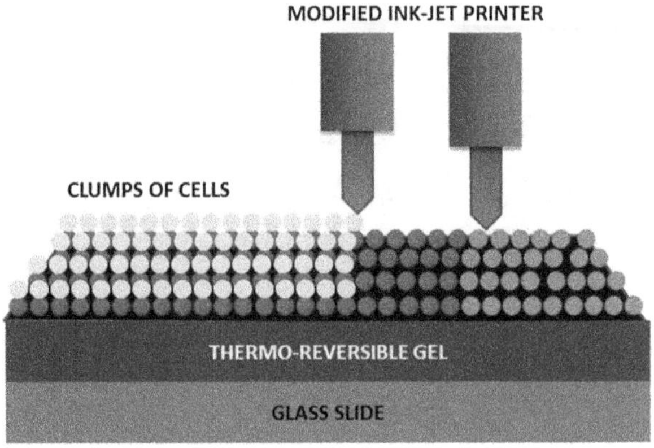

Fig. 21.1 Schematic presentation of Ink-jet printing

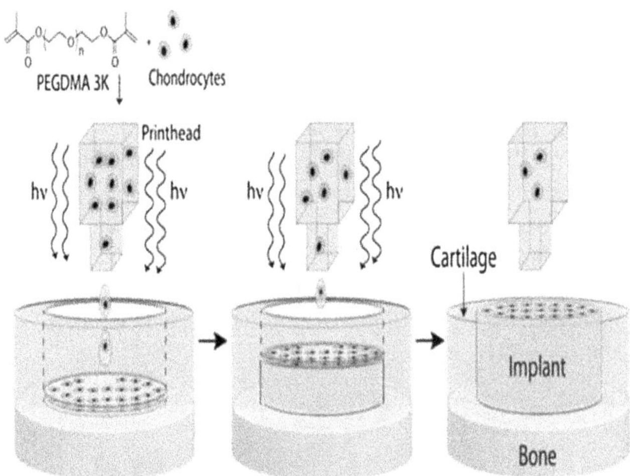

Fig. 21.2 Thermal Ink-jet printing: Schematic of bioprinting cartilage with simultaneous photo-polymerization process

Introduction

Organ printing refers to the placement of various cell types into a soft scaffold fabricated according to a computer-aided design template using a single device [1–3]. Computer aided scaffold topology design has recently gained attention as a viable option to achieve function and mass transport requirements within tissue engineering scaffolds [4–7]. An exciting advance is that of simultaneous printing of cells and biomaterials, which allows precise placement of cells and proteins within 3-D hydrogel structures. This advance raises the possibility of spatially controlling not only the scaffold structure, but also the type of tissue that can be grown within the scaffold and the thickness of the tissue as capillaries and vessels could be constructed within the scaffolds.

Combining cells with scaffolding materials to generate functional tissue constructs describes tissue engineering at its most basic level [8–11]. However, understanding and manipulating the complex relationship between the cells and the scaffolding materials, represents the great challenge for tissue engineers. What cells should be used, for example, and should the combination of cells and materials occur in vitro or in vivo? What scaffolding material will best facilitate development? How can development be guided using humoral or mechanical cues? How will the tissue construct be functionally integrated?

Cell culture in TE is three-dimensional (3-D) while classical seeding of the cells constituted monolayer (2-D). This was limitation that was bridged by inventing tissue scaffolds and using the knowledge on the cell growth parameters and influences. Scaffolds are supportive materials, biocompatible and usually biodegradable and bioresorbable after its function once is finalized [12]. They do have different surfaces and holes which enable cells to move through and establish contact activation and contact inhibition stimulating growth and differentiation supported by molecules of extracellular matrix that are in cell media [13–16].

Regenerative medicine, which consists of tissue engineering, cell therapies, and healing therapies, is still in its infancy, but is a fast-growing field.

3-D Culture

Tissue Engineered Building Blocks and creating optimal-niche tissue environment; involves triad of tissue engineering: cells, scaffolds and molecules of extracellular matrix (**EMC**) (Fig. 21.3).

In tissue engineering and regenerative medicine, "scaffolds" are typically three-dimensional (3-D) fibre-based or porous structures that are designed to carry cells and/or therapeutically active molecules. These scaffolds may be used to culture cells in 3-D for subsequent use as advanced in vitro models or implanted in a patient to encourage the regeneration of healthy, functional tissues. In the early stages after surgery, scaffolds may also provide a useful additional mechanical support for the lost or injured structures, until the recovering local host tissues have developed and been fully integrated within the patient's body.

Scaffolds for bone are commonly based on bioceramics and composites, but there is an extensive effort invested in work on the development of novel porous ceramic structures for bone tissue engineering [16, 17]. Craniofacial bone is of course vitally important for function and aesthetics in the head, neck and face, and this research has obvious implications for other specialties including orthopaedics [8]. Cartilage tissue engineering scaffolds research is directed at both polymeric systems for articular hyaline cartilage, and ceramic or composite systems for hypertrophic cartilage. In periodontal ligament scaffold research, the main interest is in understanding the role of orientation and mechanical loading in fibre-based systems.

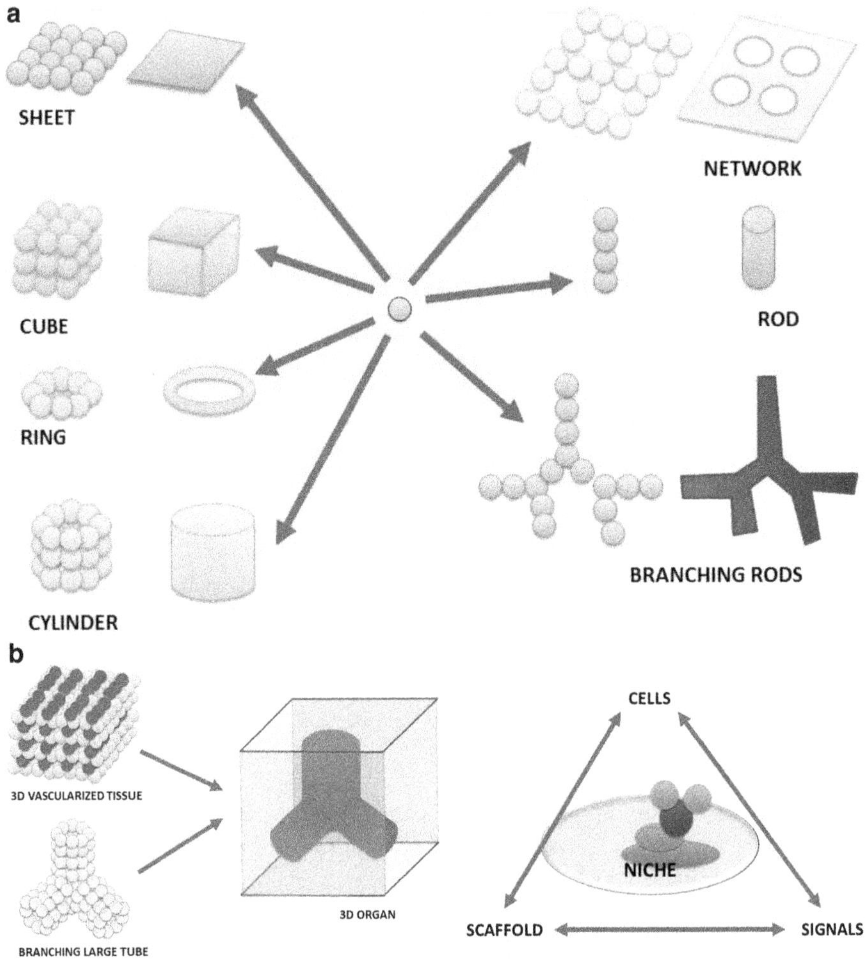

Fig. 21.3 Scaffolds for different purposes (**a**) and TRIAD of TE mimicking natural niche (**b**)

Ink-Jet Printing of the Cells and Liquid Scaffolds

Recent advances in organ printing technology (Ink-jet printing) for applications relating to medical interventions and organ replacement are the extension on the work on scaffolds. Organ printing refers to the placement of various cell types into a soft scaffold fabricated according to a computer-aided design template using a single device [6, 7]. Computer aided (CAD) scaffold topology design has gained strong attention as a viable option to achieve function and mass transport requirements within tissue engineering scaffolds. An exciting advance is that of simultaneous printing of cells and biomaterials, which allows precise placement of cells and proteins within 3-D hydrogel structures. This advance raises the possibility of spatially controlling not only the scaffold structure, but also the type of tissue that can be grown within the scaffold and the thickness of the tissue as capillaries and vessels could be constructed within the scaffolds.

Organ printing is today possible thanks to application of brilliant idea that cells can be printed onto scaffolds as it is the ink [6, 7]. This in essence, very simple technique provides a broad spectrum of TE maneuvers including the very fast recovery from large burns, the event that is extremely useful and highly necessary in those situations. Today many cell types can be printed as bio-ink using ink-jet printers: the cells survive, maintain their phenotype, differentiate and show function. They can be printed uniformly and homogeneously into confluent layers. They can be printed into 3-D structures. However, there are considerable technical barriers in the development of this emerging inkjet printing technology, such as the ability of the modified printers to deliver viable cells and the capability of the inkjet printing to fabricate functional, viable and functionally vascularized 3-D configurations [17–20]. This is the future problem which needs to be gradually solved. A new way to print living cells onto any surface and in almost any shape has been developed by researchers led by Houston Methodist Research Institute nanomedicine faculty member Lidong Qin. Unlike a similar inkjet printing process, almost all cells survive.

The new process, called Block-Cell-Printing (BloC-Printing), produces 2-D cell arrays in half an hour, prints the cells as close together as 5 μm (most animal cells are 10–30 μm wide), and allows the use of many different cell types.

Cell printing is used in so many different ways now – for drug development and in studies of tissue regeneration, cell function, and cell-cell communication. Such things can only be done when cells are alive and active. A survival rate of 50–80 % is typical as cells exit the inkjet nozzles. By comparison, we are seeing close to 100 % of cells in BloC-Printing survive the printing process.

BloC-Printing manipulates microfluidic physics to guide living cells into hook-like traps in the silicone mold. Cells flow down a column in the mold, past trapped cells to the next available slot, eventually creating a line of cells in a grid.

The position and spacing of the traps and the shape of the channel navigated by the cells is fully configurable during the mold's creation. When the mold is lifted away, the living cells remain behind, adhering to the growth medium or other substrate, in prescribed formation (Fig. 21.4, Table 21.1) [22, 23].

We have already mentioned that stem cells according to the functionality can be divided into two categories:

Normal stem cells are building blocks for our body (embryonic, fetal, cord blood and adult from different sources) [1, 21, 24, 25].

Cancer stem cells are defined as those cells within a tumour that can self-renew and drive tumorigenesis. Rare cancer stem cells have been isolated from a number of human tumours, including haematopoietic, brain, colon and breast cancers. The cancer stem-cell concept has important implications for cancer therapy. However, the generality of the cancer stem-cell hypothesis has also been challenged, most recently in a paper by Quintana et al. [26].

Cancers originally develop from normal cells that gain the ability to proliferate aberrantly and eventually turn malignant. These cancerous cells then grow clonally

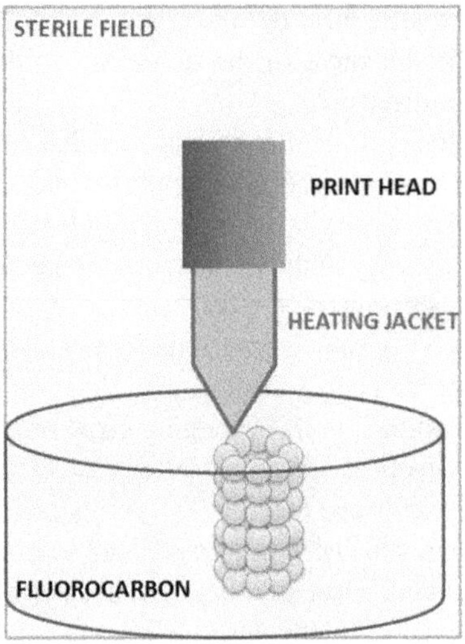

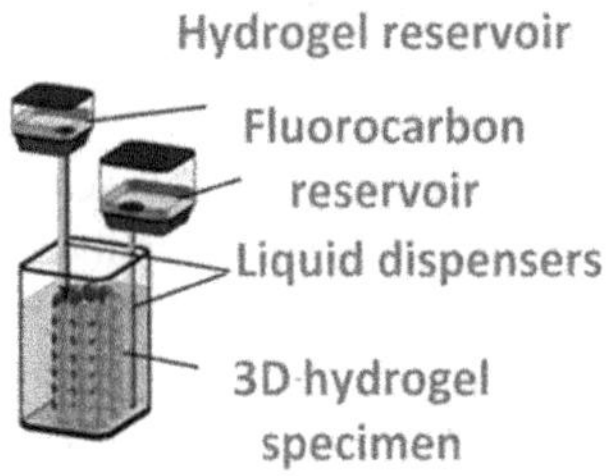

Fig. 21.4 B-Bridge has brought recently two distinct hydrogels for 3D cell culture: Cellendes, a life science company in Germany, developed a biomimetic dextran-based 3D hydrogel while Menicon Life Science in Japan manufactures a peptide-based 3-D hydrogel. Each hydrogel offers unique advantages for a variety applications like drug discovery and tissue engineering

Table 21.1 Distinction between differentiation, regeneration, dedifferentiation and degeneration: definitions and examples

Terms	Examples
Differentiation: The process by which a less specialized cell becomes a more specialized cell type	Epithelial cells will differentiate to give rise to all of the parenchymal cells (those cells which perform the function of the particular organ) of all glands, whether exocrine or endocrine; Mesenchymal cells will differentiate to give rise to cells that manufacture bone (osteoblast), cartilage (chondroblasts), muscle (myoblasts), fat (adipocytes), tendons and ligaments (fibroblasts)
Regeneration: Property to regrow whole limbs, tails, other body parts or organs if they are lost in an accident	Regeneration of a severed finger with a collagen powder derived from pigs bladder, both liver and red blood cells are able to renew, salamanders are able to regenerate limbs, nerve cell regeneration
Dedifferentiation: The loss of specialization in form or function; a reversal of cell development, esp. in plants, so that the differentiation that had occurred previously is lost and the cell becomes more generalized in structure	Pluripotency, giving rise to cells reminiscent of stem cells. Cellular dedifferentiation has also been implicated in cancer. As cancer can only be established from cells that have the potential to divide, and not terminally differentiated cells, one theory suggests that tumors may arise from the unrestrained growth of dedifferentiated cells that resemble embryonic cells
Degeneration: Progressive deterioration of physical characters from a level representing the norm of earlier generations or forms; deterioration of a tissue or an organ in which its function is diminished or its structure is impaired	Macular degeneration (Wet and Dry)

into tumors and eventually have the potential to metastasize. A central question in cancer biology is, which cells can be transformed to form tumors? Recent studies elucidated the presence of cancer stem cells that have the exclusive ability to regenerate tumors [26]. These cancer stem cells share many characteristics with normal stem cells, including self-renewal and differentiation [26]. With the growing evidence that cancer stem cells exist in a wide array of tumors, it is becoming increasingly important to understand the molecular mechanisms that regulate self-renewal and differentiation because corruption of genes involved in these pathways likely participates in tumor growth. This new paradigm of oncogenesis has been validated in a growing list of tumors. Studies of normal and cancer stem cells from the same tissue have shed light on the ontogeny of tumors. That signaling pathways such as Bmi1 and Wnt have similar effects in normal and cancer stem cell self-renewal suggests that common molecular pathways regulate both populations. Understanding the biology of cancer stem cells will contribute to the identification of molecular targets important for future therapies.

References

1. Vunjak-Novakovic, G., Scadden, D.T.: Biomimetic platforms for human stem cell research. Cell Stem Cell **8**, 252–261 (2011). PMC3048803
2. Vunjak-Novakovic, G.: Tissue engineering. J. Serbian Soc. Comput. Mech. **5**(2), 29–36 (2011)
3. Campos, D.F.D., Blaeser, A., Weber, M., Jäkel, J., Neuss, S., Jahnen-Dechent, W., Fischer, H.: Three-dimensional printing of stem cell-laden hydrogels submerged in a hydrophobic high-density fluid. Biofabrication **5**, 015003 (2013), 11pp
4. Bhumiratana, S., Vunjak-Novakovic, G.: Engineering functional bone and cartilage grafts. In: Bernstein, H.S. (ed.) Tissue Engineering in Regenerative Medicine. Stem Cell Biology and Regenerative Medicine Series. Springer, New York (in press)
5. Vunjak-Novakovic, G., Lui, K.O., Tandon, N., Chien, K.: Bioengineering heart muscle: a paradigm for regenerative medicine. Ann Rev Biomed Eng **13**, 245–267 (2011)
6. Mhashilkar, A., Atala, A.: Advent and maturation of regenerative Medicine. Curr. Stem Cell Res. Ther. **7**, 430–445 (2012)
7. Mhashilkar A: Bioengineering and Disease correction (2012). World Stem Cell Summit, December 4, 2012 West Palm Beach
8. Coreia, C., Grayson, W.L., Park, M., Hutton, D., Zhou, B., Guo, E., Niklason, L., Sousa, R.A., Reis, R.L., Vunjak-Novakovic, G.: In vitro model of vascularized bone: synergizing vascular development and osteogenesis. PLoS One **6**(12), e28352 (2011). PMID: 22164277
9. Bhumiratana, S., Vunjak-Novakovic, G.: Personalized human bone grafts for reconstructing head and face. Stem Cells Trans Med **1**(1), 64–69 (2012)
10. Gadjanski, I., Spiller, K., Vunjak-Novakovic, G.: Time-dependent processes in tissue engineering of articular cartilage. Stem Cell Rev. **8**(3), 863–881 (2012). PMID: 22016073
11. Vunjak-Novakovic G: Tissue engineering strategies for skeletal repair. Musculoskel. J. (in press)
12. Rockwood, D.N., Gil, E.-S., Park, S.-H., Kluge, J.A., Grayson, W., Bhumiratana, S., Rajkhowa, R., Wang, L., Kim, S.J., Vunjak-Novakovic, G., Kaplan, D.L.: Silk particle reinforced silk composite scaffolds for bone tissue engineering. Acta Biomater. **7**(1), 144–151 (2011). PMC2967589
13. Maidhof, R., Tandon, N., Lee, E.J., Luo, J., Duan, Y., Yeager, K., Vunjak-Novakovic, G.: Biomimetic perfusion and electrical stimulation applied in concert improved the assembly of engineered cardiac tissue. J. Tissue Eng. Regen. Med. **6**(10), e12–e23 (2012). doi:10.1002/term.525. PMID: 22170772

14. Borenstein, J.T., Vunjak-Novakovic, G.: BioMEMS and tissue engineering. IEEE Pulse **2**(6), 28–34 (2011). PMID: 22147066
15. Wan, L.Q., Vunjak-Novakovic, G.: Micropatterning chiral morphogenesis. Commun. Integr. Biol. **4**(6), 745–748 (2011)
16. Zhang, T., Wan, L.Q., Xiong, Z., Marsanno, A., Maidhof, R., Park, M., Yan, Y., Vunjak-Novakovic, G.: Chitosan-collagen based channeled scaffold for engineering functional myocardial patch with mechanical stimulation. J. Tissue Eng. Regen. Med. **6**(9), 748–756 (2012). doi:10.1002/term.481
17. Duan, Y., Liu, Z., O'Neill, J., Wan, L., Freytes, D.O., Vunjak-Novakovic, G.: Hydrogel derived from native heart matrix induces cardiac differentiation of human embryonic stem cells without supplemental growth factors. J. Cardiovasc. Transl. Res. **4**(5), 605–615 (2011). PMCID:3196310
18. Wan, L.Q., Ronaldson, K., Park, M., Taylor, G., Zhang, Y., Gimble, J.M., Vunjak-Novakovic, G. Micropatterned mammalian cells exhibit chiral morphogenesis. PNAS. **108**(30), 12295–12300 (2011); cover article, commentary in the same issue: McSheene, J.C., Burdine, R.D.: Examining the establishment of cellular axes using intrinsic chirality. PNAS. **108**(30), 12191–12192. PMCID: PMC3145729
19. Godier-Fournemont, A., Martens, T., Koeckert, M., Wan, L.Q., Parks, J., Zhang, G., Hudson, J., Vunjak-Novakovic, G.: Composite scaffold provides a cell delivery platform for cardiovascular repair. Proc. Natl. Acad. Sci. U. S. A. **108**(19), 7974–7979 (2011). PMCID:PMC393484
20. Marolt, D., Cozin, M., Vunjak-Novakovic, G., Cremers, S., Landesberg, R.: Effects of pamidronate on human alveolar osteoblasts. J. Oral Maxillofac. Surg. **70**(5), 1081–1092 (2012). PMC3223542
21. Gavrilov, S., Marolt, D., Douglas, N.C., Prosser, R.W., Khalid, I., Sauer, M.V., Landry, D.W., Vunjak-Novakovic, G., Papaionnaou, V.: Derivation of two new human embryonic stem cell lines. Stem Cells Int. **2011**, 765378 (2011). PMCID: PMC3118293
22. Singh, G., Javidfar, J., Costa, J., Guarrera, J., Miller, J., Henry, S., Jallerat, Q., Freytes, D.O., Vunjak-Novakovic, G., Sonett, J.R., Bacchetta, M.D.: Perfusion/decellularization of large animal lungs. J Heart Lung Transplantation **30**(4), S184–S185 (2011)
23. Rouwkema, J., Gibbs, S., Lutolf, M., Martin, I., Vunjak-Novakovic, G., Malda, J.: In vitro platforms for tissue engineering: implications to basic research and clinical translation. Opinion article. J. Tissue Eng. Regen. Med. **5**(8), e164–167 (2011)
24. Bhumiratana, S., Grayson, W.L., Castaneda, A., Rockwood, D., Gil, E.-S., Kaplan, D.L., Vunjak Novakovic, G.: Silk-hydroxyapatite composite provides an osteogenic scaffold for bone tissue engineering using human mesenchymal stem cells. Biomaterials **32**(11), 2812–2820 (2011)
25. Grayson, W.L., Marolt, D., Bhumiratana, S., Frohlich, M., Guo, X.E., Vunjak-Novakovic, G.: Optimizing the medium perfusion rate in bone tissue engineering bioreactors. Biotechnol. Bioeng. **108**(5), 1159–1170 (2011). PMID: 21125596
26. Quintana, E., Shackleton, M., Sabel, M.S., Fullen, D.R., Johnson, T.M., Morrison, S.J.: Efficient tumour formation by single human melanoma cells. Nature **456**(7222), 593–598 (2008). doi:10.1038/nature07567

Chapter 22
Magnetism and Magnetobiology: New Undiscovered Horizons?

I often say that when you can measure what you are speaking about, and express it in numbers, you know something about it; but when you cannot measure it, when you cannot express it in numbers, your knowledge is of a meagre and unsatisfactory kind.

Lord Kelvin (1824–1907)

Magnetism is a spectrum of physical phenomena involving forces exerted by magnets on other magnets. There are different types of magnetism as well as different sources. We shall not take it in consideration in this chapter. One can find the hierarchy flow chart for magnetism type and go deeper into phenomenology, if interested. However, it is necessary to keep in mind that magnetism and electricity are tightly linked phenomena. One of strong impacts of magnetism upon bioengineering is in biomedical engineering. There is the entire field of studies known as magneto-biology which involves new studies and concepts of magnetic fields produced in, or applied to biological systems as a medical treatment or diagnostic approach.

David Cohen (-present) first pioneering measurements of magnetic fields produced by the body, John Wikswo (1949–), development and application of microdevices (SQUID magnetometry) for instrumenting and controlling single living cells, Samuel Williamson (1950–2005) developed one of the first biomagnetometers together with Douglas Brenner, made first recordings of magnetoencephalography (MEG) signals from the visual, somatosensory and motor cortices and identified a tonotopic mapping in the auditory cortex, helped develop multi-channel and whole-head MEG systems and a new type of biomagnetometer called cryoSQUID. Tapash Chakraborty(1947–present) contributed to the theory of electronic properties of various nanoscale systems, quantum dots, double dots and rings, graphene single and double layers, and electronic transport in DNA.

M. Pavlovic, *Bioengineering*, DOI 10.1007/978-3-319-10798-1_22

Introduction

The word "magnetism" originates from Latin term for iron loadstones found close to Greek place in province Thessaly known as Magnesia—where magnetized iron ores were obtained (magnetite). That magnetic state (or phase) of the matter is considered to depend on temperature and other variables, so that the material can express more than one form of magnetism dependent on that [1]. So, one can see the Earth as a giant magnet with two poles and at the same time understand the basics of the function of compass which is the needle in a magnetic field showing us North and South (Fig. 22.1).

Biomagnetism as a Phenomenon in the Nature and Possibilities for its Measurement

Magnetism is a spectrum of physical phenomena involving forces exerted by magnets on other magnets [1, 2]. There are different types of magnetism as well as different sources. We shall not go into that in this chapter. One can find the hierarchy flow chart for magnetism type and go deeper into phenomenology, if interested. However, it is necessary to keep in mind that magnetism and electricity are tightly linked phenomena as well as that both of them exist in living tissues (cells).

There is a new area of investigations opening a new horizons in biological way of thinking known as biomagnetism. Biomagnetism involves a broad spectrum of phenomenology which is not quite explained, but is detected in life, measured and continues to attract the interest of scientists [2–7]. Living organisms have polarized membranes (cellular and inner mitochondrial), polarized DNA due to flow of electrons down the loops, and therefore, they are magnetically polarized as well.

Fig. 22.1 The Earth as a magnet and reason why we can determine N and S poles by using magnetic needle in the compass

Living matter expresses electromagnetism. And not only that. Some organisms can detect magnetic fields. This phenomenon is known as *magnetoception* [2]. Fields naturally produced by an organism are known as *biomagnetism* [2]. Some bacteria have magnets of nanoparticle's size in their bodies (*magneto-bacteria*) [2]. Some birds have miniature magnets in their retina and can "see" where to fly. Meanwhile, scientists like David Cohen have established the approach for measurement of the magnetic field of humans in shielded rooms with a superconducting magnetometer [2]. A SQUID (for superconducting quantum interference device) is a very sensitive magnetometer used to measure extremely subtle magnetic fields, based on superconducting loops containing Josephson junctions [2, 6–8]. SQUIDs are sensitive enough to measure fields as low as 5 aT (5×10^{-18} T) within a few days of averaged measurements [2].

No wonder that one of strong impacts of magnetism upon bioengineering is in biomedical engineering. There is the entire field of studies known as *magneto-biology* which involves new studies and concepts of magnetic fields produced in, or applied to biological systems as a *diagnostic approach* or a *medical treatment*, for example. Despite a lot of controversies this field is progressing in development and methodological evolution and seems to be a potential source of knew understanding and knowledge that can benefit to human health and well-being.

Currently, a very interesting from bioengineering point of view, are *magnetic labeling of stem cells* [3] and the *new concept of cancer stem cell therapy using principles of magnetism* [4].

In Vivo Imaging of Intravascularly Injected Magnetically Labeled Stem Cells

One of the most interesting and significant innovations from biomedical engineering is for sure tracking through In vivo Imaging of Intravascularly Injected Magnetically Labeled Stem Cells of different origin (Mesenchymal stem cells, Human Neural stem cells, Embryonic stem cells). It has been shown that various, synthesized magnetic particles can serve for tracking stem cells into damaged tissues (brain, heart, etc.) and their engraftment in those tissues, which is a very promising tool in cellular treatment of the diseases such as stroke, acute myocardial infarction (AMI), and probably many others [8–20].

Possibilities of Engineering Targeted Cancer Stem Cell Therapy Using Principles of Magnetism

The ultimate goal of cancer therapy lies in a few key ideas: (1) create as little side-effect of the treatment to the host's tissues, (2) treat as non-invasively as possible, and (3) have long-term viability of treatment as stem cells vary in their genotypic expression.

Fig. 22.2 Stem-line Therapy

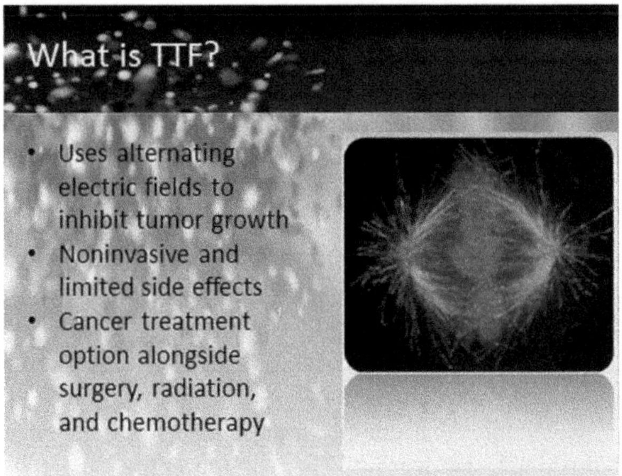

The third point may be moot if the cancer stem cells can be obliterated in the first treatment, however.

As technological advancement creates new opportunity in other realms of science, so too does it in the world of medicine. Two new treatment modalities are on the forefront of oncological intervention: *nanoparticle therapy* (already described in Drug delivery section) and *alternating magnetic fields* on replicating cancer stem cells. Although in the early phases of testing, the two show promise of accomplishing limited-to-no side-effects as well as being as non-invasive as possible. These novel treatments could be used in combination with chemotherapy or radiotherapy, as indicated (Fig. 22.2).

The other novel therapy that has shown great result is the use of **alternating magnetic fields**. As dividing cells undergo the various stages of cell replication, a developmental stage known as mitosis is the target of this therapy. During mitosis, all of the sister chromatids are lined up along the midline of the cell and still adjoined to one another by a centromere, which then become the target of spindles emergent from the centrioles at opposite poles of the dividing cell. These spindles have a *polarity in charge* due to their molecular composition. As this transient treatment field, or TTF, is applied via an external array, the spindles are disrupted by the alternating fields and a resultant disruption of cancer cell replication is accomplished The first clinical trial was in 2003 for patients with glioblastoma (GBM), the most aggressive and most common form of primary brain tumor in the United States. Two years later, three of the original ten patients were still alive, two of which had no progression of the cancer whatsoever. In 2011, the FDA approved TTF as a viable treatment for GBM. Currently, clinical trials are being run for the utilization of TTF with lung cancer, as well as in vitro research for many other types of cancer, including cervical. Some devices already are produced and utilized in the market, making the treatment more readily available. Novocure™, a commercial stage private oncology company, manufactures the device, NovoTTFTM-100A, a wearable device weighing around 6 lb that can fit into a shoulder bag for easy handling. Using non-invasive, insulated transducer arrays that are placed directly on the skin

Fig. 22.3 Electric fields

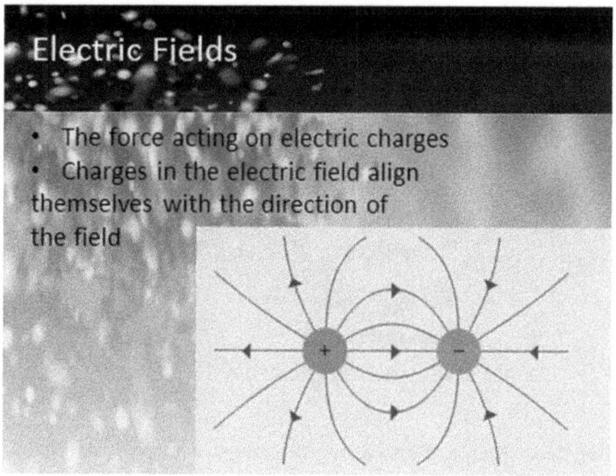

in the region surrounding the tumor, TTF therapy is unlike previous applications of electricity in medicine (Fig. 22.3).

The other novel therapy targeting cancer stem cells that has shown great result is the use of alternating magnetic fields. As dividing cells undergo the various stages of cell replication, *a developmental stage known as mitosis is the target of this therapy*. During mitosis, all of the sister chromatids are lined up along the midline of the cell and still adjoined to one another by a **centromere**, which then become the target of spindles emergent from the **centrioles** at opposite poles of the dividing cell. *These spindles have a polarity in charge due to their molecular composition.* As this transient treatment field (TTF), is applied via an external array, the spindles of cancer stem cells (which are smaller than other progenitors and normal cells) are disrupted by the alternating fields and a resultant disruption of cancer cell replication is accomplished. It is important to note, however, that **cancer stem cells are smaller than typical, normal-state mitotic cells and can therefore be targeted with specific frequencies in order to minimalize damage to healthy, non-cancerous cells** (Fig. 22.4).

The first clinical trial was in 2003 for patients with glioblastoma (GBM), the most aggressive and most common form of primary brain tumor in the United States. Two years later, three of the original ten patients were still alive, two of which had no progression of the cancer whatsoever. In 2011, the FDA approved TTF as a viable treatment for GBM.

Currently, clinical trials are being run for the utilization of TTF with lung cancer, as well as in vitro research for many other types of cancer, including cervical. Some devices already are produced and utilized in the market, making the treatment more readily available. NovocureTM, a commercial stage private oncology company, manufactures the device, NovoTTFTM-100A, a wearable device weighing around 6 lb that can fit into a shoulder bag for easy handling. Using non-invasive, insulated transducer arrays that are placed directly on the skin in the region surrounding the tumor, TTF therapy is unlike previous applications of electricity in medicine (38) (Fig. 22.5).

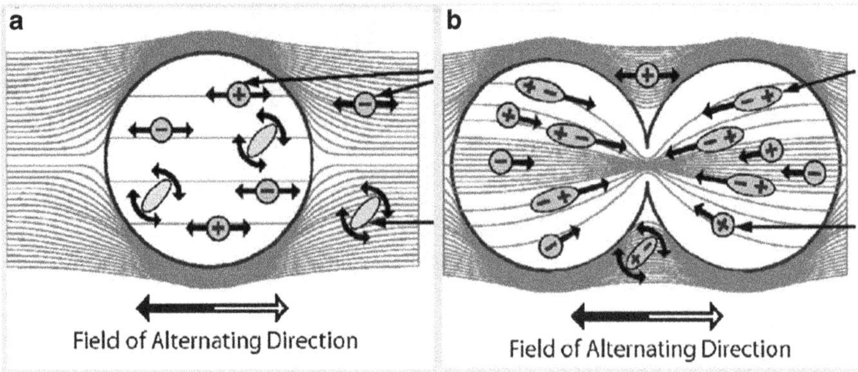

Fig. 22.4 Fields of alternating directions with polarization and difference in size of normal and cancer cells

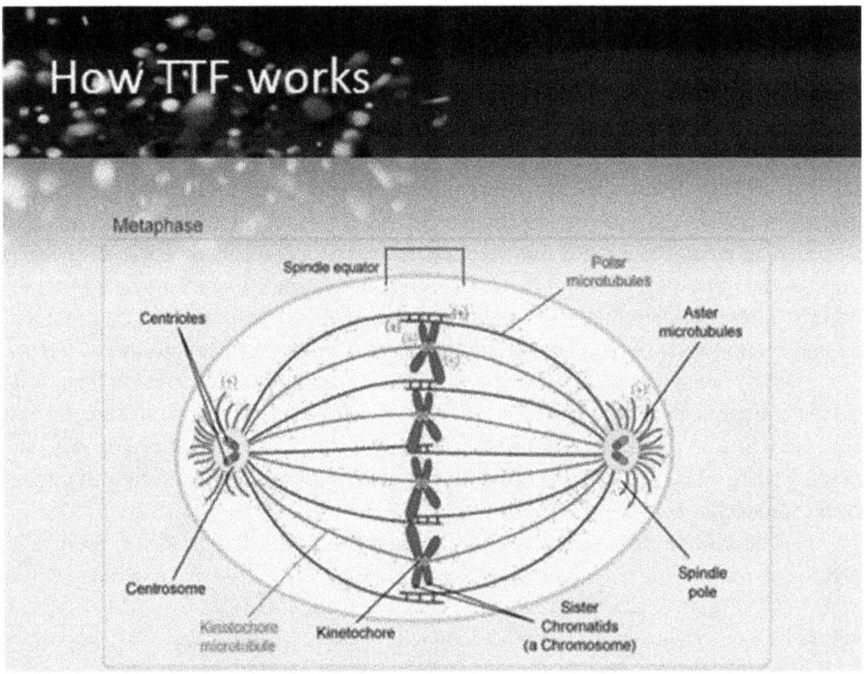

Fig. 22.5 Transient Treatment Field (TTF) Inducing Mitotic Spindle Rearrangement in Polarity. *The induced magnetic field specific to the frequency of CSCs disrupts the spindle formation and subsequent continuance of mitosis

Increased apoptosis of CSCs will result in a significantly smaller number of matured cancer cells, which can then be addressed successfully with common anticancer therapies. Thus, anticancer therapy that only results in apoptosis of the matured cancer cells and/or only inhibits the proliferation of CSCs provides a potential window of opportunity for new and more aggressive CSC mutants to occur and might be unsuccessful if not dangerous. It is expected that the elimination of cancer should target the CSC pool, and successful treatment regimens would need to be the result of an orchestrated 'target and destroy" effect. TTF therapy is a locally or regionally delivered treatment that uses electric fields within the human body that disrupt the rapid cell division exhibited by cancer cells. TTF therapy was developed to provide physicians and patients with a fourth treatment option for cancer in addition to surgery, radiation therapy and chemotherapy. Novocure developed TTF therapy from Prof. Yoram Palti's novel concept that a cell's physical properties can serve as targets for an anti-cancer therapy. Specifically, TTF therapy takes advantage of the special characteristics, geometrical shape, and rate of dividing cancer cells, all of which make them susceptible to the effects of alternating electric fields by altering the tumor cell polarity. The frequency used for a particular treatment is specific to the cell type being treated. TT Fields have been shown to disrupt mitotic spindle microtubule assembly and to lead to dielectrophoretic dislocation of intracellular macromolecules and organelles during cytokinesis. These processes lead to physical disruption of the cell membrane and to programmed cell death (apoptosis). The above mechanisms of action are consistent with the extensive research regarding the effects of TTF therapy. These results demonstrate both disruption of cancer cell division up to complete cessation of the process, as well as complete destruction of the dividing cancer cells.

While the promise of reduced cross-effect oncological treatment via CSC targeting gives great hope to the future of oncology and the patients that suffer from cancer, a great deal of work still remains. CSCs are a moving target and exist in such a small population that effective use of treatment modalities, although more promising than some miRNA studies and the like, still does not, in its current state, exist as a viable treatment option for all cancers. Much as microbiologists have difficulty in the world of fighting an ever-adapting organism, so to shall the oncologist and researchers that pursue this path. In combination with other therapies, however, it does appear that a reduction in risk associated with current treatment modalities would be evident. In light of the difficulty of the manipulation of the CSC model, the research that has been done thus far is providing a solid framework upon which a new, improved paradigm of oncological treatment will be.

1. TTF therapy is tuned to affect only one cell type at a time. TTF therapy has not been shown to affect cells that are not undergoing division.
2. TTF therapy is not expected to affect the normal functions of bone marrow in creating red and white blood cells, since the bone marrow is naturally shielded from the fields.
3. TTF therapy is delivered locally through a physical, non-chemical pathway. This allows TTF therapy to treat brain tumors, whereas other mitotic inhibitor treatments such as taxanes and vinca alkaloids have poor diffusion across the blood-brain barrier and are rarely used to treat brain tumors.
4. There is no evidence of cumulative damage to healthy tissues in the body when exposed to TTF therapy. Since the fields alternate so rapidly, they have no effect on normal quiescent cells nor do they stimulate nerves and muscles.

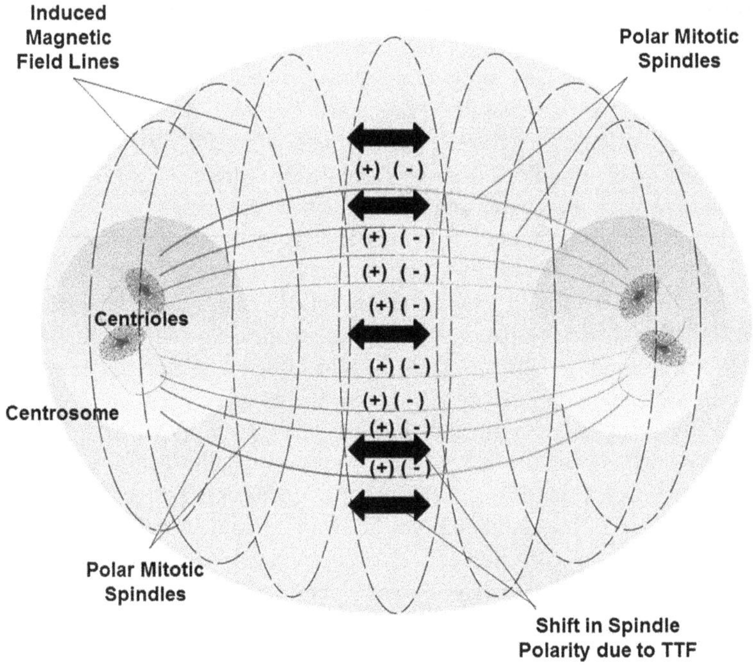

Fig. 22.6 (**a**, **b**) Transient Treatment Field (TTF) Inducing Mitotic Spindle Rearrangement in Polarity. *The induced magnetic field specific to the frequency of CSCs disrupts the spindle formation and subsequent continuance of mitosis

Taken together, these properties will potentially allow patients to receive TTF treatment for as long as necessary with minimal side effects while maintaining a high quality of life established (Fig. 22.6).

Emphasizing Bioengineering Aspect to Biomagnetism

Dynamic effects of therapeutic strategies directed against cancer stem cells (CSCs). A tumor tissue is a complex mix of cancer cells at various stages of differentiation, from uncommitted CSCs through various stages of cancer progenitor cells to matured cancer cells, with a concomitant decrease in the levels of proliferative and/or metastatic potential. Both the CSC niche with supporting cell types and the matured cancer cell compartment create an intricate network of inter-dependency [21]. Cancer therapy should ideally address both the CSCs and the matured cancer cells by slowing down proliferation and production of differentiated cancer cells and increasing apoptosis in both CSCs and matured cancer cells. In a fast-growing cancer, tumor therapy might come too late and/or be ineffective, or reduce tumor mass by killing matured cancer cells without targeting the CSC niche. The latter effect might stimulate CSC proliferation and increase the CSC pool, which would

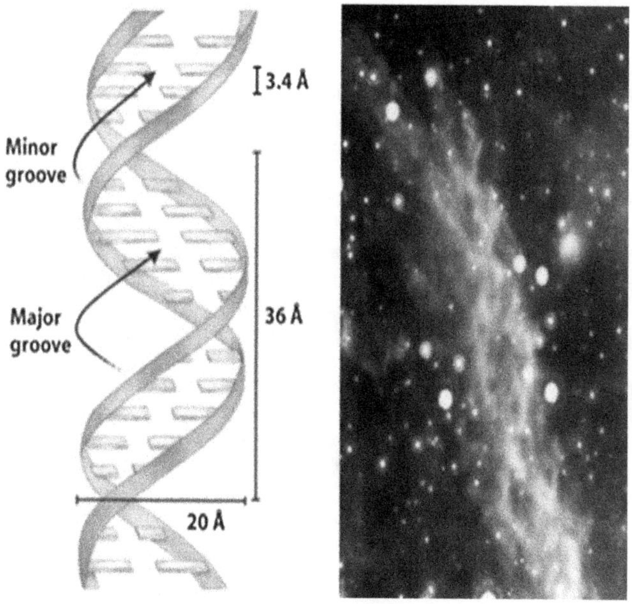

Fig. 22.7 DNA-Milky way α-helix-comparison

consequently result in a resurgence of even larger numbers of matured cancer cells [21]. In another scenario, therapeutic intervention itself might provoke an enlargement of the CSC pool by selecting for more radio- and chemoresistant CSC clones. These CSCs will have a superior ability to repair DNA damage upon radiotherapy and/or overexpress members of the ABC transmembrane pumps, resulting in the swift efflux of certain chemotherapeutics. Over time, this new generation of CSCs could also include new mutant CSCs with even more aggressive signatures. CSC therapy targets the CSC niche itself by attenuating the self-replicating potential of CSCs and disturbing cellular crosstalk within the CSC niche [21].

Increasing interest within past decade has been expressed for DNA electro/magneto-polarization in regard with α-helix conformation, nucleotide bonding, and raising still not defined possibilities based on that [16]. Just to make a comparison between micro and macro-world let us look in the α-helix of our DNA and α-helix of our Milkey Way in order to make a connections and imagine the world interconnected harder or more subtile than it was thought to be (Fig. 22.7).

References

1. Baule, G.M., McFee, R.: Detection of the magnetic field of the heart. Am. Heart J. **66**, 95–96 (1963)
2. Cohen, D., Edelsack, E.A., Zimmerman, J.E.: Magnetocardiograms taken inside a shielded room with a superconducting point contact magnetometer. Appl. Phys. Lett. **16**(7), 278–280 (1970)

3. Becker, R.O., Selden, G.: Dogs are sensitive to small variations of the Earth's magnetic field. Study: Dogs Relieve Themselves In-Line With Earth's Magnetic Field The Body Electric: Electromagnetism and the Foundation of Life. Quill, William Morrow, New York, NY (1985)
4. Williamson, S.H., Romani, G.L., Kaufman, L., Modena, I. (eds.): Biomagnetism: An Interdisciplinary Approach. NATO ASI Series. Plenum, New York (1983)
5. Cohen, D.: Boston and the history of biomagnetism. Neurol. Clin. Neurophysiol. **2004**, 114 (2004)
6. Cohen, D.: Magnetoencephalography: detection of the brain's electrical activity with a superconducting magnetometer. Science **175**, 664–666 (1972)
7. Science (Research News story by R. Crease), Sept.8, 1989; 1041-3.
8. Dubrov, A.P.: The Geomagnetic Field and Life: Geo-magnetobiology. Plenum, New York, NY (1978)
9. Hurtak, J.J.: The Power of Healing. Lecture given to the members of the Bioenergetics Institute, Johannesburg (1986)
10. Zimmerman, J.E., Theine, P., Harding, J.T.: Design and operation of stable rf-biased superconducting point-contact quantum devices, etc. J. Appl. Phys. **41**, 1572–1580 (1970)
11. Science (Research News story by R. Crease), July 26, 1991; 374-
12. Jacobson, J.I.: Exploring the potential of magneto-crystallization of genes and associated structures with respect to nerve regeneration and cancer. Int. J. Neurosci. **64**, 153–165 (1992)
13. Kirschvink, J.L.: Magnetite biomineralization and geomagnetic sensitivity in higher animals: an update and recommendations for future study. Bioelectromagnetics **10**, 239–259 (1989)
14. Kirschvink, J.L., et al.: Magnetite biomineralization in the human brain. Proc. Natl. Acad. Sci. **89**, 7683–7687 (1992)
15. Nordenstrom, B.W.W.: Impact of biological closed electric circuits (BCEC) on structure and function. Integr. Physiol. Behav. Sci. **27**, 285–303 (1992)
16. Chakraborty T: Finding Physics in DNA (2014) http://www.fysik.su.se/~hansson/nordforsk0803/Talks/tapash_dna.pdf
17. Zhao, C., Tian, M., Zhang, H.: In vivo stem cell imaging. Nucl. Med. J. **2**, 171–177 (2010)
18. Jiang, H., Cheng, Z., Tian, M.: In vivo imaging of embryonic stem cell therapy. Eur. J. Nucl. Med. Mol. Imaging **38**(4), 774–84 (2011). doi:10.1007/s00259-010-1667-y
19. Lu, S.S., Liu, S., Zu, Q.Q., Xu, X.Q., Yu, J., Wang, J.W., Zhang, Y., Shi, H.B.: In vivo MR imaging of intraarterially delivered magnetically labeled mesenchymal stem cells in a canine stroke model. PLoS One **8**(2), e54963 (2013)
20. Park, J.W., Ku, S.H., Moon, H.-H., Lee, M., Cho, D., Yang, J., Huh, Y.-M., Jeong, J.-H., Park, T.G., Mok, H., Kim, S.H.: Cross-linked iron oxide nanoparticles for therapeutic engineering and in vivo monitoring of mesenchymal stem cells in cerebral ischemia model. Macromol. Biosci. **14**, 380–389 (2014)
21. Lobo, N.A., Shimono, Y., Qian, D., Clarke, M.F.: The biology of cancer stem cells. Annu. Rev. Cell Dev. Biol. **23**, 675–699 (2007)